Air quality guidelines for Europe

Cover design: G. Gudmundsson

Air quality guidelines for Europe

WHO Regional Publications, European Series No. 23

ISBN 92 890 1114 9
ISSN 0378-2255

The designations employed and the presentation of the material in this publication do not imply the expression of any opinion whatsoever on the part of the Secretariat of the World Health Organization concerning the legal status of any country, territory, city or area or of its authorities, or concerning the delimitation of its frontiers or boundaries.

The mention of specific companies or of certain manufacturers' products does not imply that they are endorsed or recommended by the World Health Organization in preference to others of a similar nature that are not mentioned. Errors and omissions excepted, the names of proprietary products are distinguished by initial capital letters.

The views expressed in this publication are those of the participants in the meetings and do not necessarily represent the decisions or the stated policy of the World Health Organization.

PRINTED IN DENMARK

CONTENTS

PART III. INORGANIC SUBSTANCES

PART IV. EFFECTS OF INORGANIC SUBSTANCES ON VEGETATION

Foreword

Following the successful introduction of WHO guidelines for drinking-water quality, the Government of the Netherlands approached the WHO Regional Office for Europe in 1983 to suggest that air quality guidelines should be developed, using the same general philosophy and approach. We realized from the outset that this would be a difficult task and that, for example, sampling procedures were very much more complicated than in the case of water supplies. As was also the case for the drinking-water programme, we agreed that it would be inappropriate to try to formulate "standards", these being for governments and regulating agencies to decide on in the context of prevailing exposure levels and environmental, social, economic and cultural conditions. The approach has been to develop guideline values that in the opinion of the experts are appropriate for the safeguarding of public health and will guide national and local authorities in their risk management decisions. It is important that the numerical values are taken in the context of the descriptive sections of the guidelines: in many cases only an order of magnitude can be given, based on available data, but this is more useful to public health officials and regulators than no figure at all. In some cases, it has not been found useful to give a guideline value, but instead to provide a risk estimate.

Great efforts have been made during recent years in many countries of the European Region to reduce air pollution, and the intense smogs that were frequently experienced in London and other large cities up until 30 years ago no longer occur. Progress has, however, been uneven and in particular the burning of low-quality soft coals has caused increasing problems in some parts of Europe, especially when associated with atmospheric inversions.

In addition to the major pollutants such as sulfur dioxide and particulates, there is now increasing interest in the emission of small concentrations of potentially toxic inorganic and organic micropollutants.

The vast majority of Europeans spend over 75% of their lives indoors and the guidelines appropriately include consideration of indoor air quality, though not of occupational exposure. Although this publication is primarily intended to cover public health considerations, it was, I think, a useful decision to include some ecotoxicological dimensions. Most certainly, those involved in environmental management and public health practices increasingly need to view the effects of pollution on man within the context of the health of the entire biosphere. This is particularly important in the case of air pollution, which transcends national frontiers and can have marked effects at great distances from the source.

It is hoped that the guidelines will have a wide application in environmental decision-making throughout the Region as well as in other parts of the world. They were developed to a very demanding time scale and all those concerned in their preparation are aware that the publication does not represent any final judgement on the subject. In future editions new parameters will have to be addressed and the effect of combined exposures may have to be taken into account.

The farsightedness of the initiative by the Government of the Netherlands and the encouragement received from their representatives during the preparation of the guidelines are warmly acknowledged. The work involved great dedication and enthusiasm on the part of the 150 experts who participated in the various meetings and who were often under very great pressure to meet apparently impossible deadlines. It is appropriate that the unstinting efforts of the Editorial Consultation Group that met in Copenhagen in March 1986 should be specially recognized.

That the final product was delivered on time is a great tribute to all concerned and I would like to thank the secretaries and other members of the Regional Office staff for their great contribution and to mention in particular Dr Reiner Türck, project coordinator, and Dr Dinko Kello, project consultant, without whose combined qualities of scientific excellence, power of persuasion, tenacity and organizational ability, the work could not have been completed within an acceptable time frame.

J.I. Waddington
Director
Environmental Health Service

Preface

The World Health Organization has been concerned with air pollution and, in particular, its dangers to human health for 30 years. In 1957, a WHO conference was held in Milan dealing with the public health aspects of air pollution in Europe. In the same year a special report on air pollution by the WHO Expert Committee on Environmental Sanitation was published.

Since then, many activities have been undertaken covering different areas of the subject, such as measurement techniques, the compilation of air quality data from different urban areas in the world and the effects of air pollution on health. A *Manual on urban air quality management* published in 1976, a *Glossary on air pollution* published in 1980, and a *Manual on industrial air pollutants* published on behalf of WHO by Elsevier in 1983, have been part of the work of the Regional Office for Europe in this field.

With regard to air quality criteria and guides, the report of a WHO Expert Committee was published in 1972. Furthermore, the Environmental Health Criteria Programme, which started under the joint sponsorship of WHO and UNEP and now comes under the trilateral responsibility of WHO, UNEP and ILO, within the framework of the International Programme on Chemical Safety, has so far resulted in some 60 environmental health criteria documents, a significant number of which deal with air pollutants.

When the industrialized countries of the European Region began to establish environmental policies, it soon became apparent that a yardstick was needed for the evaluation of air quality. Air quality standards or objectives for major urban air pollutants were accordingly developed in many European countries. In this context, it was gradually recognized that international cooperation would be needed in certain areas, one of them being the joint assessment by scientists from various countries of the adverse effects of air pollutants. This perception may be demonstrated by a statement in the WHO Global Medium-term Programme for the Promotion of Environmental Health, 1983:

> With respect to environmental hazards assessment and control, much can be accomplished through international cooperation. The collection and assessment of information on human exposures to pollutants and on their effects on health is a very time-consuming and expensive effort, and very few countries, even in the industrialized world, have the necessary resources and expertise to do this alone. The pooling of resources to conduct these assessments can provide the governments with the necessary information which they need to take action.

It is obvious why the WHO Regional Office for Europe therefore felt fully justified in embarking on its project of establishing air quality guidelines for the Region. The recent successful development of guidelines for drinking-water quality was regarded as a further incentive.

From the very beginning of the project, it was clear that some basic principles would have to be followed.

- The guidelines should describe the latest state of scientific knowledge. They would have to be developed together in a short time period in order to guarantee this objective.

- The information provided would have to be condensed, describing only the essential factors leading to the final conclusions. To create a one-volume book on 28 pollutants or groups of pollutants, all encyclopaedic types of information would have to be ignored.

- The description of scientific findings would have to be of a kind that would be understandable to a broad and rather heterogeneous readership. The flow of the arguments would have to be clear.

- The rationale for the guideline recommendations should also contain a description of uncertainties in the evaluation process due to missing, inadequate or equivocal data. Any illusion should be avoided that it is possible to condense a very complex situation in reality down to a simple figure without making assumptions. In particular, questions of safety or protection factors, combined effects and high-risk groups would have to be discussed in this context.

- Another important goal would be to create a basic common structure for the description of pollutants and the rationale for guidelines, without deleting too much of the "handwriting" of specific authors and working groups.

- Finally, it was a prerequisite that the draft guidelines would have to undergo several intensive reviews, giving everybody involved the chance to look at the whole document at various stages of development.

It can easily be understood that the achievement of these goals required a high spirit of cooperation among all the experts involved in this publication. It is for the reader to decide how successfully their task has been accomplished.

The term "guidelines" should be understood literally, meaning that the main objective is to provide guidance to those interested in air pollution problems as well as to those directly involved in air quality management.

The guidelines consider various toxic (carcinogenic and noncarcinogenic) substances, and for a few substances also their ecological effects. No differentiation is made in terms of guidelines for outdoor or indoor pollution. The guidelines address concentration and exposure times regardless of whether the exposure is inside or outside buildings. However, typical occupational exposures are not the main focus of attention as these guidelines relate to the general population.

For those compounds that were not reported to induce carcinogenic effects and on which data regarding carcinogenic effects were lacking or insufficient, a threshold assumption was made and guideline values were proposed. For carcinogenic substances, the guidelines provide an estimate of lifetime cancer risk arising from exposure to those substances.

The guidelines are intended to provide background information and guidance to governments in making risk management decisions, particularly in setting standards. It should be strongly emphasized that the guideline values are not to be regarded as standards in themselves. It is obvious that there is a variety of ways to control air pollution and to protect our health and environment: it is advisable to use all the tools available. Air quality guidelines alone may not be very effective in the fight against pollution, but they should prove extremely useful in the framework of an overall environmental policy.

R. Türck
Project Coordinator
Air Quality Guidelines

Part I
General

1

Introduction

Human beings need a continuous supply of food, air and water to exist. The requirements for air and water are relatively constant (10–20 m^3 and 1–2 litres per day respectively). Air and water are also used in industrial processes of energy conversion, in manufacturing, and for the removal of waste products, some of which may be injurious to human health. In a comprehensive set of guidelines for drinking-water developed by WHO *(1)*, guideline values were recommended for specified contaminants, a consistent process of assessment being used.

The WHO Regional Office for Europe has subsequently developed the present air quality guidelines for the European Region. The task of developing such guidelines is more difficult than that of drawing up drinking-water guidelines, since for air, unlike water, there is no centrally supplied and controlled source. The development of consistent rules for assessing 28 chemical air contaminants also posed a challenge.

These air quality guidelines should be seen as a contribution to the target on air pollution contained in WHO's regional strategy for health for all. This target states that "by 1995, all people of the Region should be effectively protected against recognized health risks from air pollution" *(2)*. Accordingly, "the achievement of this target will require the introduction of effective legislative, administrative and technical measures for the surveillance and control of both outdoor and indoor air pollution, in order to comply with criteria to safeguard human health" *(2)*.

Various chemicals are emitted into the air from both natural and man-made (anthropogenic) sources. The quantities may range from hundreds to millions of tonnes annually. Natural air pollution stems from various biotic and abiotic sources (e.g. plants, radiological decomposition, forest fires, volcanoes and other geothermal sources, emissions from land and water), leading to a natural background concentration that varies according to local sources or specific weather conditions. Anthropogenic air pollution has existed at least since people learned to use fire, but it has increased rapidly since industrialization began. The increase in air pollution as a consequence of the expanding use of fossil energy sources and the growth in the manufacture and use of chemicals has been accompanied by mounting public awareness of and concern about its detrimental effects on health and the environment. Moreover, knowledge of the nature, quantity, physicochemical behaviour and effects of air pollutants has greatly increased in recent years. Nevertheless, more needs to be known. Certain aspects of the health effects of air pollutants require further assessment; these include newer scientific areas such as developmental toxicity. The proposed guideline values will undoubtedly be changed as future studies lead to new information.

The impact of air pollution is broad. In man, the pulmonary deposition and absorption of inhaled chemicals can have direct consequences for health. However, public health can also be indirectly affected by the deposition of air pollutants in plants, animals and the other environmental media, resulting in chemicals entering the food chain or being present in drinking-water and thereby constituting additional sources of human exposure. Furthermore, the direct effects of air pollutants on plants, animals and soil can influence the structure and function of ecosystems, including their self-regulation ability, thereby affecting the quality of life.

Although in recent decades major efforts have been made to reduce air pollution, the situation in the European Region is still not satisfactory. While air pollution has decreased and peak concentrations have been reduced in many larger cities and urban areas, the overall pollution in terms of the amounts of pollutants released into the atmosphere has often been only slightly reduced or has remained unchanged, and concentrations have even increased in some areas and for some pollutants *(2–5)*.

Many countries of the European Region encounter rather similar air pollution problems, partly because pollution sources are comparable, and in any case air pollution does not respect national frontiers. The subject of the transboundary medium- and long-range transport of air pollution has received increasing attention in Europe in recent years. International efforts to combat its consequences are under way, for instance within the framework of the Convention on Long-range Transboundary Air Pollution established by the United Nations Economic Commission for Europe *(6)*.

The task of reducing levels of exposure to air pollutants is a complex one. It begins with an analysis to determine which chemicals are present in the air, at what levels, and whether these levels of exposure are hazardous to human health and the environment. It must then be decided whether an unacceptable risk is present. When a problem is identified, mitigation strategies are developed and implemented so as to prevent excessive risk to public health in the most efficient way.

Analyses of air pollution problems are exceedingly complicated. Some are national in scope (e.g. definition of actual levels of exposure of the population, determination of acceptable risk, identification of the most efficient control strategies), while others are of a more basic character and are applicable in all countries (e.g. analysis of the relationships between chemical exposure levels, doses and their effects). The latter form the basis of the present guidelines.

The most direct and important source of air pollution affecting the health of many people is tobacco smoke. Even those who do not smoke may inhale the smoke produced by others ("passive smoking"). Indoor pollution in general and occupational exposure in particular also contribute substantially to overall human exposure: indoor concentrations of nitrogen dioxide, carbon monoxide, respirable particulates, formaldehyde and radon are often higher than outdoor concentrations *(7)*.

Outdoor air pollution can originate from a single point source which may affect only a relatively small area. More often, outdoor air pollution is caused by a mixture of pollutants from a variety of diffuse sources, such as traffic and heating, and from point sources. Finally, in addition to those emitted by local sources, pollutants transported over medium and long distances contribute further to the overall level of air pollution.

The relative contribution of emission sources to human exposure to air pollution may vary according to regional and lifestyle factors. Although indoor air pollution will be of higher relevance than outdoor pollution as far as certain air pollutants are concerned, this does not diminish the importance of outdoor pollution. In terms of the amounts of substances released, the latter is far more important and may have deleterious effects on animals, plants and materials as well as adverse effects on human health.

Nature of the Guidelines

The primary aim of the air quality guidelines[a] is to provide a basis for protecting public health from adverse effects of air pollution and for eliminating, or reducing to a minimum, those contaminants of air that are known or likely to be hazardous to human health and wellbeing.

The guidelines are intended to provide background information and guidance to governments in making risk management decisions, particularly in setting standards, but their use is not restricted to this. They also provide information for all who deal with air pollution. The guidelines may be used in planning processes and various kinds of management decision at community or regional level. When guideline values are indicated, this does not necessarily mean that they must take the form of general countrywide standards, monitored by a comprehensive network of control stations. In the case of some agents, guideline values may be of use mainly for carrying out local control measures around point sources.

It should be emphasized that when air quality guideline values are given, these values are not standards in themselves. Before standards are adopted, the guideline values must be considered in the context of prevailing exposure levels and environmental, social, economic and cultural conditions *(1)*. In certain circumstances there may be valid reason to pursue policies which will result in pollutant concentrations above or below the guideline values.

Ambient air pollutants can cause several significant effects which require attention: irritation, odour annoyance, acute and long-term toxic effects (including carcinogenic effects). Air quality guidelines either indicate levels combined with exposure times at which no adverse effect is expected concerning noncarcinogenic endpoints, or they provide an estimate of lifetime

[a] Guidelines in the present context are not restricted to suggested numerical values, but also include any kind of recommendation or guidance in the relevant field.

cancer risk arising from those substances which are proven human carcinogens or carcinogens with at least limited evidence of human carcinogenicity (see p. 12).

The guidelines represent the current best scientific judgement, but there is a need for periodic revision, since much remains to be determined regarding the toxicity of air pollutants for humans.

It is believed that inhalation of an air pollutant in concentrations and for exposure times below a guideline value will not have adverse effects on health and, in the case of odorous compounds, will not create a nuisance of indirect health significance (see definition of health, *Constitution of the World Health Organization*). Compliance with recommendations regarding guideline values does not guarantee the absolute exclusion of effects at levels below such values. For example, highly sensitive groups especially impaired by concurrent disease or other physiological limitations may be affected at or near concentrations referred to in the guideline values. Health effects at or below guideline values can also result from combined exposure to various chemicals or from exposure to the same chemical by multiple routes.

It is important to note that guidelines have been established for single chemicals. Chemicals, in mixture, can have additive, synergistic or antagonistic effects; however, knowledge of these interactions is still rudimentary. With a few exceptions, such as the combined effect of sulfur dioxide and particulates, there is insufficient information at present to establish guidelines for mixtures. An adequate margin of safety should exist between the guideline values and concentrations at which toxic effects will occur.

Risk estimates for carcinogens do not indicate a safe level; they are presented so that the carcinogenic potencies of different carcinogens can be compared and an assessment of overall risk made.

Although health effects were the major consideration in establishing the guidelines, ecologically based guidelines for preventing adverse effects on terrestrial vegetation were also considered and guideline values were recommended for a few substances. These ecological guidelines for vegetation have been established because, in the long term, only a healthy total environment can guarantee human health and wellbeing (see p. 17). Ecological effects on species other than plants have not been discussed, since they are outside the scope of this book.

The guidelines do not differentiate between indoor and outdoor exposure (with the exception of exposure to mercury) because, although the sites influence the type and concentration of chemicals, they do not directly affect the basic exposure–effect relationship. Occupational exposure has been considered in the evaluation process, but it was not a main focus of attention as these guidelines relate to the general population. However, it should be noted that occupational exposure may add to the effects of environmental exposure.

The guidelines do not apply to very high short-term concentrations which may result from accidents or natural disasters.

The health effects of tobacco smoking have not been assessed here, the carcinogenic effects of smoking having recently been evaluated by IARC *(8* and see Annex 1*)*. Neither have the effects of air pollutants on climate been considered, since too many uncertainties remain to allow an evaluation of the possible adverse health and environmental effects. However, possible changes of climate have to be investigated very seriously by the appropriate bodies because their overall consequences, for example the "greenhouse effect", may go beyond direct adverse effects on human health or ecosystems.

Procedures used in Establishing the Guidelines

The first step in the process of establishing air quality guidelines was the selection of pollutants. Air pollutants of special environmental and health significance to countries of the European Region were identified and selected on the basis of the following criteria suggested by a WHO Scientific Group *(9)*:

(*a*) severity and frequency of observed or suspected adverse effects on human health, where irreversible effects are of special concern;

(*b*) ubiquity and abundance of the agent in man's environment, with emphasis on air pollutants;

(*c*) environmental transformations or metabolic alterations, as these alterations may lead to the production of chemicals with greater toxic potential;

(*d*) persistence in the environment, particularly if the pollutant would resist environmental degradation and accumulate in humans, the environment or food chains; and

(*e*) population exposed (size of exposed population and special groups at risk).

Other factors affecting the selection were the timetable of the project and the fact that only those substances could be considered for which sufficient documentation was available (such as the WHO *Environmental health criteria* documents). On the basis of these criteria, the following 28 pollutants were selected for evaluation.

Organic air pollutants

Acrylonitrile
Benzene
Carbon disulfide
1,2-Dichloroethane
Dichloromethane
Formaldehyde
Polynuclear aromatic hydrocarbons (carcinogenic fraction)
Styrene
Tetrachloroethylene
Toluene
Trichloroethylene
Vinyl chloride

Inorganic air pollutants

Arsenic
Asbestos
Cadmium
Carbon monoxide
Chromium
Hydrogen sulfide
Lead
Manganese
Mercury
Nickel
Nitrogen oxides
Ozone/photochemical oxidants
Particulate matter
Radon
Sulfur oxides
Vanadium

After a planning meeting in early 1984 that offered suggestions on content, format, workplan and timetables for the air quality guidelines project, a series of nine meetings involving more than 130 experts took place to evaluate various air pollutants (see Annex 2).

Before the meeting of each working group, scientific background documents were prepared as a basis for discussion and for establishing the guidelines. After each meeting, a text on the individual pollutant or pollutant group was drafted on the basis of the amended background documents, incorporating the working group's conclusions and recommendations. The drafts were then circulated to all participants of the meetings for their comments and corrections. An editorial consultation group of scientists was then convened to review the documents for clarity of presentation, adequacy of description of the rationale supporting each guideline and consistency in the application of criteria. Certain sections in which inconsistencies were noted were again submitted for review, whereupon the final draft was prepared and submitted for extramural review; it was sent to the governments of Member States of the Region, and to organizations and individuals engaged in air quality research or management. The process concluded with a review in a final meeting, at which the recommendations and conclusions of all the working groups were submitted for final appraisal.

References

1. *Guidelines for drinking-water quality. Vol. 1. Recommendations.* Geneva, World Health Organization, 1984.
2. *Targets for health for all.* Copenhagen, WHO Regional Office for Europe, 1985.
3. **United Nations Economic Commission for Europe.** *Airborne sulphur pollution: effects and control.* New York, United Nations, 1984 (Air Pollution Series, No. 1).
4. **International Register of Potentially Toxic Chemicals.** *List of environmentally dangerous chemical substances and processes of global significance: scientific monographs.* Geneva, United Nations Environment Programme, 1984 (UNEP Report No. 2).
5. *Pilot compendium of environmental data.* Paris, Organisation for Economic Co-operation and Development, 1984.
6. **United Nations Economic Commission for Europe.** Convention on Long-range Transboundary Air Pollution. *International digest of health legislation,* **30**(4): 965 (1979).
7. *Estimating human exposure to air pollutants. GEMS: Global Environmental Monitoring System.* Geneva, World Health Organization, 1982 (WHO Offset Publication, No. 69).
8. *Tobacco smoking.* Lyon, International Agency for Research on Cancer, 1986 (IARC Monographs on the Evaluation of the Carcinogenic Risk of Chemicals to Humans, Vol. 38).
9. Background and purpose of the WHO environmental health criteria programme. *In: Mercury.* Geneva, World Health Organization, 1976 (Environmental Health Criteria, No. 1).

2

Criteria used in establishing guideline values

Relevant information on the pollutants was carefully considered during the process of establishing guideline values. Ideally, guideline values should represent concentrations of chemical compounds in air that would not pose any hazard to the human population. However, the realistic assessment of human health hazards necessitates a distinction between absolute safety and acceptable risk. To aim at achieving absolute safety, one would need a detailed knowledge of dose–response relationships in individuals in relation to all sources of exposure, the types of toxic effect elicited by specific pollutants or their mixtures, the existence or nonexistence of "thresholds" for specified toxic effects, the significance of interactions and the variation in sensitivity and exposure levels within the human population. However, such comprehensive and conclusive data on environmental contaminants are not always available. Very often the relevant data are scarce and the quantitative relationships uncertain; scientific judgement and consensus therefore play an important role in establishing acceptable levels of population exposure. Value judgements are unavoidable, because terms such as "adverse" and "sufficient evidence" are not in themselves totally objective, their meaning being based on generally agreed judgements.

Although it may be accepted that a certain risk can be tolerated or is simply unavoidable, the risk within a population may not be equally distributed. There may be subpopulations which are at considerably higher risk from the same exposure. Therefore, groups at special risk in the general population must be taken specifically into account in the risk management process. Even if knowledge about groups with specific sensitivity is available, unknown factors may exist that change the risk in an unpredictable manner.

Criteria Common to Carcinogens and Noncarcinogens

Sources, levels and routes of exposure

Available data are provided on the current levels of human exposure to pollutants from all sources, including the air. Special attention is given to atmospheric concentrations in urban and in nonpolluted rural areas and in the indoor environment. Where appropriate, concentrations in the workplace are also indicated for comparison with environmental levels. To provide information on the contribution from air in relation to all other sources, data on uptake by inhalation, ingestion from water and food, and dermal contact are given where relevant. However, for most chemicals, data on total human exposure are lacking to some extent.

Kinetics and metabolism

Available data on the toxicokinetics of distribution in humans and experimental animals are indicated for inter- and intraspecies extrapolation, especially to assess the magnitude of body-burden from long-term, low-level exposures and to characterize better the mode of toxic action. Data concerning the distribution of an agent in the body are important in determining the molecular or tissue dose to target organs. High-to-low-dose and interspecies extrapolations are more easily carried out using equivalent tissue doses. Metabolites are mentioned, particularly if they are known or believed to exert a greater toxic potential than the original agent. Additional data of interest include the rate of excretion and the biological half-life.

Criteria for Endpoints other than Carcinogenicity

For those compounds reportedly without carcinogenic effects (or for which data on carcinogenicity were lacking or insufficient), the starting-point for the derivation of guideline values was to define the lowest concentration at which effects are observed in humans, animals and plants. In doing so, an attempt was made to define a lowest-observed-adverse-effect level. The question whether the lowest-observed-effect level or the no-observed-effect level should be used instead is mainly a matter of availability of data. If a series of data fixes the lowest-observed-effect level and the no-observed-effect level, either of those levels might be used. The gap between the lowest-observed-effect level and the no-observed-effect level is among the factors included in judgements concerning the appropriate margin of protection. However, a single, free-standing no-observed-effect level which is not defined in reference to a lowest-observed-effect level or a lowest-observed-adverse-effect level is not conclusive. Opinions on this subject differ, but the working consensus was that the level of concern in terms of human health is more relatable to the lowest-observed-adverse-effect level; this level was therefore used whenever possible. In the case of irritant and sensory effects on humans, it is desirable where possible to determine the no-observed-effect level.

On the basis of the evidence concerning adverse effects, judgements about the protection factors (safety or uncertainty factors) needed to minimize health risks were made. Averaging times were included, since the time of exposure is critical in determining toxicity. Criteria applied to each of these key factors are described below.

Criteria for selection of a lowest-observed-adverse-effect level

The distinction between adverse and nonadverse effects poses considerable difficulties *(1)*. Any observable biological change may be considered an adverse effect under certain circumstances. The definition of an adverse effect has been given as "any effect resulting in functional impairment and/or pathological lesions that may affect the performance of the whole organism or which contributes to a reduced ability to respond to an additional challenge" *(2)*. Even with such a definition, a significant degree of subjectivity and uncertainty remains. Ambient levels of major air pollutants frequently cause subtle effects that are typically detected only by sensitive methods. This makes it exceedingly difficult, if not impossible, to achieve a broad consensus as to which effects are adverse. To resolve this difficulty, data should be ranked in three categories.

1. Observations, even of potential health concern, which are single findings that have not been verified by other groups. Because of the lack of verification by other investigators, such data could not readily be used as a basis for guideline values. They do, however, indicate the need for further research and may be considered in evaluating a margin of protection.

2. A lowest-observed-effect level: such a level is represented by data which have been supported by other scientific information. When the results are in a direction that might result in pathological change, there is a higher degree of health concern. Scientific judgement based on all available health information is used to determine how effects in this category can be used in determining the pollutant level that is to be avoided so that excessive risk can be prevented.

3. A substantial change in the direction of pathological effects: these findings have had a major influence on guideline considerations.

Criteria for selection of protection factors

In previous evaluations by WHO, protection factors, usually called safety factors, have been applied to derive guidelines from accepted criteria for adverse effects on health *(3,4)*. The rationale has been that such a factor allows for a variety of uncertainties, for example, about possibly undetected effects on particularly sensitive members of the population, synergistic effects of multiple exposures, and the adequacy of existing data. Traditionally, the safety factor has been used to allow for uncertainties in extrapolation from animals to humans and from a small group of individuals to a large population *(1)*.

In these guidelines, the terms "protection factor" and "margin of protection" have been used in preference to "safety factor" or "margin of safety" because the word "safety" may convey to the public the impression of absolute freedom from risk; this goes beyond what is intended by scientists when they refer to safety factors. These factors are applied in guidelines for the protection of human health. They are not applied to ecological guidelines, because they already include a kind of protection factor with regard to the variety of species covered.

A wide range of factors for protecting human health is used in this book, based on scientific judgements concerning the interplay of various criteria. The decision process for developing protection factors has been complex, involving the transformation of mainly nonquantitative information into a single number expressing the judgement of a group of scientists.

Some of the factors which are taken into account in deciding the margin of protection can be grouped under the heading of scientific uncertainty. Uncertainty occurs because of limitations in the extent or quality of the data base. One can confidently set a lower margin of protection (i.e. use of a smaller number), when a large number of high quality, mutually supportive scientific experiments in different laboratories using different approaches clearly demonstrate the dose–response, including a lowest-observed-effect level and a no-observed-effect level. In reality, difficulties inherent in studying air pollutants and the failure to perform much needed and very specific research usually preclude this situation.

Where a protection factor was adopted in the air quality guidelines, the reasoning behind this factor is given in the scientific background information. As previously mentioned, exceeding a guideline value with an incorporated protection factor does not necessarily mean that adverse effects will result. However, the risk to public health will increase, particularly in situations where the most sensitive population group is exposed to several pollutants simultaneously. It is therefore necessary to exercise some kind of judgement regarding the size of the protection factor.

Groups within a population respond differently to pollutants *(5)*. Individuals with pre-existing lung disease, for instance, can be at higher risk from exposure to air pollutants than healthy people. Differences in response can be due to factors other than pre-existing health factors, such as age, sex, level of exercise taken, or to unknown factors. Thus, the population must be considered very heterogeneous in respect of response to air pollutants. Existing information does not allow adequate assessment of the proportion of the population that has enhanced response. However, an estimate of even a few per cent of the total population entails a large number of people.

Effects observed in laboratory animals in the absence of human studies generally require a larger protection factor, because humans may be more susceptible than laboratory animal species. Negative data from human studies will tend to reduce the magnitude of the protection factor. Also of importance are the nature and reversibility of the reported effect. A pollutant level producing slight alterations in physiological parameters requires a smaller protection factor than a pollutant level producing a clearly adverse effect. Scientific judgement about protection factors will also take into account the toxicology of pollutants, including the type of metabolites formed, variability in metabolism or response in humans suggesting hypersusceptible groups, and the likelihood that the compound or its metabolites will accumulate in the body. It is also important to consider the exposure level used in health studies and to make appropriate conversions to environmental situations.

It is obvious, therefore, that diverse factors must be taken into account in proposing a margin of protection. The protection factor cannot be assigned by a simple mathematical formula; it requires experience, wisdom and judgement.

Criteria for selection of averaging times

The development of toxicity is a complex function of the interaction between concentration and time of exposure. A chemical may cause acute, minor, reversible effects after brief exposure and irreversible or incapacitating effects after prolonged exposure. Our knowledge is usually insufficient to delineate these concentration–time interrelationships. Therefore, expert judgement must be applied, based on the weight of the evidence available *(6)*. Generally, when short-term exposures lead to adverse effects, short-term averaging times are recommended. The use of a long-term average under such conditions would be misleading, since the typical pattern of repeated peak exposures is averaged over time and the risk manager has difficulty in deciding upon effective strategies. In other cases, exposure–response knowledge is sufficient to recommend a long-term average. This frequently occurs for chemicals that accumulate in the body over time, thereby resulting in adverse effects. In such cases, the integral of exposure can have more impact than the pattern of peak exposure.

It should be noted that these averaging times are based on effects. Therefore, if the guidelines are used as a basis for regulation, the regulator needs to select the most appropriate and practical standards in relation to the guidelines, without necessarily using the guidelines directly.

A similar situation occurs for effects on vegetation. Plants are generally damaged by short-term exposures to high concentration as well as by long-term exposures to low concentration. Therefore, both short- and long-term guidelines to protect plants are proposed.

Criteria for consideration of sensory effects

Some of the substances selected for evaluation have malodorous properties at concentrations far below those at which toxic effects occur. Although odour annoyance cannot be regarded as an adverse health effect in a strict sense, it affects the quality of life *(7)*. Therefore, odour threshold levels for such chemicals have been indicated where relevant and used as a basis for separate guideline values.

For practical purposes, the following aspects and respective levels were considered in the evaluation of sensory effects:

(*a*) intensity, where the *detection threshold level* is defined as the lower limit of the perceived intensity range (by convention the lowest concentration that can be detected in 50% of the cases in which it is present);

(*b*) quality, where the *recognition threshold level* is defined as the lowest concentration at which the sensory effect, e.g. odour, can be recognized correctly in 50% of the cases;

(*c*) acceptability and annoyance, where the *nuisance threshold level* is defined as the concentration at which not more than a small proportion of the population (less than 5%) experiences annoyance for a small part of the time (less than 2%); since annoyance will be influenced by a number of psychological and socioeconomic factors, a nuisance threshold level cannot be defined on the basis of concentration alone.

Criteria for Carcinogenic Endpoint

Cancer risk assessment is basically a two-step procedure, involving a qualitative assessment of how likely it is that an agent is a human carcinogen, and a quantitative assessment of the cancer rate the agent is likely to cause at given levels and durations of exposure *(8)*.

Qualitative assessment of carcinogenicity

The decision to consider a substance as a carcinogen is based on the qualitative evaluation of all available information on carcinogenicity, ensuring that the association is unlikely to be due to chance alone. Here the classification criteria of the International Agency for Research on Cancer have been applied *(9)*. These classify chemicals for carcinogenicity in the following way.

Group 1 — Proven human carcinogens. This category includes chemicals or groups of chemicals for which there is sufficient evidence from epidemiological studies to support a causal association between the exposure and cancer.

Group 2 — Probable human carcinogens. This category includes chemicals and groups of chemicals for which, at one extreme, the evidence of human carcinogenicity is almost sufficient, and those for which, at the other extreme, it is inadequate. To reflect this range, the category is divided into two subgroups according to higher (Group 2A) and lower (Group 2B) degrees of evidence.

Group 2A. This group is usually used for chemicals for which there is at least limited evidence of carcinogenicity in humans and sufficient evidence for carcinogenicity in animals.

Group 2B. This group is usually used for chemicals for which there is inadequate evidence of carcinogenicity in humans and sufficient evidence of carcinogenicity in animals. In some cases the known chemical properties of a compound and the results of short-term tests have allowed its transfer from Group 3 to Group 2B, or from Group 2B to Group 2A.

Group 3 — Unclassified chemicals. This group includes chemicals or groups of chemicals which cannot be classified as to their carcinogenicity in humans.

With regard to this classification scheme, it is likely that in the future some chemicals in Group 3 can be classified as noncarcinogenic.

It was concluded that the qualitative evaluation applied by IARC should serve as the baseline in establishing the air quality guidelines for carcinogens, and that the IARC categorization scheme should be used if no new divergent evidence was available. In this respect, it was decided that all chemicals categorized in Groups 1 and 2A, i.e. proven human carcinogens and carcinogens with at least limited evidence of human carcinogenicity, should be treated as human carcinogens, and guidelines should be formulated accordingly, indicating only risk estimates. For chemicals classified in Group 2B (inadequate evidence in humans, sufficient evidence in animals) it was decided that, until new evidence appeared, guidelines would point out the carcinogenic effects in laboratory animals and cite risk estimates (if such estimates could be reasonably obtained) in the health risk evaluation part of the scientific background information. However, these risk estimates, based on animal data only, would not be incorporated in the guideline recommendations because of various uncertainties in this connection. Guideline values based on noncarcinogenic endpoints would be given for these pollutants.

Quantitative assessment of carcinogenic potency

The aim of risk assessment is to apply information available from very specific study situations (mainly occupational studies) to the general population in order to calculate the possible risk to the latter. Therefore, quantitative risk assessment or, more specifically, dose–response assessment generally includes the extrapolation of risk from relatively high dose levels (characteristic of animal experiments or occupational exposures), where cancer responses can be measured, to relatively low dose levels, which are of concern in environmental protection and where such risks are too small to be measured directly, either in animal studies or in epidemiological studies *(10)*.

The choice of the extrapolation model depends on the current understanding of the mechanisms of carcinogenesis *(11)*. No single mathematical procedure can be regarded as fully appropriate for low dose extrapolation. Methods based on a linear, nonthreshold assumption have been used at the international level *(10,12)* and the national level (various health assessment documents produced by the US Environmental Protection Agency (EPA) and the National Institute of Public Health, Netherlands) more frequently than models which assume a safe or virtually safe threshold.

In these guidelines the risk associated with lifetime exposure to a certain concentration of a carcinogen in the air has generally been estimated by linear extrapolation and the carcinogenic potency expressed as the incremental unit risk estimate. The incremental unit risk estimate for an air pollutant is defined as "the additional lifetime cancer risk occurring in a hypothetical population in which all individuals are exposed continuously from birth throughout their lifetimes to a concentration of $1 \mu g/m^3$ of the agent in the air they breathe" *(13)*.

Calculations expressed in unit risk estimates provide the opportunity to compare the carcinogenic potency of different agents and can help to set priorities in pollution control according to the existing exposure situation. By using unit risk estimates, any reference to the "acceptability" of risk is avoided. The decision on the acceptability of a risk should be made by national authorities in the framework of risk management.

For those substances for which appropriate human studies are available, the method known as the "average relative risk model" *(14)* has been generally used and is therefore described in more detail below.

For animal cancer bioassays several methods have been used to estimate the incremental risks. Two general approaches have been proposed. A strictly linearized estimate has been used by US EPA generally *(11)*. Nonlinear relations have been proposed by others where either the concentration–tumour response was found experimentally or where metabolism is of limited capacity. Accordingly, risk estimates based on animal bioassays are considered separately.

Quantitative assessment of carcinogenicity based on human data

The quantitative assessment using the average relative risk model includes four steps: (*a*) selection of studies; (*b*) standardized description of study results in terms of relative risk, exposure level and duration of exposure; (*c*) extrapolation towards zero dose; and (*d*) application to a general (hypothetical) population.

First, a reliable human study must be identified, where the exposure of the study population can be estimated and the excess of cancer incidence is statistically significant. If several studies exist, the best representative study should be selected or several risk estimates evaluated.

When a study is identified, the relative risk (R) as a measure of response must be calculated. It is important to note that the 95% confidence limits around the central value for the relative risk can be wide and should be specifically stated and evaluated. The relative risk is then introduced in the following formula (average relative risk model) which combines steps (*c*) and (*d*) and allows the unit lifetime risk (UR) (i.e. risk associated with a lifetime exposure to 1 $\mu g/m^3$) to be calculated:

$$UR = \frac{P_o(R - 1)}{X}$$

where: P_o = background lifetime risk; this is taken from age/cause-specific death or incidence rates found in national vital statistics tables using the life table methodology, or it is available from a matched control population

R = relative risk, being the ratio between the observed (O) and expected (E) number of cancer cases in the exposed population; the relative risk is sometimes expressed as the standardized mortality ratio SMR = (O/E) × 100

X = lifetime average exposure (standardized lifetime exposure for the study population on a lifetime continuous exposure basis); in the case of occupational studies, X represents a conversion from the occupational 8-hour, 240-day exposure over a specific number of working years and can be calculated as X = 8-hour TWA × 8/24 × 240/365 × (average exposure duration [in years])/(life expectancy [70 years]), where TWA is the time-weighted average ($\mu g/m^3$).

It should be noted that the unit lifetime risk depends on P_0 (background lifetime risk), which is determined from national age-specific cancer incidence or mortality rates. Since these rates are also determined by exposures other than the one of interest and may vary from country to country, it follows that the UR may also vary from one country to another.

Necessary assumptions for average relative risk method

Before any attempt is made to assess the risk in the general population, numerous assumptions are needed at each phase of the risk assessment process to fill in various gaps in the underlying scientific data base. Therefore, as a first step in any given risk assessment, an attempt should be made to identify the major assumptions that have to be made, indicating their probable consequences. These assumptions are as follows.

1. *The response (measured as relative risk) is some function of cumulative dose or exposure.*

2. *There is no threshold dose for carcinogens.*

Many stages in the basic mechanism of carcinogenesis are not yet known or are only partly understood. However, taking available scientific findings into consideration, several scientific bodies *(3,10,12,15–17)* have concluded that there is no scientific basis for assuming a threshold or no-effect level for chemical carcinogens. This view is based on the fact that most agents that cause cancer also cause irreversible damage to deoxyribonucleic acid (DNA). The assumption applies for all nonthreshold models.

3. *The linear extrapolation of the dose–response curve towards zero gives an upper-bound conservative estimate of the true risk function if the unknown (true) dose–response curve has a sigmoidal shape.*

The scientific justification for the use of a linear nonthreshold extrapolation model stems from several sources: the similarity between carcinogenesis and mutagenesis as processes which both have DNA as target molecules; the strong evidence of the linearity of dose–response relationships for mutagenesis; the evidence for the linearity of the DNA binding of chemical carcinogens in the liver and skin; the evidence for the linearity in the dose–response relationship in the initiation stage of the mouse 2-stage tumorigenesis model; and the rough consistency with the linearity of the dose–response relationships for several epidemiological studies *(10)*. This assumption applies for all linear models.

4. *There is constancy of the relative risk in the specific study situation.*

In a strict sense, constancy of the relative risk means that the background age/cause-specific rate at any time is increased by a constant factor. The advantage of the average relative risk method is that this needs to be true only for the average.

Advantages of the method

The average relative risk method was selected in preference to many other more sophisticated extrapolation models because it has several advantages, the main one being that it seems to be appropriate for a fairly large class of different carcinogens, as well as for different human studies. This is possible because averaging doses, i.e. averaging done over concentration and duration of exposure, give a reasonable measure of exposure when dose rates are not constant in time. This may be illustrated by the fact that the use of more sophisticated models *(13, 14, 18, 19)* results in risk estimates very similar to those obtained by the average relative risk method.

Another advantage of the method is that the carcinogenic potency can be calculated when estimates of the average level and duration of exposure are the only known parameters besides the relative risk. Furthermore, the method has the advantage of being simple to apply, allowing non-experts in the field of risk models to calculate a lifetime risk from exposure to the carcinogens.

Limitations of the method

As pointed out earlier, the average relative risk method is based on several assumptions which appear to be valid in a wide variety of situations. However, there are specific situations in which the method cannot be recommended, mainly because the assumptions do not hold true.

The cumulative dose concept, for instance, is inappropriate when the mechanism of the carcinogen suggests that it cannot produce cancer throughout all stages of the cancer development process. Also, specific toxicokinetic properties, such as a higher excretion rate of a carcinogen at higher doses or a relatively lower production rate of carcinogenic metabolites at lower doses, may diminish the usefulness of the method in estimating cancer risk. Furthermore, supralinearity of the dose–response curve or irregular variations in the relative risk over time which cannot be eliminated would reduce the value of the model. However, evidence concerning these limitations either does not exist or is still too preliminary to make the average relative risk method inappropriate for carcinogens evaluated here.

A factor of uncertainty, rather than of methodological limitation, is that data on past exposure are nearly always incomplete *(12, 17)*. Although it is generally assumed that in the majority of studies the historical dose rate can be determined within an order of magnitude, there are possibly greater uncertainties, even of more than two orders of magnitude, in some studies. In the risk assessment process it is of crucial importance that this degree of uncertainty be clearly stated. This is often done simply by citing upper and lower limits of risk estimates *(12)*. Duration of exposure and the age- and time-dependence of cancer caused by a particular substance are less uncertain parameters, although the mechanisms of relationship are not so well understood *(8)*.

Risk estimates from animal cancer bioassays

Animal bioassays of chemicals provide important information on the human risk of cancer from exposure to chemicals. These data enhance our confidence in assessing human cancer risks on the basis of epidemiological data.

Several chemicals considered in this volume have been studied using animal cancer bioassays. The process is continuing and new information on the potential carcinogenicity of chemicals is rapidly appearing. Consequently, the status of chemicals is constantly being reassessed.

During the preparation of this book, dichloromethane was classified by IARC as showing sufficient evidence of carcinogenicity in animals, on the basis of studies in rats and mice *(20)*. Detailed studies of the kinetics of metabolism of dichloromethane have also recently been completed, indicating that the capacity of mammals to metabolize dichloromethane is limited. Thus, the extent of metabolism of dichloromethane at high doses where cancer bioassays are conducted is less than at levels of environmental exposure. The linearized cancer risk models may therefore represent considerable overestimates of the carcinogenic potential of dichloromethane in humans at levels likely to occur in the environment. As with dichloromethane, there are also considerable uncertainties in establishing human risk estimates derived from animal data for formaldehyde and 1,2-dichloroethane. Therefore, the significance of such estimates is still very problematic.

There is little doubt of the importance of animal bioassay data in reaching an informed decision on a chemical. The collection and use of data such as those on saturation mechanisms, absorption, deposition and metabolic pathways, as well as on interaction with other chemicals, is important and should be continued. Regrettably, these data were not available for the above-mentioned chemicals at the time of this evaluation of guidelines for air pollutants. The process of evaluating guidelines and the impact of human exposure to these chemicals should continue and be revised as new information becomes available.

Interpretation of risk estimates

The risk estimates presented in this book should *not* be regarded as being equivalent to the true cancer risk. Quantitative risk estimates can provide policy-makers with rough estimates of risk which may serve well as a basis for setting priorities, balancing risks and benefits, and establishing the degree of urgency of public health problems among subpopulations inadvertently exposed to carcinogens *(11)*.

Ecological Effects

The importance of taking an integrated view of both health and ecological effects in air quality management was recognized from the beginning of the project. Ecological effects may have a significant indirect influence on human health and wellbeing. For example, most of the major urban air pollutants are known to have adverse effects at low levels on plants, including food crops. A consultation group was therefore convened to consider ecological effects of sulfur oxides, nitrogen oxides and ozone/photochemical oxidants on terrestrial vegetation. These substances are important both because of the high anthropogenic amounts produced and because of their wide distribution. They deserve special attention because of significant adverse effects on ecological systems in concentrations far below those known to be harmful to humans.

The pollutants selected for consideration here form only part of the vast range of air pollutants that have ecological effects. The project timetable permitted only an evaluation of adverse effects on terrestrial plant life, although effects on animal and aquatic ecosystems are also of great concern in parts of Europe. Nevertheless, even this limited evaluation clearly indicates the importance attached to the ecological effects of such pollutants in the European Region.

References

1. *Principles and methods for evaluating the toxicity of chemicals. Part I.* Geneva, World Health Organization, 1978 (Environmental Health Criteria, No. 6).
2. **US Environmental Protection Agency.** Guidelines and methodology used in the preparation of health effect assessment chapters of the consent decree water quality criteria. *Federal register,* **45**: 79347-79357 (1980).
3. *Guidelines for drinking-water quality. Vol. 1. Recommendations.* Geneva, World Health Organization, 1984.
4. **Vettorazzi, G.** *Handbook of international food regulatory toxicology. Vol.1. Evaluations.* New York, SP Medical and Scientific Books, 1980.
5. *Guidelines on studies in environmental epidemiology.* Geneva, World Health Organization, 1983 (Environmental Health Criteria, No. 27).
6. WHO Technical Report Series, No. 506, 1972 (*Air quality criteria and guides for urban air pollutants:* report of a WHO Expert Committee).
7. *Sulfur oxides and suspended particulate matter.* Geneva, World Health Organization, 1979 (Environmental Health Criteria, No. 8).
8. **Peakall, D.B. et al.** Methods for quantitative estimation of risk from exposure to chemicals. *In:* Vouk, V.B. et al., ed. *Methods for estimating risk of chemical injury: human and non-human biota and ecosystems.* New York, John Wiley & Sons, 1985.
9. *Chemicals, industrial processes and industries associated with cancer in humans. IARC Monographs, Volumes 1 to 29.* Lyon, International Agency for Research on Cancer, 1982 (IARC Monographs on the Evaluation of the Carcinogenic Risk of Chemicals to Humans, Supplement 4).
10. *Arsenic.* Geneva, World Health Organization, 1981 (Environmental Health Criteria, No. 18).
11. **Anderson, E.L.** Quantitative approaches in use in the United States to assess cancer risk. *In:* Vouk, V.B. et al., ed. *Methods for estimating risk of chemical injury: human and non-human biota and ecosystems.* New York, John Wiley & Sons, 1985.
12. *Some industrial chemicals and dyestuffs.* Lyon, International Agency for Research on Cancer, 1982 (IARC Monographs on the Evaluation of the Carcinogenic Risk of Chemicals to Humans, Vol. 29).
13. *Health assessment document for nickel.* Research Triangle Park, NC, US Environmental Protection Agency, 1985 (Final report No. EPA-600/8-83-012F).

14. *Health assessment document for acrylonitrile.* Washington, DC, US Environmental Protection Agency, 1983 (Final report No. EPA-600-8-82-007F).
15. **National Research Council.** *Drinking water and health.* Washington, DC, National Academy of Sciences, 1977.
16. **Anderson, E.L. et al.** Quantitative approaches in use to assess cancer risk. *Risk analysis,* **3**: 277–295 (1983).
17. *Risk assessment and risk management of toxic substances. A report to the Secretary, Department of Health and Human Services.* Washington, DC, US Department of Health and Human Services, 1985.
18. *Health assessment document for chromium.* Research Triangle Park, NC, US Environmental Protection Agency, 1984 (Final report No. EPA-600-8-83-014F).
19. *Health assessment document for inorganic arsenic.* Research Triangle Park, NC, US Environmental Protection Agency, 1984 (Final report No. EPA-600-8-83-021F).
20. *Some halogenated hydrocarbons and pesticide exposures.* Lyon, International Agency for Research on Cancer, 1986 (IARC Monographs on the Evaluation of the Carcinogenic Risk of Chemicals to Humans, Vol. 41).

3

Summary of the guidelines

The term "guidelines" in the context of this book implies not only numerical values (guideline values), but also any kind of guidance given. Accordingly, for some substances the guidelines encompass recommendations of a more general nature that will help to reduce human exposure to harmful levels of air pollutants. For some pollutants no guideline values are recommended, but risk estimates are indicated instead. Table 1 summarizes the different endpoints on which guideline values and carcinogenic risk estimates have been based for organic and inorganic substances, showing that all relevant biological effects (endpoints) were evaluated and sometimes more than one endpoint was considered for guideline recommendations.

The numerical guideline values and the risk estimates for carcinogens (Tables 2–5) should be regarded as the shortest possible summary of a complex scientific evaluation process. Scientific results are an abstraction of real life situations, and this is even more true for numerical values and estimates based on such results. Numerical guideline values, therefore, are not to be regarded as separating the acceptable from the unacceptable, but rather as indications. They are proposed in order to help avoid major discrepancies in reaching the goal of effective protection against recognized hazards. Moreover, numerical guidelines for different substances are not directly comparable. Variations in the quality and extent of the scientific information and in the nature of critical effects result in guideline values which are only comparable between pollutants to a limited extent.

Owing to the different bases for evaluation, the numerical values for the various air pollutants should be considered in the context of the accompanying scientific documentation giving the derivation and scientific considerations. Any *isolated* interpretation of numerical data should therefore be avoided and guideline values should be used and interpreted in conjunction with the information contained in the appropriate sections.

It is important to note that guidelines are for individual chemicals. Pollutant mixtures can yield differing toxicities, but data are at present insufficient for guidelines relating to mixtures (except that of sulfur dioxide and suspended particulates) to be laid down.

Guideline Values based on Effects other than Cancer

The guideline values for individual substances based on effects other than cancer and odour are given in Table 2. Guideline values for combined exposure to sulfur dioxide and particulate matter are indicated in Table 3.

The emphasis in the guidelines is placed on exposure, since this is the element that can be controlled to lessen dose and hence lessen response. As stated earlier, the starting-point for the derivation of guideline values was to define the lowest concentration at which adverse effects are observed. On the basis of the body of scientific evidence and judgements of protection (safety) factors, the guideline values were established.

However, compliance with the guideline values does not guarantee the absolute exclusion of undesired effects at levels below the guideline values. It means only that guideline values have been established in the light of current knowledge and that protection factors based on the best scientific judgements have been incorporated, though some uncertainty cannot be avoided.

For some of the substances, a direct relationship between concentrations in air and possible toxic effects is very difficult to establish. This is especially true of those metals for which a greater body-burden results from ingestion than from inhalation. For instance, available data show that the food chain is, for most people, the critical route of nonoccupational exposure to lead and cadmium. On the other hand, airborne lead and cadmium may contribute significantly to the contamination of food by these metals. Complications of this kind were taken into consideration and an attempt was made to develop air quality guidelines which would also prevent those toxic effects of air pollutants that resulted from uptake through both ingestion and inhalation.

For certain compounds, such as organic solvents, the proposed health-related guidelines are orders of magnitude higher than current ambient levels. The fact that existing environmental levels for some substances are much lower than the guideline levels by no means implies that pollutant burdens may be increased up to the guideline values. Any level of air pollution is a matter of concern, and the existence of guideline values never means a licence to pollute.

The approach taken in the preparation of the air quality guidelines was to use expert panels to evaluate data on the health effects of individual compounds. As part of this approach, each chemical is considered in isolation. Inevitably, there is little emphasis on such factors as interaction between pollutants that might lead to additive or synergistic effects and on the environmental fate of pollutants (e.g. the role of solvents in atmospheric photochemical processes leading to the formation or degradation of ozone, the formation of acid rain and the propensity of metals and trace elements to accumulate in environmental niches). These factors militate strongly against allowing a rise in ambient pollutant levels. Many uncertainties still remain, particularly regarding the ecological effects of pollutants, and therefore efforts should be continued to maintain air quality at the best possible level.

Unfortunately, the situation with regard to actual environmental levels and proposed guideline values for some substances is just the opposite,

Table 1. Established guideline values and risk estimates

Substance	IARC Group classification	Risk estimate based on carcinogenic endpoint	Guideline value(s) based on:		
			toxicological endpoint	sensory effects or annoyance reaction	ecological effects
Organic substances					
Acrylonitrile	2A	×			
Benzene	1	×			
Carbon disulfide	—		×	×	
1,2-Dichloroethane	—[a]		×		
Dichloromethane	—[a]		×		
Formaldehyde	2B		×		
Polynuclear aromatic hydrocarbons (Benzo[*a*]pyrene)	—[b]	×			
Styrene	3		×	×	
Tetrachloroethylene	3		×	×	
Toluene	—		×	×	
Trichloroethylene	3		×		
Vinyl chloride	1	×			
Inorganic substances					
Arsenic	1	×			
Asbestos	1	×			

Inorganic substances (contd)					
Cadmium	2B		x		
Carbon monoxide	—		x		
Chromium (VI)	1	x			
Hydrogen sulfide	—		x	x	
Lead	3		x		
Manganese	—		x		
Mercury	—		x		
Nickel	2A[c]	x			
Nitrogen dioxide	—		x		x
Ozone/photochemical oxidants	—		x		x
Radon	—	x			
Sulfur dioxide and particulate matter	—		x		x
Vanadium	—		x		

[a] Not classified, but sufficient evidence of carcinogenicity in experimental animals.

[b] Not classified, but sufficient evidence of carcinogenicity of PAH in humans in some occupational exposures (*IARC Monographs on the Evaluation of the Carcinogenic Risk of Chemicals to Humans*, Vol. 34). Sufficient evidence of carcinogenicity for benzo[*a*]pyrene in animal studies. Benzo[*a*]pyrene is present as a component of the total content of polycyclic aromatic hydrocarbons in the environment (*IARC Monographs on the Evaluation of the Carcinogenic Risk of Chemicals to Humans*, Vol. 32).

[c] Exposures from nickel refineries are classified in Group 1.

Table 2. Guideline values for individual substances based on effects other than cancer or odour/annoyance[a]

Substance	Time-weighted average	Averaging time	Chapter
Cadmium	1– 5 ng/m³	1 year (rural areas)	19
	10–20 ng/m³	1 year (urban areas)	
Carbon disulfide	100 μg/m³	24 hours	7
Carbon monoxide	100 mg/m³[b]	15 minutes	20
	60 mg/m³[b]	30 minutes	
	30 mg/m³[b]	1 hour	
	10 mg/m³	8 hours	
1,2-Dichloroethane	0.7 mg/m³	24 hours	8
Dichloromethane (Methylene chloride)	3 mg/m³	24 hours	9
Formaldehyde	100 μg/m³	30 minutes	10
Hydrogen sulfide	150 μg/m³	24 hours	22
Lead	0.5–1.0 μg/m³	1 year	23
Manganese	1 μg/m³	1 year[c]	24
Mercury	1 μg/m³[d] (indoor air)	1 year	25
Nitrogen dioxide	400 μg/m³	1 hour	27
	150 μg/m³	24 hours	
Ozone	150–200 μg/m³	1 hour	28
	100–120 μg/m³	8 hours	
Styrene	800 μg/m³	24 hours	12
Sulfur dioxide	500 μg/m³	10 minutes	30
	350 μg/m³	1 hour	
Sulfuric acid	—[e]	—	30
Tetrachloroethylene	5 mg/m³	24 hours	13
Toluene	8 mg/m³	24 hours	14
Trichloroethylene	1 mg/m³	24 hours	15
Vanadium	1 μg/m³	24 hours	31

[a] Information from this table should *not* be used without reference to the rationale given in the chapters indicated.

[b] Exposure at these concentrations should be for no longer than the indicated times and should not be repeated within 8 hours.

[c] Due to respiratory irritancy, it would be desirable to have a short-term guideline, but the present data base does not permit such estimations.

[d] The guideline value is given only for indoor pollution; no guidance is given on outdoor concentrations (via deposition and entry into the food chain) that might be of indirect relevance.

[e] See Chapter 30.

Note. When air levels in the general environment are orders of magnitude lower than the guideline values, present exposures are unlikely to present a health concern. Guideline values in those cases are directed only to specific release episodes or specific indoor pollution problems.

Table 3. Guideline values for combined exposure to sulfur dioxide and particulate matter[a]

	Averaging time	Sulfur dioxide ($\mu g/m^3$)	Reflectance assessment: black smoke[b] ($\mu g/m^3$)	Gravimetic assessment: Total suspended particulates (TSP)[c] ($\mu g/m^3$)	Gravimetic assessment: Thoracic particles (TP)[d] ($\mu g/m^3$)
Short term	24 hours	125	125	120[e]	70[e]
Long term	1 year	50	50	—	—

[a] No direct comparisons can be made between values for particulate matter in the right- and left-hand sections of this table, since both the health indicators and the measurement methods differ. While numerically TSP/TP values are generally greater than those of black smoke, there is no consistent relationship between them, the ratio of one to the other varying widely from time to time and place to place, depending on the nature of the sources.

[b] Nominal $\mu g/m^3$ units, assessed by reflectance. Application of the black smoke value is recommended only in areas where coal smoke from domestic fires is the dominant component of the particulates. It does not necessarily apply where diesel smoke is an important contributor.

[c] TSP: measurement by high volume sampler, without any size selection.

[d] TP: equivalent values as for a sampler with ISO-TP characteristics (having 50% cut-off point at 10 μm); estimated from TSP values using site-specific TSP/ISO-TP ratios.

[e] Values to be regarded as tentative at this stage, being based on a single study (involving sulfur dioxide exposure also).

i.e. guideline values are below the existing levels in some parts of Europe. For instance, the guideline values recommended for major urban air pollutants such as nitrogen oxides, ozone and sulfur oxides point to the need for a significant reduction of emissions in some areas.

For substances with malodorous properties at concentrations below those where toxic effects occur, guideline values likely to protect the public from odour nuisance were established; these were based on data provided by expert panels and field studies (Table 4). In contrast to other air pollutants, odorous substances in ambient air often cannot be determined easily and systematically by analytical methods because the concentration is usually very low. Furthermore, odours in the ambient air frequently result from a complex mixture of substances and it is difficult to identify individual ones; future work may have to concentrate on odours as perceived by individuals rather than on separate odorous substances.

Table 4. Rationale and guideline values based on sensory effects or annoyance reactions, using an averaging time of 30 minutes

Substance	Detection threshold	Recognition threshold	Guideline value
Carbon disulfide in viscose emissions			20 μg/m³
Hydrogen sulfide	0.2–2.0 μg/m³	0.6–6.0 μg/m³	7 μg/m³
Styrene	70 μg/m³	210–280 μg/m³	70 μg/m³
Tetrachloroethylene	8 mg/m³	24–32 mg/m³	8 mg/m³
Toluene	1 mg/m³	10 mg/m³	1 mg/m³

Guidelines based on Carcinogenic Effects

In establishing criteria upon which guidelines could be based, it became apparent that carcinogens and noncarcinogens would require different approaches. These are determined by theories of carcinogenesis which postulate that there is no threshold for effects (i.e. that there is no safe level). Therefore, risk managers are faced with two decisions: either to prohibit a chemical or to regulate it at levels that result in an acceptable degree of risk. Indicative figures for risk and exposure assist the risk manager to reach the latter decision. Therefore, air quality guidelines are indicated in terms of incremental unit risks in respect of those carcinogens for which at least limited evidence of carcinogenicity in humans exists (Table 5).

Table 5. Carcinogenic risk estimates based on human studies[a]

Substance	IARC Group classification	Unit risk[b]	Site of tumour
Acrylonitrile	2A	2×10^{-5}	lung
Arsenic	1	4×10^{-3}	lung
Benzene	1	4×10^{-6}	blood (leukaemia)
Chromium (VI)	1	4×10^{-2}	lung
Nickel	2A	4×10^{-4}	lung
Polynuclear aromatic hydrocarbons (carcinogenic fraction)[c]		9×10^{-2}	lung
Vinyl chloride	1	1×10^{-6}	liver and other sites

[a] Calculated with average relative risk model.

[b] Cancer risk estimates for lifetime exposure to a concentration of 1 μg/m³.

[c] Expressed as benzo[*a*]pyrene (based on benzo[*a*]pyrene concentration of 1 μg/m³ in air as a component of benzene-soluble coke-oven emissions).

Separate consideration is given to risk estimates for asbestos (Table 6) and radon daughters (Table 7) because they refer to different physical units and are indicated in the form of ranges.

Unfortunately, the recent reclassification of dichloromethane by IARC has not allowed sufficient time to publish a detailed risk estimate which takes into account important information on the metabolism of the compound. The risk estimate for cancer from the animal bioassay is not used for this reason in the guidelines.

Table 6. Risk estimates for asbestos[a]

Concentration	Range of lifetime risk estimates	
500 F*/m³ (0.0005 F/ml)	$10^{-6} - 10^{-5}$	(lung cancer in a population where 30% are smokers)
	$10^{-5} - 10^{-4}$	(mesothelioma)

[a] See Chapter 18 for an explanation of these figures.

Note. F* = fibres measured by optical methods.

Table 7. Risk estimates and recommended action level[a] for radon daughters

Exposure	Lung cancer excess lifetime risk estimate	Recommended level for remedial action in buildings
1 Bq/m³ EER	$(0.7 \times 10^{-4})-(2.1 \times 10^{-4})$	≧ 100 Bq/m³ EER (annual average)

[a] See Chapter 29 for an explanation of these figures and for further information.

Formaldehyde represents a chemical for which cancer bioassays in rats have resulted in nonlinear exposure response curves. The nonlinearity of the tumour incidence with exposure concentrations led Starr & Buck[a] to introduce the "delivered dose" (amount of formaldehyde covalently bound to respiratory mucosal DNA) as the measure of exposure into several low-dose extrapolation models. Results showed considerable differences in the ratio between risk estimates based on the administered dose and those based on the delivered dose, with a great variance of ratios between models. Since estimates vary because of the inherent differences in approach, cancer risk estimates are referred to but not used for the guidelines. In addition, such estimates should be compared with human epidemiological data when an informed judgement has to be made.

The evidence for carcinogenicity of 1,2-dichloroethane in experimental animals is sufficient, being based on ingestion data. No positive inhalation bioassays are available. Consequently, an extrapolation from the ingestion route to the inhalation route is needed to provide a cancer risk estimate from the bioassay data. Such extrapolations are best conducted when detailed information is available on the kinetics of metabolism, distribution and excretion. Two estimates calculated from data on oral studies are provided for the risk of cancer through inhalation of 1,2-dichloroethane, but they lack detailed data for the route-to-route extrapolation and are not used in the guidelines.

It is important to note that quantitative risk estimates may give an impression of accuracy which in fact they do not have. An excess of cancer in a population is a biological effect and not a mathematical function, and uncertainties of risk estimation are caused not only by inadequate exposure data but also, for instance, by the fact that specific metabolic properties of agents are not reflected in the models. Therefore, the guidelines do not indicate that a specified lifetime risk is virtually safe or acceptable.

[a] **Starr, T.B. & Buck, R.D.** The importance of delivered dose in estimating low-dose cancer risk from inhalation exposure to formaldehyde. *Fundamental and applied toxicology*, **4**: 740–753 (1984).

Table 8. Guideline values for individual substances based on effects on terrestrial vegetation

Substance	Guideline value	Averaging time	Remarks
Nitrogen dioxide	95 μg/m³ 30 μg/m³	4 hours 1 year	In the presence of SO_2 and O_3 levels which are not higher than 30 μg/m³ (arithmetic annual average) and 60 μg/m³ (average during growing season), respectively
Total nitrogen deposition	3 g/m²	1 year	Sensitive ecosystems are endangered above this level
Sulfur dioxide	30 μg/m³ 100 μg/m³	1 year 24 hours	Insufficient protection in the case of extreme climatic and topographic conditions
Ozone	200 μg/m³ 65 μg/m³ 60 μg/m³	1 hour 24 hours averaged over growing season	
Peroxyacetylnitrate	300 μg/m³ 80 μg/m³	1 hour 8 hours	

The decision on the acceptability of a certain risk should be taken by the national authorities in the context of a broader risk management process. Risk estimate figures should not be applied in isolation when regulatory decisions are being made; combined with data on exposure levels and individuals exposed, they may be a useful contribution to risk assessment. Risk assessment can then be used together with technological, economic and other considerations in the risk management process.

Guidelines based on Ecological Effects on Vegetation

Although the main objective of the air quality guidelines is the direct protection of human health, it was decided that ecological effects of air pollutants on vegetation should also be considered. The effects of air pollutants on the natural environment are of special concern when they occur at concentrations lower than those that damage human health. In such cases, air quality guidelines based only on effects on human health would allow for environmental damage that might indirectly affect human wellbeing.

It should be understood that the pollutants selected (SO_x, NO_x and ozone/photochemical oxidants) (Table 8) are only a few of a larger category of air pollutants that may adversely affect the ecosystem. Furthermore, the effects which were considered are only part of the spectrum of ecological effects. Effects on aquatic ecosystems were not evaluated, nor were effects on animals taken into account. Nevertheless, the available information indicates the importance of these pollutants and of their effects on terrestrial vegetation in the context of the European Region.

4

Use of the guidelines in protecting public health

When strategies to protect public health are under consideration, the air quality guidelines need to be placed in the perspective of man's total chemical exposure. The interaction of man and the biosphere is complex. Individuals can be exposed briefly or throughout life to chemicals in air, water and food; exposures may be environmental and occupational. A further complication is the wide diversity of individuals with regard to dose–response relationships; each person has a pre-existing status (e.g. age, sex, pregnancy, pulmonary disease, cardiovascular disease, genetic differences) and a lifestyle in which such factors as exercise and nutrition play a key part. All these different elements may influence a person's susceptibility to chemicals. At present, the full effects of the chemical biosphere on an individual cannot be assessed. Various sensitivities also exist within the plant kingdom and need to be considered in protecting the environment. However, the problem can be approached more confidently once basic causes are understood.

To be fully effective, the regulatory approach to controlling air pollution will probably differ somewhat from country to country. Several sources of air pollutants have unique national components that are best subject to national control procedures. For example, indoor ventilation and heating systems, major determinants of indoor air pollution levels, are heavily influenced by the geographical location of the building.

The importance of international collaboration and the sharing of control technologies cannot be overemphasized, since many pollution sources are common to a number of countries. Such collaboration can be very cost-effective. Furthermore, when air pollutants are not transnational, differences related to societal opinions on degrees of acceptable risk are likely in controls.

In devising methods to mitigate human exposure, care must be taken to avoid creating new problems. For example, some types of emission control have the potential for generating more waste and sewage sludge or creating new chemicals in transformation processes that may prove to be more toxic than the original ones. If a new product is introduced, consideration must be given to its toxicity relative to that of the old product.

Any kind of air quality control that is undertaken must be based on knowledge about the hazard potential of air pollutants. Therefore, regardless of whether technological and emission control measures are performed in an air quality management system using air quality standards or whether other strategies are used, effect-related guidelines are a prerequisite for successful pollution control.

Part II

Organic substances

5

Acrylonitrile

General Description

Acrylonitrile ($CH_2=CH—C\equiv N$) is a volatile, flammable, colourless liquid with a characteristic odour. It is somewhat soluble in water and miscible with most organic solvents. Technical-grade acrylonitrile (AN) is more than 99% pure, with minor quantities of impurities and stabilizers *(1,2)*.

Sources
Acrylonitrile does not occur as a natural product. It is used in the production of acrylic and modacrylic fibres, resins and rubbers, and as a chemical intermediate. It may enter the environment during production, processing (manufacture), handling and storage, transportation and disposal of wastes. As well as emissions during manufacture, there are also losses to the atmosphere due to accidents and inadequate maintenance. Acrylonitrile emissions from plants in the USA in 1974 were estimated to be 14 100 tonnes, 2.2% of the total production. More recent estimates, following the introduction of stricter emission controls in the USA, indicate an overall reduction in emissions and a change in pattern: in 1981, 800 tonnes for AN production and 3000 tonnes for end-product manufacture *(1)*. Total emissions from 41 factories in 10 western European countries in 1981 were estimated to be 4970 tonnes, the proposed emission factor for Europe being 0.05% for AN production and 0.2–0.5% for AN processing *(2)*.

Occurrence in air
The dispersion of AN is closely related to wind patterns; the highest levels in the ambient air were found close to plants, especially downwind *(3)*, and they rapidly declined with distance. Dry deposition and wet deposition (by rainfall) are believed to play a negligible role *(2)*.

The degradation processes of AN in the air are primarily chemical; they result in the formation of hydrocyanic acid in 50% of the mass and (based upon measurement of reaction rates between AN and hydroxyl radicals at concentrations present in the air) are believed to be mainly responsible for the half-time (9–32 hours) of AN in ambient air *(1,2)*. In the soil, AN is degraded by microorganisms *(1)*. Its half-time in water was found to be 5–7 days *(1)*. However, an accidental spillage of 91 000 litres of AN from a tanker resulted in contamination of the soil and groundwater for more than a year, despite a cleaning process lasting 108 days after the accident *(1)*.

In the Netherlands, national emissions are estimated to be about 230 tonnes/year and, with a similar contribution from foreign sources, make up an average AN concentration in the ambient air of unpolluted areas of 0.01 $\mu g/m^3$. This exposure was estimated to concern 12.91 million of the country's 14 million population. Apart from these large-scale calculations, small-scale calculations relevant to the largest Dutch sources were also made *(2)*. Neither of the two types of calculation was confirmed by measurements. Of the total population of the Netherlands, an estimated 2100 are exposed to an annual average concentration of 3 $\mu g/m^3$, 42 100 to 1 $\mu g/m^3$ or more, and 1.1 million to 0.1 $\mu g/m^3$ or more.

In the vicinity of two AN-producing plants in Japan, AN concentrations in the range of 390–608 mg/m^3 were found near the exhausts of both ships and storage tanks *(1)*. In a study of AN concentrations in different media near 11 industrial sites in the United States, the highest air concentrations were found close to the plants and downwind, with levels decreasing from, for instance, 249.4 mg/m^3 at 0.5 km to 0.3 mg/m^3 at 1.3 km from the plant *(3)*. From a comparison of measured and calculated concentrations, estimates based on models have been validated *(1)*.

Workplace levels are mostly far higher than concentrations in the ambient air. Concentrations vary considerably in different countries, factories, and even types of occupation/operation. The ranges of concentration at some workplaces may be extremely wide but, with a short duration of peak levels, the average concentrations may be markedly less than the maxima. The highest occupational concentrations reported have been up to 600 mg/m^3. In Polish factories, concentrations of up to 20 mg/m^3 were measured, while maxima of 11 mg/m^3 were described for the USSR. Workplace concentrations in the fabrication of articles from polymers containing AN and in the handling of polyacrylic fibres were found to be very low and well below 2.2 mg/m^3 (1 ppm) *(1)*. The mean workplace concentration in the Dutch industries was reported to be less than 4.4 mg/m^3 in 90–95% of the cases *(2)*.

Numerous nonfatal and fatal cases of poisoning occurred in Florida in homes that had been fumigated with mixtures of AN and carbon tetrachloride or dichloromethane and considered safe for occupancy by subjective evaluation *(1)*. Similar fatalities were reported in the Federal Republic of Germany *(1)*. Indoor AN concentrations, with outdoor AN in ambient air being the source of AN contamination, may follow a pattern known with sulfur dioxide; due to its reactivity, AN may be present indoors at lower concentrations. AN in cigarettes might contribute to AN concentrations in the indoor air, but no measurements are available. Nevertheless, according to one study, cigarettes contain 1–2 mg AN per 100 cigarettes *(1)*. It is not known if AN results from the combustion process itself or from the fumigation of tobacco. Acrylonitrile has been used as a fumigant for stored tobacco *(4)*.

Conversion factors

1 ppm in air = 2.205 mg/m^3
1 mg/m^3 = 0.4535 ppm (20 °C, 101.3 kPa)

Routes of Exposure

Air

Using calculated concentration isopleths of yearly averages near the largest Dutch sources of AN, regional maps and population density data, together with an estimate of foreign contribution to ambient air concentrations of AN, the average daily intake for the Dutch population (14 million) was estimated to be 0.2 μg for 12.91 million, 2 μg for 0.9 million, 20 μg for 30 000 and 60 μg for 2100 persons. If respiratory ventilation at rest (6–7 litres/minute) and 46% pulmonary retention of AN *(1)* were considered, a 24-hour exposure to an AN concentration of 1 $\mu g/m^3$ would result in an average daily uptake of 4.0–4.6 μg *(2)*. The actual exposure at this concentration may be lower because of exposure to lower indoor air concentrations. Furthermore, the calculation of a relative emission per km^2 in 10 European countries indicated that, as far as domestic sources are concerned, 8 out of 10 countries should have ambient concentrations of AN lower or markedly lower than those in the Netherlands.

Occupational exposure

If one applies a pulmonary retention of 46% AN observed in three human volunteers exposed to 20 mg/m^3 *(1)* to people occupationally exposed to 10 mg/m^3 and assumes a ventilation of 20 litres/minute, there would be an intake of 44 mg AN during an 8-hour working shift. People occupationally exposed to 1 mg/m^3 would have an intake of 4.4 mg during a working shift.

People exposed at work may be expected to live near the factory, and such areas would have higher AN concentrations in ambient air than the national average, but such additional exposure would only be a fraction of the occupational exposure.

Smoking

Cigarette smoking can be a significant source of AN indoor air pollution. If only 10% of the 1–2 mg AN per 100 cigarettes *(1)* were absorbed, smoking 20 cigarettes per day would result in a daily intake of 20–40 μg. Therefore, cigarettes may be the most important source of nonoccupational AN exposure.

Drinking-water

No data are available on AN concentrations in drinking-water. Two studies in the USA determined AN concentrations in effluents from a chemical plant at 100 mg/litre and 3.5–4.3 mg/litre near an acrylic-modacrylic fibre plant *(1)*. AN was not detected in the soil or sediments near industrial sites *(3)*.

The industry is mostly situated in densely populated areas where tapwater is available. Rain washout of AN from ambient air is believed to play a minor role in removing AN from the atmosphere *(2)*. The accidental spillage of transported AN caused contamination of wells lasting more than a year, but the affected area was small and the probability of such accidents occurring is low (in the USA it was calculated that, with transport by rail, one accident would occur approximately every 6 years). Acrylonitrile intake from drinking-water thus appears doubtful.

Food

The use of AN copolymers for making beverage bottles was banned in the USA in 1977 *(1)* owing to the risk that remaining monomers might enter the contents. Such copolymers are still used in other products: analysis of AN in soft margarines, concentrated butter and shortening showed contamination in the range of 0.01–0.04 mg/kg *(1)*. The AN content of beverages in nitrile resin bottles was usually 2–3 μg/kg, but levels of up to 9 μg/kg were found *(1)*. Acrylonitrile concentrations of 0–19 mg/kg were found in dry food experimentally fumigated with AN at a concentration of 10 g/m^3 *(1)*; the level decreased by 30–70% over a period of 2 months. Shelled walnuts contained 0–8.5 mg/kg *(1)*. Some goods (e.g. walnuts) have been fumigated with mixtures containing AN, a practice now discontinued in some countries *(4)*.

The reported contamination with AN of soft margarine and butter (up to 40 μg/kg) by containers made of AN copolymers would, on the basis of a consumption of 100 g/day, result in a daily AN intake of 4 μg (probably less due to lower AN level and consumption). A government survey of the AN content in food suggested that the average daily intake of AN in the United Kingdom was likely to be less than 0.3 μg/person *(1)*. On the basis of the reported AN content of beverages in Sweden *(1)*, a 0.33-litre bottle would give an intake of 1–3 μg.

Other routes of exposure

Free AN monomers have been found in commercial AN polymers at levels of less than 1 mg/kg (acrylic and modacrylic fibres), 15 mg/kg (ABC resin) and 0–750 mg/kg (nitrile rubbers and latex materials) *(1)*. There is no record or estimate of human intake from these sources.

Relative significance of different routes of exposure

Uptake by nonoccupationally exposed persons living in the vicinity of plants, especially downwind, may be close to 20 μg AN per day, and 20–40 μg AN per day by people smoking AN-fumigated cigarettes. Consumption of average amounts of butter or soft margarine in AN copolymer containers could result in an AN intake of 1–4 μg/day, while the drinking of beverages distributed in such containers could represent an AN intake of 1–10 μg/day. As data on both these sources are as yet uncertain, due to the introduction of new regulations in some countries, it cannot be estimated conclusively which of these routes is the more important in a particular country. However, if cigarettes or tobacco are fumigated with AN, smoking may be an important source of nonoccupational exposure.

Kinetics and Metabolism

Absorption

The retention of AN in the respiratory tract of three volunteers exposed for up to 4 hours to a concentration of 20 mg/m^3 was 46 $\pm$ 1.6% and did not change throughout the inhalation period *(1)*. Dermal absorption of AN vapours in rabbits was estimated to be 100 times less efficient than respiratory absorption *(1)*. Absorption from the gastrointestinal tract of rats was complete but much slower, especially from the stomach, than from other routes *(5)*.

Distribution

Acrylonitrile is apparently uniformly distributed in the organism. It is rapidly eliminated from the blood of rats *(5)*, the elimination half-time being 15–19 minutes.

Biotransformation

Acrylonitrile is metabolized essentially along three pathways. First, an oxidative pathway, catalysed by microsomal monooxygenases, forms glycidonitrile, with subsequent reactions leading to several final products: thiocyanate via cyanide, cyanoacetic and acetic acid, and 2-cyanoethanol. Second, AN and glycidonitrile are metabolized to glucuronide(s). Third, AN and glycidonitrile react with glutathione (catalysed by GSH S-transferases) to form mercapturic acids, the final metabolites excreted in urine. A cyclic metabolite has also been reported *(1)*. Glycidonitrile is suspected to be the reactive intermediate responsible for the mutagenic properties of AN *(1)*, and the cyanide formed may significantly contribute to acute AN toxicity by inhibiting cytochrome oxidase *(1)*.

Elimination

Acrylonitrile mercapturic acids represent the most promising metabolites for biological monitoring. Methods for determining them have been elaborated *(1)*, and these metabolites have been identified in the urine of humans occupationally or experimentally exposed to AN. In animals, the chemobiokinetics and metabolism of AN were significantly modified by the simultaneous administration of organic solvents, and low doses of these chemicals also significantly increased the acute toxicity of AN *(5)*. Fatalities observed after home fumigation with AN mixed with carbon tetrachloride or dichloromethane may have been due to such metabolic interference *(1)*.

Health Effects

Effects on experimental animals and *in vitro* test systems

Toxicological effects

The range of LD_{50} values for AN given by various routes to different animal species is between 25 and 186 mg/kg *(1)*. The range of LC_{50} for 4-hour inhalation of AN is between 300 and 990 mg/m^3 in several species *(1)*. Acute toxicity studies showed effects on respiration, circulation, adrenals and brain, but only at high concentrations or with lethal doses. Irritation of mucous membranes and the skin was also reported *(1)*.

Ninety-day inhalation studies in dogs, mice and rats at 58 mg/m^3 showed no lethal effects. At 117 mg/m^3 some dogs died and at 234 mg/m^3 mice and rats died. Pulmonary inflammation was reported in rats exposed to 240 mg/m^3 for 6 months. Studies in which AN was administered in drinking-water at concentrations up to 300 mg/litre failed to show specific changes other than those mentioned above. Histological lesions were minimal or absent in the liver of rats that had drunk water with AN concentrations of 100–300 mg/litre for 12 months *(1)*. Daily exposure of rats to 22 mg/m^3 for 7 weeks resulted in enlargement of the liver, kidneys, heart and spleen. In a conditioned food reflex test, rats given AN intraperitoneally at 20 mg/kg per day for 6 weeks were affected *(1)*.

Histological liver changes were observed in cats, but not in rats, guinea pigs, rabbits, dogs and rhesus monkeys exposed for 8 weeks to 220–230 mg/m^3, while kidney and lung changes were found in all the species. The weights of the liver, kidneys, spleen, pituitary gland, lungs, gonads, thyroid gland, adrenals, heart and brain of rats, mice and dogs exposed to 234 mg/m^3 for 13 weeks were within normal limits, while in another study the exposure of rats to 22–100 mg/m^3 for 13 weeks caused enlargement of the liver, kidneys, heart and spleen *(1)*.

Animal experiments have shown that the mechanism(s) of acute AN toxicity may be related to the liberated cyanide and reaction with sulfhydryls and sulfhydryl-dependent processes; interactions with microsomal monooxygenases and induction of lipid peroxidation may also play a role in AN toxicity *(1)*. The relevance of these studies to human toxicity has yet to be established.

Exposure to low doses of toluene or styrene markedly increases the lethal effects of AN in animals and markedly influences the metabolism of AN.

Teratogenic effects were observed in rats exposed to 174 mg/m^3 for 6 hours per day or receiving 25 mg acrylonitrile per kg body weight orally during days 6–15 of gestation, whereas embryotoxic effects were absent at a dose of 10 mg/kg *(1)*.

Mutagenic and carcinogenic effects

Acrylonitrile was mutagenic in several strains of *Salmonella typhimurium (1,2)*. Positive results were observed with 0.5–1.5 mg AN per plate, microsomal metabolic activation was required and the effect was dose-related in the TA 1535 strain. Urine from AN-treated animals was mutagenic. Glycidonitrile (AN epoxide) did not require metabolic activation to exert mutagenic effects in *S. typhimurium (1)*. In *Escherichia coli* and *Saccharomyces cerevisiae* the mutagenic activity of AN did not require metabolic activation while, in the former microorganisms, 10 μg AN per plate was not mutagenic *(1)*.

Several *in vitro* tests have indicated effects of AN on DNA: induced strand breakage, cell transformation and a shift in the sedimentation pattern of DNA reminiscent of that observed in treatment with carcinogens were seen in Syrian golden hamster embryo cells. Sister chromatid exchange frequency was increased in Chinese hamster ovary cells and human lymphocytes with an activation system. Chromosome aberrations in hamster lung fibroblast cells were also observed *(1,6)*.

Three studies in rats and mice, using doses of up to 40 mg/kg per day for 16 days, up to 21 mg/kg per day for 30 days and up to 500 ppm (mg/litre) in drinking-water for 90 days, did not reveal chromatic or chromosomal aberrations or bone marrow abnormalities, nor chromosomal aberrations in somatic or germ cells; the dominant lethal assay in mice was also negative *(1,2)*. The only positive study on bone marrow chromosome aberrations gave no details on the exposure *(2)*. A comparison with vinyl chloride revealed that DNA alkylation occurs to a much lesser extent with AN *(1)*.

In carcinogenicity studies in rats, oral administration of AN corresponding to 10 mg/kg body weight for 20 months, or drinking-water containing 10, 30 or 100 mg/litre for 23–26 months, or water containing 0, 35, 100 or 300 mg/litre (mean dosage levels about 0, 4, 9 or 22 mg/kg body weight daily in both males and females) induced increased occurrence of tumours of the CNS, Zymbal gland (ear canal), nonglandular portion of the stomach, tongue, mammary glands and small intestine, while in the last study the incidence of tumours of the pituitary, thyroid, adrenals, pancreas and uterus was lower than in the controls *(1,7–9)*. In rats, inhalation of 11, 22, 44 or 88 mg/m^3 for 52 weeks (and observation for another year) resulted in an increased incidence of some tumours, e.g. mammary tumours, forestomach papillomas and acanthomas and encephalic gliomas *(10)*. In rats, inhalation of 0, 44 or 176 mg/m^3 of AN for 2 years *(11)* resulted in an increased incidence of tumours of the CNS, Zymbal gland, tongue, stomach, small intestine, mammary glands and nasal turbinates, and an apparent decrease in tumours of the pituitary gland, adrenals, thyroid, pancreas and testes. These results, obtained only with rats, have been considered sufficient evidence that AN is a carcinogen in rats *(1,2,12,13)*.

Effects on humans

Toxicological effects

Workmen exposed to AN concentrations varying from 35 to 220 mg/m^3 for 20–45 minutes complained of dull headaches, fullness of the chest, and irritation of the eyes, nose and throat; some reported "intolerable" itching of the skin and nervous irritability. In some cases higher concentrations also caused vertigo, nausea, vomiting, tremors, uncoordinated movements, convulsion, diarrhoea and jaundice. The symptoms were reversible. Some of the symptoms were reported on chronic exposure to concentrations ranging from 0.6 to 11 mg/m^3 *(1)*.

Brief skin contact with liquid AN caused dermatitis lasting up to 3 months (24 months according to other reports). Dermatitis caused by chronic exposure to AN has frequently been observed *(1)*. Contact with liquid AN is the most likely source of this dermatitis.

Haematological changes have been reported in employees exposed to AN at 2.5–5 mg/m^3, but not observed in other investigations in humans, nor in animals exposed to markedly higher AN concentrations for long periods of time *(1)*.

Workmen occupationally exposed to 35–220 mg/m³ for 20–45 minutes with unspecified frequency did not show signs of liver damage. Some abnormal liver function tests were reported at AN occupational exposures of up to 44 mg/m³, but were later absent after the concentrations fell below 9 mg/m³.

Considering the results of animal and human studies, there is little probability that exposures below 9 mg/m³ cause impairment of the liver or other parenchymal organs.

In one study, blepharoconjunctivitis was seen in all employees examined; some of them had severe alterations ascribed to AN, but the exposure concentration was not cited *(1)*.

Carcinogenic effects

Seven of the 12 available carcinogenicity studies present no evidence of carcinogenic risk from exposure to AN *(1,2)*; the other five studies give some indication that such exposure is associated with cancer *(14–18)*. All these studies, and particularly the seven that show no relationship between exposure to AN and cancer, have some deficiencies with regard to methodology, size of population and exposure to other chemicals, including carcinogens, and they reveal insufficient consideration of smoking habits.

In a retrospective cohort study, O'Berg *(14)* investigated 1345 male employees of a textile fibre plant in the USA up to the end of 1976. The employees were identified as having had potential exposure to AN between the commissioning of the plant in 1950 and 1966. Estimates for exposure were made retrospectively, assuming 44 mg/m³ (20 ppm) for high and 22 mg/m³ (10 ppm) for medium exposure *(4)*. Twenty-five cancer cases occurred versus 20.5 expected, based on company rates; however, 25.5 would have been expected according to the data of the National Cancer Institute. In particular, an excess of lung cancer (8 cases versus 4.4 expected) was found, but 7 of these 8 employees were known to have been smokers. Among the employees who had been exposed when the plant started production, 23 cancer cases were found versus 12.9 expected. A latency of 20 years was postulated. In an occupational study *(15)*, Thiess et al. found an excess of mortality from lung cancer, but exposure to other substances, some of them known carcinogens, makes interpretation difficult. Delzell & Monson *(16)* reported a higher mortality from lung cancer among 322 employees of a chemical rubber plant (9 versus 5.9 for white males and 4.7 for other rubber workers from the same city), with the greatest risk among men who had worked for 5–14 years and had started working at least 15 years before death. The lack of significant differences in "all cancer" and "lung cancer" mortality despite marked differences in the found versus expected rates is due to the low number of employees followed. Workers employed in the polymerization of AN and the spinning of acrylic fibres were followed by Werner & Carter *(17)*, who considered the excess of lung cancer in workers younger than 44 years to be particularly relevant; however, owing to several insufficiencies, their own interpretation was that the results were inconclusive.

There was a marked "healthy worker effect" in total mortality (found/expected deaths) in the studies of O'Berg (89/121), Thiess et al. (89/99), Werner & Carter (68/72.4) and Delzell & Monson (74/89.5) *(6)*.

No excess of lung cancer was found in 9525 Japanese workers exposed to AN *(18)*. There was no increase in the total number of deaths due to cancer, nor in the numbers of deaths from colon cancer, while 7 deaths due to liver, gall bladder or cystic duct cancer were found versus 5 expected.

Evaluation of Human Health Risks

Exposure

On the basis of large-scale calculations using dispersion models, the average annual ambient air concentration of AN in the Netherlands was estimated to be about $0.01 \mu g/m^3$ *(2)*, which is below the present detection limit of $0.3 \mu g/m^3$ *(2,3)*. Production figures *(2)* indicate that, in 8 out of 10 European countries for which data are available, ambient concentrations of AN are lower or markedly lower than this. Near industrial sites, AN air concentrations can exceed $100 \mu g/m^3$ over a 24-hour period, but are usually less than $10 \mu g/m^3$ at a distance of about 1 km. AN concentrations in the air at the workplace have exceeded $100 mg/m^3$, but shift averages are usually in the range of $1–10 mg/m^3$. Exposure from smoking is possible, if AN is used for tobacco fumigation, and could amount to $20–40 \mu g$ daily for an average smoker.

A more sensitive method of determination, with a detection limit below $0.1 \mu g/m^3$, is required in order to examine concentrations in the ambient air and to allow populations at possible risk to be identified.

Health risk evaluation

Acute and noncancer chronic toxicity may occur at concentrations still reported in some industries. Subjective complaints were reported in acute exposure to $35 mg/m^3$, and in chronic exposure to $11 mg/m^3$, $4.2–7.2 mg/m^3$ or $0.6–6 mg/m^3$. Teratogenic effects in animals were observed at $174 mg/m^3$ and carcinogenicity was shown in rats exposed for 2 years to $44 mg/m^3$.

Twelve epidemiological studies investigating the relationship between AN exposure and cancer are available; only five indicate a carcinogenic risk from exposure to AN *(2)*. Negative studies suffered from small cohort size, insufficient characterization of exposure, short follow-up times and relatively youthful cohorts. Although four of the remaining five epidemiological studies indicate a higher risk of lung cancer and one study showed a higher mortality rate for liver, gall bladder and cystic duct cancer, all have problems with regard to methodology, definition and/or size of the population, existence of exposure to other carcinogens, and duration of follow-up period.

In laboratory animals an increased incidence of tumours of the CNS, Zymbal gland, stomach, tongue, small intestine and mammary glands was observed at all doses tested *(12)*. However, there is a clear difference between animal and human studies concerning the tumorigenic response to AN: no lung tumours have been produced in animals and no brain tumours have been observed in humans.

Acrylonitrile was categorized in Group 2A by IARC *(12)* on the basis of the sufficient evidence of its carcinogenicity in experimental animals and the limited evidence of its carcinogenicity in humans.

The epidemiological study by O'Berg *(14)* presents the clearest available evidence of AN as a human lung carcinogen. Furthermore, in this study there were no confounding exposures to other carcinogenic chemicals during exposure to AN. It was therefore used to make an estimate of the incremental unit risk. As this study has now been updated to the end of 1983 for cancer incidence and to the end of 1981 for overall mortality, the most recent data are used here *(19)*. Out of 1345 workers exposed to AN, a total of 43 cases of cancer occurred versus 37.1 expected. Ten cases of lung cancer were observed versus 7.2 expected, based on the company rates. Lung cancer, which had been the focus of the previous report *(14)*, remained in excess, but not as high as before; two new cases occurred since 1976, with 2.8 expected. This means that the relative risk (R) would be 10/7.2 = 1.4, significantly lower than in the previous report. On the assumption made by the US Environmental Protection Agency *(4)* for the first O'Berg study *(14)* that the 8-hour time-weighted average exposure was 33 mg/m^3 (15 ppm), and with an estimated work duration of 9 years, the average lifetime daily exposure (X) is estimated to be 930 μg/m^3 [X = 33 mg/m^3 $\times$ 8/24 $\times$ 240/365 $\times$ 9/70].

Using the average relative risk model, the lifetime unit risk (UR) for exposure to 1 μg/m^3 can be calculated to be 1.7×10^{-5} [UR = P_0(R – 1)/X = 0.04(1.4 – 1)/930].

Using animal data, an upper-bound risk of cancer associated with a lifetime inhalation exposure to AN was calculated from a rat inhalation study *(11)* to be 1.5×10^{-5} *(4)*.

The calculated unit risk based on the human study is consistent with that of the animal study, although the human estimate is uncertain, particularly because of the lack of documentation on exposure.

Guidelines

Because AN is carcinogenic in animals and there is limited evidence of its carcinogenicity in humans, it is treated as if it were a human carcinogen. Therefore, no safe level for AN can be recommended. At an air concentration of 1 μg AN per m^3, the lifetime risk is estimated to be 2×10^{-5}.

References

1. *Acrylonitrile.* Geneva, World Health Organization, 1983 (Environmental Health Criteria, No. 28).
2. *Criteriadocument over acrylonitril* [Acrylonitrile criteria document]. The Hague, Ministerie van Volkshuisvesting, Ruimtelijke Ordening en Milieubeheer, 1984 (Publikatiereeks Lucht, No. 29).
3. **Going, J.E. et al.** *Environmental monitoring near industrial sites: acrylonitrile.* Washington, DC, US Environmental Protection Agency, 1979 (Report No. EPA-560/6-79-003).

4. *Health assessment document for acrylonitrile.* Washington, DC, US Environmental Protection Agency, 1983 (Final report No. EPA-600/8-82-007F).

5. **Gut, I. et al.** Relationship between acrylonitrile biotransformation, pharmacokinetics and acute toxicity. A short review. *Giornale italiano di medicina del lavoro,* **3**: 131–136 (1981).

6. **Koerselman, W. & van der Graaf, M.** Acrylonitrile: a suspected human carcinogen. *International archives of occupational and environmental health,* **54**: 317–324 (1984).

7. **Quast, J.F. et al.** *A two-year toxicity and oncogenicity study with acrylonitrile incorporated in the drinking water of rats. Final report.* Midland, MI, Dow Toxicology Research Laboratory, 1980.

8. **Bio/dynamics Inc.** *A twenty-four month oral toxicity/carcinogenicity study of acrylonitrile administered to Spartan rats in the drinking water.* St Louis, MO, Monsanto, 1980, Vol. 1–2 (Project No. 77-1745).

9. **Bio/dynamics Inc.** *A twenty-four month oral toxicity/carcinogenicity study of acrylonitrile administered in the drinking water to Fischer 344 rats.* St Louis, MO, Monsanto, 1980, Vol. 1–4 (Project No. 77-1744).

10. **Maltoni, C. et al.** Experimental contributions in identifying brain potential carcinogens in the petrochemical industry. *Annals of the New York Academy of Sciences,* **381**: 216–249 (1982).

11. **Quast, J.F. et al.** *A two-year toxicity and oncogenicity study with acrylonitrile following inhalation exposure of rats. Final report.* Midland, MI, Dow Toxicology Research Laboratory, 1980.

12. *Chemicals, industrial processes and industries associated with cancer in humans. IARC Monographs, Volumes 1 to 29.* Lyon, International Agency for Research on Cancer, 1982, pp. 25–27 (IARC Monographs on the Evaluation of the Carcinogenic Risk of Chemicals to Humans, Supplement 4).

13. **Ministry of Agriculture, Fisheries and Food.** *Survey of acrylonitrile and methacrylonitrile in food contact materials and in food.* London, H.M. Stationery Office, 1982 (Food Surveillance Paper No. 6).

14. **O'Berg, M.T.** Epidemiologic study of workers exposed to acrylonitrile. *Journal of occupational medicine,* **22**: 245–252 (1980).

15. **Thiess, A.M. et al.** Mortalitätsstudie bei Chemiefacharbeitern verschiedener Produktionsbetriebe mit Exposition auch gegenüber Acrylnitril [Mortality study of chemical workers in different production plants also exposed to acrylonitrile]. *Zentralblatt für Arbeitsmedizin Arbeitsschutz, Prophylaxe und Ergonomie,* **30**: 259–267 (1980).

16. **Delzell, E. & Monson, R.R.** Mortality among rubber workers. VI. Men with potential exposure to acrylonitrile. *Journal of occupational medicine,* **24**: 767–769 (1982).

17. **Werner, J.B. & Carter, J.T.** Mortality of United Kingdom acrylonitrile polymerisation workers. *British journal of industrial medicine,* **38**: 247–253 (1981).

18. **Nakamura K.** [Mortality study of acrylonitrile workers]. *In: Annual report of the National Institute of Industrial Health for 1980.* Tokyo, 1981, p. 31 (in Japanese).
19. **O'Berg, M.T. et al.** Epidemiologic study of workers exposed to acrylonitrile: an update. *Journal of occupational medicine,* **27**: 835–840 (1985).

6

Benzene

General Description

Benzene (C_6H_6) is a colourless, clear liquid with a density of 0.87 g/cm^3 (at 20 °C) and a boiling-point of 80.1 °C. It has a melting-point of 5.5 °C and a vapour pressure of 9.95 kPa at 20 °C.

Benzene is slightly soluble in water (1.8 g/litre at 25 °C) and miscible with alcohol, chloroform, diethyl ether, acetone, acetic acid and carbon tetrachloride *(1,2)*.

Chemically, benzene is fairly stable and undergoes substitution and addition reactions. Ring cleavage can also occur *(2)*.

Benzene is available commercially in three standardized grades containing varying concentrations of toluene, xylene and phenol, and traces of carbon disulfide, thiophene and naphthalene (and related compounds).

Sources

The annual worldwide production of benzene has been estimated at 14 million tonnes *(3)*. Production in western Europe in 1979 was estimated to be 4.8 million tonnes and the annual production capacity in 1980 at least 6.9 million tonnes, the major producers being the United Kingdom, the Federal Republic of Germany and the Netherlands. The yearly production of benzene in CMEA countries during 1977–1979 has been reckoned at 2 million tonnes *(2)*. In western Europe, benzene is produced in about equal quantities from catalytic reforming of naphtha, from pyrolysis of petrol and from toluene hydrodealkylation *(2)*.

Benzene is mainly used as a raw material for the production of substituted aromatic hydrocarbons *(4)*. It is a constituent of crude oil and in Europe is present in petrol in a proportion of around 5%, occasionally up to 16%. In the past, benzene was used as a solvent; this use has now been abandoned in most industrialized countries owing to the hazards involved. However, it is still used as a laboratory reagent and in sample collection, preparation and extraction.

The total global annual cycle of benzene, including benzene in fossil fuels, is estimated to be 32 million tonnes *(5)*. Of this amount, 4 million tonnes are thought to be lost to the environment *(5)*. The major source is emission from motor vehicles and evaporation losses during the handling, distribution and storage of petrol *(6)*. Benzene emitted to the atmosphere appears to have a half-time of less than one day *(3)*. Atmospheric washout and dilution of benzene by rain may occur, but re-evaporation will take place owing to the high vapour pressure of benzene. Absorption and biodegradation by vegetation have also been documented *(7)*. It is believed that plant and animal matter release benzene to the environment *(1)*. The burning of wood and organic material also results in an appreciable release of benzene.

Occurrence in air

Benzene concentrations in ambient air are generally between 3 and 160 μg/m^3 (0.001–0.05 ppm) *(1)*. Higher ambient benzene levels are found in metropolitan areas *(2)*. In the vicinity of petrol stations, petrol storage tanks and benzene-producing/handling industries, concentrations of up to several hundred μg/m^3 have been reported *(2,4)*. In a limited survey of levels inside houses, concentrations of benzene averaged 20 μg/m^3, with concentrations in outdoor air of 30 μg/m^3 (measured over a period of one to several days) *(4)*.

Conversion factors

1 ppm = 3.19 mg/m^3

1 mg/m^3 = 0.313 ppm

Routes of Exposure

Air

This is the primary route of entry of benzene into the body. About 50% of inhaled benzene in the air is absorbed. Benzene intake, based on a 24-hour respiratory volume of 20 m^3 at rest, will be 10 μg per day for each 1 μg/m^3 (0.0003 ppm) in the air. The daily adult intake at a typical ambient benzene level of 16 μg/m^3 (0.005 ppm) will therefore be about 160 μg. Together with other pollutants, benzene participates in the photochemical process that results in oxidant smog.

Tobacco and tobacco smoke

Cigarette smoke contains relatively high benzene concentrations (150–204 mg/m^3) and represents an important source of exposure for smokers *(8)*. Estimates of the uptake of benzene from smoking range from 10 to 30 μg per cigarette *(4)*, which would constitute an additional intake of up to 600 μg per day for persons smoking 20 cigarettes per day.

Household solvents and glues

Materials containing small amounts of benzene, such as glues, adhesives, solvents, and some household cleaning products, also contribute to benzene exposures, especially under conditions of inadequate handling or when there is no proper ventilation. Solvent abuse has recently been reported *(4)*.

Drinking-water

Benzene has been identified as a contaminant in drinking-water at levels of 0.1–0.3 μg/litre, the highest reported concentration being 20 μg/litre *(9)*.

Food

Benzene has been detected in several foods: eggs (500–1900 μg/kg, or 25–100 μg per egg), irradiated beef (19 μg/kg) and canned beef (2 μg/kg). It has also been detected in fish, cooked chicken, roasted nuts, and various fruits, vegetables, and dairy produce (no levels reported) *(2)*. Dietary intake of benzene may be as high as 250 μg/day *(1)*, and conventional cooking may produce an increase in the benzene content of food.

Relative significance of different routes of exposure

Table 1 summarizes the range of benzene intake in smoking and non-smoking adults. Nonsmokers living in rural areas have an estimated intake of about 0.3 mg benzene per day, whereas heavy smokers living in urban areas may receive as much as 5 times this amount. Occupational benzene exposure would contribute an intake additional to these values. Benzene levels of 26–64 μg/m³ (0.008–0.02 ppm) have been found in the breath of individuals without specific benzene exposure, suggesting that diet may be the source *(2)*. The intake values in Table 1 do not take into account the excretion (mainly exhalation) of absorbed benzene. The retention of absorbed benzene (50% of inhaled benzene is absorbed) has been measured at about 70% for both men and women following exposure to benzene concentrations of 160–200 mg/m³ *(2)*.

Table 1. Estimated daily intake of benzene from the major sources of exposure

Diet and drink (range in μg)		Respiratory sources (range in μg)		Approximate range of total intake from all sources (in μg)
Food[a]	Water[a]	Ambient air[b]	Smoking (packs/day[c])	
100–250	1–5	30–300 (residential)	(0) 0	130– 550
			(1) 600	700–1200
			(2) 1200	1300–1800

[a] Assuming complete absorption.

[b] Allowing for 50% absorption and assuming a daily respiratory volume of 20 m³.

[c] Figures in brackets indicate number of packs of 20 cigarettes (assuming an intake of 30 μg/cigarette).

Kinetics and Metabolism

Absorption

Benzene is efficiently absorbed by inhalation, the absorption level being about 50% *(2)*. The available information concerning absorption of benzene via the skin is limited, but indicates a low level of penetration in comparison with other solvents *(10)* at a rate of 0.4 mg/cm^2 per hour *(11)*.

Distribution

The high lipophilicity and low water solubility of benzene favour its distribution to fat-rich tissues. Given a knowledge of the fat content of different tissues, the partition coefficient between tissue and blood can be calculated *(12)*. Such calculations show that benzene distributed by the blood should accumulate mainly in fat-rich tissues like adipose tissue and bone marrow. This has been confirmed in animal experiments *(13)*. As fatty tissue typically has a low blood flow, a considerable time may elapse before equilibration occurs between benzene levels in blood and air. Experimental exposure of human volunteers to concentrations of benzene less than 32 mg/m^3 for 8 hours per day has shown that the body-burden of benzene doubles between day 1 and day 5, as revealed by the benzene concentration in exhaled breath *(14)*. Benzene thus accumulates in fat depots during long-term exposure. In animal experiments (mice), accumulation of benzene metabolites has been observed in bone marrow and liver, the highest concentrations being observed in the former *(13)*.

Biotransformation

Benzene is oxidized by the cytochrome P-450-dependent mixed-function oxidase system. In humans voluntarily exposed to 100 ppm (320 mg/m^3) benzene for 5 hours, 61% of absorbed benzene was metabolized to phenol, 6.4% to catechol and 2% to hydroquinone, while 26% was exhaled unmetabolized. The major part of the metabolites was excreted as sulfate or glucoronic acid conjugates. Sulfate conjugates predominated when the phenol concentration was below 400 mg/litre urine, but above that level the glucoronides constituted an increasing fraction of the excreted metabolites *(15)*.

More detailed studies, performed in laboratory animals, have shown, *inter alia*, that a metabolite of benzene is responsible for haematotoxicity *(16,17)*. The major biotransformation pathways are the formation of phenol via the epoxide and catechol via benzene dihydrodiol. Hydroquinone, catechol and hydroxyhydroquinone can be further converted to p-benzosemiquinone and p-benzoquinone, to o-benzoquinone, and to o-hydroxy-p-benzosemiquinone and hydroxy-p-benzoquinone, respectively. All these metabolites have been shown to bind covalently to microsomal proteins or mitochondrial DNA *(18–20)*. Metabolism also leads to the opening of the benzene ring, and this pathway may also play a role in haematotoxicity *(21)*.

Although the principle routes of metabolism seem to be similar in all mammalian species studied, considerable inter- and intra-species differences exist in the contribution of these pathways. Such intra-species differences have been seen in certain strains of mice *(22)*.

Elimination

Around 30% of absorbed benzene is exhaled unchanged in breath *(15,23)*. After exposure to 320 mg benzene per m^3 for 5 hours, exposed volunteers excreted 70% of the absorbed benzene in the urine as phenol, hydroquinone or catechol conjugates *(15)*. The elimination kinetics of benzene in exhaled breath reflect its release from the different depots of the body. This release is dependent on the partition coefficient between blood and organs and the degree of blood perfusion. The elimination curve can best be described as a two-compartment model, one compartment comprising a number of fast elimination phases with an average half-time of 2.5 hours, and a main compartment comprising a number of slow phases with an average half-time of 28 hours *(14,24)*.

Biological indices of exposure

At benzene concentrations above 32 mg/m^3 (10 ppm) in air, there is a correlation between phenol excretion in urine and the level of exposure *(25)*. At levels below 32 mg/m^3, benzene exposure is reflected in the amount exhaled in breath. Concentrations of benzene in breath reflect mainly the integrated benzene exposure *(14)*. The low equilibrium levels of benzene in blood preclude this biomonitor as an indicator of low-level benzene exposure (typical ambient air levels of 16 μg/m^3 result in a blood level of 0.1 μg/litre at equilibrium).

Health Effects

Effects on experimental animals and *in vitro* test systems

Toxicological effects

Haematotoxic effects like leucopenia, lymphopenia and anaemia have been produced in most laboratory animals (mice, rats, rabbits, dogs and guinea pigs) *(26)*. Fairly high levels of benzene exposure are needed to produce these effects, most laboratory animals requiring several weeks of daily exposure to 1600–3200 mg/m^3 in air. The mouse, however, seems to be an order of magnitude more sensitive to benzene exposure than rats and other studied species. Toft et al. *(27)* exposed NMRI mice to concentrations of 65 mg benzene per m^3 for 8 hours daily, 5 days per week for 2 weeks. This exposure produced increased micronuclei in bone marrow cells and depression of the *in vitro* colony-forming ability of stem cells from the tibial marrow. The effect was more pronounced in male than in female mice. At an exposure of 45 mg/m^3 a significant increase in micronuclei was still apparent, but these effects were not detected at an exposure of 32 mg/m^3. Baarson et al. *(28)* exposed C57BL/6 mice to 10 ppm (32 mg/m^3) benzene for 6 hours a day for up to 178 days. The progenitor cells showed a reduction in colony-formation ability. Spleen colony-forming unit erythroid cells (CFU-E) were depressed to within 10% of control values. Exposure to

32 mg benzene per m^3, 6 hours per day for 6 days, produced a significant depression of mitogen-induced blastogenesis in mouse lymphocytes *(29)*. Thus, in summary, several authors have demonstrated that long-term exposure of mice to benzene concentrations of 32–65 mg/m^3 results in an inhibition of early differentiating blood cell elements.

Carcinogenic effects

Long-term experimental carcinogenicity bioassays have shown that benzene is a carcinogen producing a variety of tumours (including lymphomas and leukaemias) in various organs of rats and mice. Several of these benzene-induced tumours rarely occur in untreated test animals. Carcinogenic effects were seen following exposure by ingestion and inhalation. Dose–response relationships have been observed *(30–32)*. No experimental data are available on the carcinogenic effects of doses lower than 32 mg/m^3.

Effects on humans

Toxicological effects

Exposure to high levels (more than 3200 mg/m^3) of benzene causes neurotoxic symptoms. In one lethal case, the exposure level was reported to be 63 800 mg/m^3 during an exposure period of 5–10 minutes *(33)*. At autopsy, both inflammation of the respiratory tract and haemorrhage of the lungs were seen following acute poisoning *(34)*.

Persistent exposure to toxic levels of benzene may cause injury to the human bone marrow, resulting in persistent pancytopenia. Early manifestations of toxicity are anaemia, leucocytopenia or thrombocytopenia. In severe cases, fatal aplastic anaemia develops, caused by the inhibition of bone marrow function.

A large number of cases of haematological conditions related to benzene exposure have been reported in the literature and reviewed by Goldstein *(35)*. A few epidemiological studies of benzene-exposed populations have also been performed, proving the etiological role of benzene in these conditions. Aksoy et al. *(36)* reported a study of 217 shoe-makers exposed to 95–650 mg benzene per m^3 for durations of 3 months to 17 years. Leucocytopenia, thrombocytopenia or pancytopenia were seen in 51 cases. Savilahti *(37)* investigated 147 shoe-factory workers exposed to benzene concentrations of around 1280 mg/m^3. Pathological blood values were seen in 107 cases, of which one had lethal pancytopenia. Thrombocytopenia was the most common finding. In a follow-up study *(38)*, 125 workers were re-examined 9 years after cessation of exposure. Blood cell counts revealed reduced numbers of erythrocytes and thrombocytes when compared to unexposed controls. This reduction was more pronounced in men than in women, but did not show any correlation to the severity of haematotoxicity reported in the initial study. Fishbeck et al. *(39)* reported macrocytosis in workers exposed to benzene at levels of about 80 mg/m^3 for several years, but no deviation in haemoglobin or haematocrit levels was observed. After cessation of exposure the erythrocyte size returned to normal.

Lymphocytopenia has also been reported after benzene exposure *(40–43)*. Smolik et al. reported that in a large number of workers exposed to around 16–144 mg benzene per m^3, serum complement levels and the immunoglobulins IgG and IgA were decreased and IgM was slightly increased *(42,43)*. Exposure to other solvents may have contributed to this effect.

Several studies have shown that exposure to benzene, at levels causing adverse haematotoxic effects, is associated with both stable and unstable chromosome aberrations in blood lymphocytes and bone marrow cells *(2,44–47)*. Increased chromosome aberrations of blood lymphocytes have also been reported in workers without haematotoxic effects *(47)*, and in workers exposed to less than 80 mg benzene per m^3 *(48,49)*, but other factors such as smoking and exposure to other solvents may have contributed to these effects *(50)*.

In a study that included control for smoking, Sarto et al. *(51)* found an increase in chromosomal aberrations of peripheral blood lymphocytes in workers exposed (mean time of exposure 11.4 years) to relatively low concentrations of benzene (between 0.6 and 40 mg/m^3).

No conclusive reports of fetotoxic and teratogenic effects of benzene have been found. One report refers to a woman with severe pancytopenia caused by benzene exposure in the eighth month of pregnancy. Although chromosome aberrations were seen in the blood lymphocytes of the mother, no cytogenetic abnormalities (including aberrations) were found in the child *(52)*.

Carcinogenic effects

Benzene is a known human carcinogen (IARC Group 1). A large number of cases of myeloblastic and erythroblastic leukaemia associated with benzene exposure have been reported in the literature *(2,35)*. Scattered cases of chronic myeloid and lymphoid leukaemia and of correlated malignant lymphohaemoproliferative diseases have also been reported in connection with known benzene exposure. Several epidemiological studies on benzene-exposed workers have revealed a statistically significant association between acute leukaemia and occupational exposure to benzene. In a six-year follow-up study of 44 cases of benzene pancytopenia, 6 (14%) had developed leukaemia *(53)*. The incidence of leukaemia in two other studies was 11 out of 66 (17%) and 13 out of 135 (10%) in cases of benzene haemopathy accumulated over 33 and 15 years respectively *(54)*. Aksoy *(55)* compared the incidence of different types of leukaemia in benzene-exposed cases and in nonexposed cases from a Turkish region. The ratio of acute nonlymphocytic leukaemia to chronic leukaemia was 27 : 2 in benzene-exposed cases and 13 : 23 in nonexposed cases, suggesting the relative risk of about 24 : 1 for acute nonlymphocytic leukaemia *(2)*. Vigliani *(54)* estimated the relative risk of acute leukaemia in workers heavily exposed to benzene to be at least 20 : 1 compared with the general population. Rinsky et al. *(56)* found a relative risk of 5.6 : 1 in workers exposed to benzene and other chemicals. Exposure to benzene was estimated at between 32 and 320 mg/m^3, and the average duration of exposure was 8.5 years. Ott et al. *(57)*, investigating a cohort of 594 benzene-exposed workers, found a relative risk of 3.75 : 1. Exposure levels were estimated to be in the range of 3.2–96 mg/m^3 (1–30 ppm).

Factors modifying the toxicity of benzene in man and animals

Substances that can induce benzene-metabolizing enzymes are likely to modify the haematotoxicity of benzene. It has been shown that benzene itself *(58)*, phenobarbital *(59)*, toluene *(13)* and ethanol *(60)* can modify the metabolism and haematotoxicity of benzene if animals are pretreated with these substances. Toluene *(13)* has been shown to inhibit the metabolism of benzene and decrease its haematotoxicity. Ethanol enhances the haematotoxicity of benzene in mice *(60)*.

Evaluation of Human Health Risks

Exposure

In residential areas ambient air benzene levels are generally in the range 3–30μg/m^3 and are dependent mainly on the traffic volume. The daily intake of benzene from air may therefore range from 30μg to 300μg. The daily dietary intake of benzene from food and water has been estimated to be in the order of 100–250μg. Persons smoking 20 cigarettes per day would have an increased intake of approximately 600μg. For typical daily intake values from main sources of exposure see Table 1.

Health risk evaluation

Several case reports and epidemiological studies of workers exposed to benzene clearly demonstrated excess incidences of leukaemia *(2)*. However, the possibility cannot be excluded that in some studies other chemicals have contributed to the excess risk of leukaemia reported. Carcinogenicity has also been proven in mice and rats *(2)*, suggesting a multisystem carcinogenic effect, not just leukaemia.

In attempts to estimate the carcinogenic risk of benzene, it may be asked whether a linear model is the best choice, as benzene toxicity is due to a metabolite of unknown metabolic profile and, in addition, the efficiency of DNA repair systems at low exposure levels may be increased. However, for reasons of consistency, and as the mechanism of benzene-induced leukaemia is not understood well enough to suggest another more appropriate model, the average relative risk model was used for estimation of the incremental unit risk.

Two extensively investigated study populations of workers exposed to benzene were selected for quantitative risk assessment.

The NIOSH cohort described in several papers *(56,61–63)*, comprised 748 workers who had been exposed to benzene in the production of rubber film material for 10 years and were followed up for a period of 25 years. Seven workers had died from leukaemia, compared with an expected figure of 1.25 based on the national mortality rate of 0.007 from leukaemia among adult males in the USA, giving a relative risk (R) of 5.6. The average duration of exposure was estimated to be 8.5 years *(2)*, and the average exposure levels during the critical study period were assumed to have been in the range

30–300 mg/m^3 *(56)*. Short-term area samples measured over the years indicated that some benzene levels were above 300 mg/m^3 *(63)*. Also, peak values up to 2170 mg/m^3 were recorded *(62)*.

Assuming exposure was at the upper end of the concentration range and using therefore the upper estimate of 300 mg/m^3 (about 100 ppm), the average lifetime daily exposure (X) can be calculated to be 8 mg/m^3 [$X = 300\ mg/m^3 \times 8/24 \times 240/365 \times 8.5/70$]. The unit risk (UR) associated with a lifetime exposure to 1 μg/m^3 can then be calculated to be 4×10^{-6} [$UR = P_o(R - 1)/X = 0.007(5.6 - 1)/8000$].

The second epidemiological study population comprised workers of the Dow Chemical Company. The first description of this population and an analysis of mortality and other data was made by Ott et al. in 1978 *(57)*. An update of the mortality data to the end of 1982 was made by Bond et al. *(64)*. The latter also expanded the study population from the original 594 to 965 workers. Average exposure levels reported by Ott et al. *(57)* were in the range 3–100 mg/m^3, corresponding to areas of very low up to high exposure. In the latter area peak exposure levels occurred, with a maximum of 3000 mg/m^3.

Infante et al. *(63)* estimated that the average exposure for the entire cohort was 16 mg/m^3. The average exposure duration was estimated to be 8–9 years *(2)*. Taking the higher estimate of exposure (100 mg/m^3), the average lifetime daily exposure can be calculated to be 2.66 mg/m^3 [$X = 100\ mg/m^3 \times 8/24 \times 240/365 \times 8.5/70$].

The data as updated by Bond et al. showed a nonsignificant excess of total deaths from leukaemia based on four observed cases. However, all four cases were myelogenous leukaemias and for this subcategory this was a clear excess *(64)*. The four cases of myelogenous leukaemia observed gave a relative risk of 4.4 when compared with the 0.9 cases expected. Based on the lifetime mortality rate of 0.003 (P_o) for myelogenous leukaemia, the incremental unit risk calculated using the upper exposure estimate is 3.8×10^{-6} [$UR = 0.003(4.4 - 1)/2660$].

The obtained estimates of 4×10^{-6} and 3.8×10^{-6} are very similar, and the estimate of 4×10^{-6} may be considered as a risk estimate of lifetime exposure to 1 μg benzene per m^3.

IARC conducted a quantitative cancer risk assessment for benzene *(2)*, assuming a linear relationship between the cumulative benzene dose and the relative risk of developing leukaemia. The excess risk of a working lifetime exposure (45 years) for the NIOSH cohort was estimated to be in the range of (140×10^{-3})–(170×10^{-3}). Assuming an average exposure of 300 mg/m^3 for the workers, extrapolation to a unit lifetime risk of 1 μg/m^3 (8/24 hours) would result in a value between 3×10^{-6} and 4×10^{-6}.

The US Environmental Protection Agency's Carcinogen Assessment Group (CAG) reported a unit risk of 7.5×10^{-6} obtained from the geometric mean of risk estimates from three independent epidemiological studies *(65)*. More recently, the CAG *(66)* calculated the unit risk for leukaemia due to a lifetime exposure to 1 ppm benzene using six models and combinations of assumptions and epidemiological studies. On the basis of data from all the epidemiological studies, 21 unit risks were estimated for exposures to 1 ppm, ranging from 0.9×10^{-2} to 1×10^{-1}, corresponding to the incremental unit risk from exposure to 1 μg/m³ of 2.8×10^{-6} to 3×10^{-5} (conversion factor: 3.134×10^{-4}). Using an average value over all the mathematical models, a combined unit risk estimate of 8.1×10^{-6} $(\mu g/m^3)^{-1}$ was derived as the CAG's "best-judgement" unit risk.

Guidelines

No safe level for airborne benzene can be recommended, as benzene is carcinogenic to humans and there is no known safe threshold level.

At an air concentration of 1 μg benzene per m³, the estimated lifetime risk of leukaemia is 4×10^{-6}.

References

1. **Brief, R.S. et al.** Benzene in the workplace. *American Industrial Hygiene Association journal,* **41**: 616–623 (1980).
2. Benzene and Annex. *In: Some industrial chemicals and dyestuffs.* Lyon, International Agency for Research on Cancer, 1982 (IARC Monographs on the Evaluation of the Carcinogenic Risk of Chemicals to Humans, Vol. 29).
3. **Korte, F. & Klein, W.** Degradation of benzene in the environment. *Ecotoxicology and environmental safety,* **6**: 311–327 (1982).
4. **Fishbein, L.** An overview of environmental and toxicological aspects of aromatic hydrocarbons. I. Benzene. *Science of the total environment,* **40**: 189–218 (1984).
5. **Merian, E. & Zander, M.** Volatile aromatics. *In:* Hutzinger, O., ed. *Handbook of environmental chemistry. Vol. 3, Part B, Anthropogenic compounds.* Berlin, Springer-Verlag, 1982, pp. 117–161.
6. **Howard, P.H. & Durkin, P.R.** *Sources of contamination, ambient levels and fate of benzene in the environment.* Washington, DC, US Environmental Protection Agency, 1974 (Report No. EPA-560/5-75-005).
7. **Cheremisinoff, P.N. & Morresi, A.C.** *Benzene — basic and hazardous properties.* Basle, Marcel Dekker, 1979.
8. **Lauwerys, R.** *Industrial health and safety. Human biological monitoring of industrial chemicals. 1. Benzene.* Luxembourg, Commission of the European Communities, 1979 (CEC Report EUR 6570/11979).
9. Benzene. *In: 2nd Annual report on carcinogens.* Research Triangle Park, NC, National Toxicology Program, 1981, pp. 49–52 (Report NTP81-43).
10. **Conca, G.L. & Maltagliati, A.** Studio sull'assorbimento transcutaneo del benzolo [Transcutaneous absorption of benzene]. *Medicina del lavoro,* **46**: 194–198 (1955).
11. **Hanke, J. et al.** Wchlanianie benzenu przez skòre u ludzi [The absorption of benzene through the skin in men]. *Medycyna pracy,* **12**: 413–426 (1961).
12. *Benzene.* Copenhagen, Scandinavian Expert Group for Threshold Limit Value Documentation, 1979 (in Danish).

13. **Andrews, L.S. et al.** Effects of toluene on the metabolism, disposition and haemopoietic toxicity of ^{3}H-benzene. *Biochemical pharmacology,* **26**: 293–300 (1977).
14. **Berlin, M. et al.** Breath concentration as an index of the health risk from benzene: studies on the accumulation and clearance of inhaled benzene. *Scandinavian journal of work, environment & health,* **6**: 104–111 (1980).
15. **Teisinger, J. et al.** Metabolismus benzeno u cloveca [The metabolism of benzene in man]. *Pracovni lekarstvi,* **4**: 175–188 (1952).
16. **Snyder, R. et al.** Biochemical toxicology of benzene. *In:* Hodgson, E. et al., ed. *Reviews in biochemical toxicology.* New York, Elsevier-North Holland, 1981, Vol. 3, pp. 123–153.
17. **Tunek, A.** *Metabolism and toxicity of benzene.* Thesis, Institute of Environmental Health, University of Lund, 1981.
18. **Irons, R.D. & Pfeifer, R.W.** Benzene metabolites: evidence for an epigenetic mechanism of toxicity. *In:* Tice, R. et al., ed. *Genotoxic effects of airborne agents.* New York, Plenum, 1981, pp. 241–256.
19. **Kalf, G.F. et al.** Inhibition of RNA synthesis by benzene metabolites and their covalent binding to DNA in rabbit bone marrow mitochondria *in vitro. American journal of industrial medicine,* **7**: 485–492 (1985).
20. **Gill, D.P. & Ahmed, A.E.** Covalent binding of [^{14}C] benzene to cellular organelles and bone marrow nucleic acids. *Biochemical pharmacology,* **30**: 1127–1131 (1981).
21. **Witz, G. et al.** Short-term toxicity of *trans,trans*-muconaldehyde. *Toxicology and applied pharmacology,* **80**: 511–516 (1985).
22. **Longacre, S.L. et al.** Influence of strain differences in mice on the metabolism and toxicity of benzene. *Toxicology and applied pharmacology,* **60**: 398–409 (1981).
23. **Srbova, J. et al.** Absorption and elimination of inhaled benzene in man. *Archives of industrial hygiene and occupational medicine,* **2**: 1–8 (1950).
24. **Hunter, C.G. & Blair, D.** Benzene: pharmacokinetic studies in man. *Annals of occupational hygiene,* **15**: 193–199 (1972).
25. **Lauwerys, R.** Benzene. *In:* Alessio, L. et al., ed. *Human biological monitoring of industrial chemicals series.* Luxembourg, Commission of the European Communities, Joint Research Centre, Ispra Establishment, 1983, pp. 2–22.
26. **Leong, B.K.J.** Experimental benzene intoxication. *Journal of toxicology and environmental health,* Suppl. 2: 45–62 (1977).
27. **Toft, K. et al.** Toxic effects on mouse bone marrow caused by inhalation of benzene. *Archives of toxicology,* **51**: 295 (1982).
28. **Baarson, K.A. et al.** Repeated exposure of C57BL mice to 10 ppm benzene markedly depressed erythropoietic colony formation. *Toxicology letters,* **20**: 337–342 (1984).
29. **Rosen, M.G. et al.** Depressions in B and T lymphocyte mitogen-induced blastogenesis in mice exposed to low concentrations of benzene. *Toxicology letters,* **20**: 343–349 (1984).

30. **Maltoni, C. et al.** Benzene: a multipotential carcinogen. Results of long-term bioassays performed at the Bologna Institute of Oncology. *American journal of industrial medicine,* **4**: 589–630 (1983).

31. **Maltoni, C. et al.** Experimental studies on benzene carcinogenicity at the Bologna Institute of Oncology: current results and ongoing research. *American journal of industrial medicine,* **7**: 415–466 (1985).

32. *Toxicology and carcinogenesis studies of benzene in F344/N rats and B6C3F1 mice (gavage studies).* Research Triancle Park, NC, National Toxicology Program, 1984 (Technical Report NTP-83-072).

33. **Flury, F.** Moderne gewerbliche Vergiftungen in pharmakologisch-toxikologischer Hinsicht [Modern industrial poisoning in respect to pharmacology and toxicology]. *Naunyn-Schmiedebergs Archiv für Pharmakologie und experimentelle Pathologie,* **138**: 65–82 (1928).

34. **Winek, C.L. & Collom, W.D.** Benzene and toluene fatalities. *Journal of occupational medicine,* **13**: 259–261 (1971).

35. **Goldstein, B.D.** Hematotoxicity in humans. *Journal of toxicology and environmental health,* Suppl. 2: 69–105 (1977).

36. **Aksoy, M. et al.** Haematological effects of chronic benzene poisoning in 217 workers. *British journal of industrial medicine,* **28**: 296–302 (1971).

37. **Savilahti, M.** Mehr als 100 Vergiftungsfälle durch Benzol in einer Schuhfabrik [More than 100 cases of benzene intoxication in a shoe factory]. *Archiv für Gewerbepathologie und Gewerbehygiene,* **15**: 147–157 (1956).

38. **Hernberg, S. et al.** Prognostic aspects of benzene poisoning. *British journal of industrial medicine,* **23**: 204–209 (1966).

39. **Fishbeck, W.A. et al.** Effects of chronic occupational exposure to measured concentrations of benzene. *Journal of occupational medicine,* **20**: 539–542 (1978).

40. **Greenburg, L. et al.** Benzene (benzol) poisoning in the rotogravure printing industry in New York City. *Journal of industrial hygiene,* **21**: 395–420 (1939).

41. **Goldwater, L.J.** Disturbances in the blood following exposure to benzol. *Journal of laboratory and clinical medicine,* **26**: 957–973 (1941).

42. **Lange, A. et al.** Serum immunoglobulin levels in workers exposed to benzene, toluene and xylene. *Internationales Archiv für Arbeitsmedizin,* **31**: 37–44 (1973).

43. **Smolik, R. et al.** Serum complement level in workers exposed to benzene, toluene and xylene. *Internationales Archiv für Arbeitsmedizin,* **31**: 243–247 (1973).

44. **Pollini, G. & Colombi, R.** Il danno cromosomico midollare nell'anemia aplastica benzolica [Damage to bone marrow chromosomes in benzolic aplastic anaemia]. *Medicina del lavoro,* **55**: 241–255 (1964).

45. **Pollini, G. & Colombi, R.** Il danno cromosomico dei linfociti nell'emopatia benzenica [Chromosomal damage in lymphocytes during benzene haemopathy]. *Medicina del lavoro,* **55**: 641–654 (1964).

46. **Pollini, G. et al.** Le alterazione cromosomiche dei linfociti rilevate dopo cinque anni insoggetti gia'affetti da emopatia benzolica [Chromosome changes in lymphocytes five years after benzene haemopathy]. *Medicina del lavoro,* **60**: 743–758 (1969).
47. **Forni, A.M. et al.** Chromosome changes and their evolution in subjects with past exposure to benzene. *Archives of environmental health,* **23**: 385–391 (1971).
48. **Hartwich, G. & Schwanitz, G.** Chromosomenuntersuchungen nach chronischer Benzol-Exposition [Chromosome studies after chronic exposure to benzol]. *Deutsche medizinische Wochenschrift,* **97**: 45–49 (1972).
49. **Fredga, K. et al.** Chromosome studies in workers exposed to benzene. *In: Genetic damage in man caused by environmental agents.* New York, Academic Press, 1979, pp. 187–203.
50. **Fredga, K. et al.** Chromosome changes in workers (smokers and non-smokers) exposed to automobile fuels and exhaust gases. *Scandinavian journal of work, environment & health,* **8**: 209–221 (1982).
51. **Sarto, F. et al.** A cytogenetic study on workers exposed to low concentrations of benzene. *Carcinogenesis,* **5**: 827–832 (1984).
52. **Forni, A.M. et al.** Chromosome studies in workers exposed to benzene or toluene or both. *Archives of environmental health,* **22**: 373–378 (1971).
53. **Aksoy, M.** Benzene and leukaemia. *Lancet,* **1**: 441 (1978).
54. **Vigliani, E.C.** Leukemia associated with benzene exposure. *Annals of the New York Academy of Sciences,* **271**: 143–151 (1976).
55. **Aksoy, M.** Leukemia in workers due to occupational exposure to benzene. *New Istanbul contribution to clinical science,* **12**: 3–14 (1977).
56. **Rinsky, R.A. et al.** Leukemia in benzene workers. *American journal of industrial medicine,* **2**: 217–245 (1981).
57. **Ott, M.G. et al.** Mortality among individuals occupationally exposed to benzene. *Archives of environmental health,* **33**: 3–10 (1978).
58. **Tunek, A. & Oesch, F.** Unique behaviour of benzene mono-oxygenase: activation by detergent and different properties of benzene- and phenobarbital-induced mono-oxygenase activities. *Biochemical pharmacology,* **28**: 3425–3429 (1979).
59. **Gill, D.P. et al.** Modifications of benzene myelotoxicity and metabolism by phenobarbital, SKF-525A and 3-methylcholanthrene. *Life sciences,* **25**: 1633–1640 (1979).
60. **Baarson, K.A. et al.** The hematotoxic effects of inhaled benzene on peripheral blood, bone marrow, and spleen cells are increased by ingested ethanol. *Toxicology and applied pharmacology,* **64**: 393–404 (1982).
61. **Infante, P.F. & White, M.C.** Projections of leukemia risk associated with occupational exposure to benzene. *American journal of industrial medicine,* **7**: 403–413 (1985).
62. **White, M.C. et al.** A quantitative estimate of leukemia mortality associated with occupational exposure to benzene. *Risk analysis,* **2**: 195–204 (1982).

63. **Infante, P.F. et al.** Assessment of leukemia mortality associated with occupational exposure to benzene. *Risk analysis,* **4**: 9–13 (1984).
64. **Bond, G.G. et al.** An update of mortality among chemical workers exposed to benzene. *British journal of industrial medicine,* **43**: 685–691 (1986).
65. **Carcinogen Assessment Group.** *Final report on population risk to ambient benzene exposures.* Springfield, VA, US Environmental Protection Agency, 1979 (PB 82-227372).
66. **Carcinogen Assessment Group.** *Interim quantitative cancer unit risk estimates due to inhalation of benzene.* Washington, DC, US Environmental Protection Agency, 1985 (Internal Report No. EPA-600/X-85-022).

7

Carbon disulfide

General Description

Carbon disulfide (CS_2) in its pure form is a colourless, volatile and inflammable liquid with a sweet aromatic odour. The technical product is a yellowish liquid with a disagreeable odour.

Sources

Carbon disulfide is used in large quantities as an industrial chemical for the production of viscose rayon fibres. In this technological process, for every kilogram of viscose produced, about 20–30 g of carbon disulfide and 4–6 g of hydrogen sulfide are emitted *(1)*. Additional release of carbon disulfide, carbonyl sulfide and hydrogen sulfide takes place from coal gasification plants; data on the total emission from these plants are not available.

The ventilation discharge from viscose plants can reach several millions of m^3 per hour, with a carbon disulfide content varying from 20 to 240 mg/m^3, which represents a total emission of 15–40 tonnes of carbon disulfide daily *(2)*.

Exposure to carbon disulfide is mostly confined to those engaged in technological processes in the viscose industry. However, the general population living near viscose plants may also be exposed to carbon disulfide emissions.

Occurrence in air

The primary source of carbon disulfide in the environment is emission from viscose plants, around which environmental pollution is especially great. A scientific review of Soviet literature indicates values ranging from 0.01 to 0.21 mg/m^3 around viscose plants *(2)*.

A recent Austrian study reports that concentrations of 0.05 ppm (157 $\mu g/m^3$) were often exceeded in the vicinity of viscose plants, even at a distance of several kilometres, and concentrations close to the plants could be 5–10 times higher. The highest peak concentrations were between 3 and 6 mg/m^3 *(3)*.

During soil treatment with a 50% carbon disulfide emulsion for fumigation, carbon disulfide concentration in the respiration zone was found to be as high as 0.03 mg/m^3 on the first day. This concentration decreases quickly, so that carbon disulfide is not detectable the next day *(2)*.

Carbon disulfide present in air could be partially decomposed by light. Oxidation leads to the formation of carbonyl sulfide, sulfur dioxide and carbon monoxide *(4)*. Carbonyl sulfide in particular causes an unpleasant odour.

Workplace concentrations of carbon disulfide have been found to range from less than 9 mg/m^3 to peaks exceeding 6200 mg/m^3 *(1)*. As a result of various precautions taken over a period of time, average carbon disulfide concentrations have been reduced from about 250 mg/m^3 in 1955–1965 to about 20–30 mg/m^3 *(5)*.

Conversion factors

1 ppm = 3.13 mg/m^3
1 mg/m^3 = 0.32 ppm

Routes of Exposure

The main source of environmental pollution (indoor and outdoor) by carbon disulfide is emission into the air from viscose plants. There are no data for pollution on a regional or global scale, nor is information available on dispersion and transformation.

Air
Inhalation is the main route of carbon disulfide absorption in both occupational and environmental exposure.

Drinking-water
Carbon disulfide can reach waterways via the wastewaters of viscose rayon plants. Drinking-water normally does not contain carbon disulfide.

Food
Carbon disulfide can contaminate juice and wine distilled from grapes harvested in vineyards treated with carbon disulfide *(2)*.

Other routes of exposure
Dermal absorption of carbon disulfide can represent an additional route of entry in occupational exposure. In environmental exposure it does not constitute a hazard.

Kinetics and Metabolism

Absorption
The majority of studies showed that in man an equilibrium between the carbon disulfide concentrations in inhaled and exhaled air is achieved during the first 60 minutes of exposure. In the state of equilibrium, retention is about 40–50%, depending on the amount of carbon disulfide in inhaled air and on the coefficient of its partition between blood and tissues. Higher

retention was registered in volunteers exposed to carbon disulfide for the first time, in comparison with continuously exposed workers *(4)*.

Distribution

In humans 10–30% of carbon disulfide absorbed by the body is exhaled and a further 70–90% undergoes biotransformation. The metabolites produced, together with less than 1% of unchanged carbon disulfide, are excreted in the urine *(4)*.

After absorption, carbon disulfide is transported by the blood, being distributed between blood erythrocytes and plasma in the ratio 2 : 1. It disappears relatively quickly from the blood and is distributed to various tissues and organs. The solubility of carbon disulfide in lipids and fats and its binding to amino acids and proteins govern its distribution in the body.

Biotransformation

The metabolism of carbon disulfide is basically performed by two main pathways: reaction with amino acids and protein (glutathione) and via the microsomal cytochrome P-450 monooxygenase system *(4,5)*.

Amino acids of blood plasma react with carbon disulfide, forming dithiocarbamic acid and a cyclic compound of the thiazolinone type. The "free" carbon disulfide present in the blood is distributed to various tissues and organs, where it reacts with endogenous amines to form acid-labile metabolites (dithiocarbamates and 2-thio-5-thiazolidinone) excreted in the urine. Dithiocarbamate metabolites are potential chelating agents. A part of the carbon disulfide is bound to glutathione (GSH), forming thiazolidine-2-thione-4-carboxylic acid and then 2-oxythiazolidine-4-carboxylic acid, which is excreted in the urine.

The other metabolic pathway involves the microsomal cytochrome P-450 monooxygenase system catalyzing the oxidative desulfuration of carbon disulfide to an unstable oxygen intermediate. The intermediate spontaneously degrades to atomic sulfur and carbonyl sulfide or undergoes attack by water, forming atomic sulfur and monothiocarbamate. Carbonyl sulfide is oxidized to atomic sulfur and carbon dioxide *(4,5)*.

The atomic sulfur liberated in these reactions can be covalently bound to the macromolecules or be oxidized to sulfate and excreted in urine. Formed carbonyl sulfide can be catalyzed by carbonic anhydrase to monothiocarbamate, which is spontaneously degraded to carbon dioxide and HS^-. The HS^- is oxidized to sulfate or other still unknown metabolites. Monothiocarbamate can enter the urea cycle, forming thiourea, which is excreted in urine *(4,5)*.

Elimination

The elimination of carbon disulfide via exhaled air takes place in a typical three-phase process: rapidly from the respiratory tract, more slowly from the blood and very slowly from tissues and organs *(4)*.

A small quantity (about 1%) of unchanged carbon disulfide is excreted via the urine, saliva and faeces. The excretion of the "free" carbon disulfide through the skin is about three times greater than that via the urine. The majority of metabolized carbon disulfide is excreted in the form of "ethereal sulfates" and the other part in the form of the metabolites mentioned earlier *(4,5)*.

Health Effects

Effects on experimental animals and *in vitro* test systems

Toxicological effects

An extensive literature is available on the toxic effects of carbon disulfide on experimental animals; these are reviewed in some publications *(1–5)*.

In acute and subacute exposure of experimental animals (dogs, cats, mice and rats) to carbon disulfide, effects on the CNS are the first signs to be noted. In experimental animals carbon disulfide produces destruction of the myelin sheath and axonal changes in both central and peripheral neurons. Degenerative changes have been observed in the cortex, basal ganglia, thalamus, brain stem and spinal cord *(4,5)*.

Neuropathy and myelopathy were extensively studied in rats and rabbits. In the muscle fibres atrophy of the denervation type occurred secondary to the polyneuropathy. The slowing down of nerve conduction velocity in the sciatic nerves preceded clinical symptoms *(5)*.

Some studies have shown that carbon disulfide causes vascular changes in various organs of animals as well as myocardial lesions *(4)*. Enhancing effects of carbon disulfide on atherosclerosis (elevated serum cholesterol, phospholipids and triglycerides) were observed in experimental animals and confirmed by biochemical studies *(5)*.

Although many studies on the hepatic effects of carbon disulfide in animals pretreated with phenobarbital have been performed, limited information is available on animals exposed to carbon disulfide only. Subchronic inhalation studies on rabbits indicated a hepatotoxic effect *(5)*.

Several studies on animals have demonstrated renal effects of carbon disulfide, but have failed to correlate them with the renal morphology: reduced urinary output, increased urinary protein and amino acids *(5)*.

Damage to the endocrine structures and functional alterations (increased thyroid activity, depressed adrenal functions) were described in animals exposed to carbon disulfide. These changes may play a part in atherogenic and reproductive effects of carbon disulfide *(5)*.

Histological studies on the ophthalmological effects of carbon disulfide in animals have demonstrated the following: changes in retinal ganglia, vascular degeneration, and tyrolysis of the retina and atrophy of the optic nerve *(4,5)*.

The teratogenic and embryotoxic potential of carbon disulfide was examined in inhalation studies of several animal species, but with no positive effects *(5)*. However, some studies indicated impairment of embryogenesis in female rats *(2,4)*.

Mutagenic and carcinogenic effects

Comprehensive testing of the mutagenic potential of carbon disulfide was performed on several types of bacteria (Ames test) and *Drosophila,* with no positive results *(5)*.

Effects on humans

Toxicological effects

Numerous epidemiological studies on carbon disulfide exposure among workers in viscose rayon plants have been reviewed *(1–5)*.

Acute and subacute poisoning appear due to exposure to carbon disulfide concentrations of 500–3000 mg/m³ and are characterized by predominantly neurological and psychiatric symptoms, "encephalopathia sulfocarbonica" such as irritability, anger, mood changes, manic delirium and hallucinations, paranoic ideas, loss of appetite, gastrointestinal disturbances and sexual disorders *(1–7)*.

More subtle neurological changes at lower carbon disulfide concentrations have been reported; the symptoms are a reduction of nerve conduction velocities and psychological disturbances *(8–10)*. In workers exposed for 10–15 years to carbon disulfide in concentrations of around 10 mg/m³, sensory polyneuritis and increased pain threshold were reported *(11)*. These neurological disturbances were accompanied by psychological and neurobehavioural disorders *(1,4,5)*.

At an exposure of between 100 and 500 mg/m³ carbon disulfide has neurological, vascular and other effects on the eye: focal haemorrhage, exudative changes, optic atrophy, retrobulbar neuritis, microaneurisms and vascular sclerosis. These morphological changes are accompanied by altered functions (colour vision, adaptation to dark, pupillary light reaction, convergence, accommodation and visual acuity), altered motility (eyelids, bulb) and sensitivity (cornea, conjunctiva) *(1,4,5,12,13)*. Some changes in the eye may be influenced by ethnic differences. For example, "microangiopathia sulfocarbonica" is reported in young Japanese and Yugoslav workers *(4,13)*, but not in Finnish workers *(4)*.

In chronic exposure to medium and lower carbon disulfide concentrations (20–300 mg/m³), the effects of carbon disulfide on blood vessels in various organs and tissues, especially cerebral and renal arteries, producing encephalopathy and nephropathy, are well established *(1,4,5)*.

Reports from Finland and the United Kingdom have drawn attention to the association between occupational exposure to carbon disulfide and coronary heart disease (CHD), even at lower exposures (30–120 mg/m³) *(1,4,5,14–17)*. At low carbon disulfide concentrations (under 30 mg/m³) the question of CHD risk remains to be resolved. However, ethnic differences are noted here too. Reports from Japan, the Netherlands and Yugoslavia *(4,5)* showed that there was no evidence that carbon disulfide exposure affected CHD incidence. These contradictory observations suggest that carbon disulfide has coronary effects when recognized factors predisposing to the development of CHD are present.

Chronic exposure to carbon disulfide concentrations of 50–150 mg/m³ affected the endocrine system by causing decreased excretion of some hormones in the urine of young exposed workers (17-ketosteroids, 17-hydroxycorticosteroids, androsterone and etiocholanolone) *(4,18)*. Decreased thyroxine levels in the serum of exposed workers imply a possible reduction in thyroid activity, which could influence vascular changes *(4,5)*.

A report on hypospermia, asthenospermia and teratospermia in young workers exposed to 40–80 mg/m^3 of carbon disulfide confirmed gonadal injury *(19)*.

Some conflicting reports have been published on disturbances of reproductive functions in women occupationally exposed to carbon disulfide. According to some reports *(20,21)*, exposure of women to carbon disulfide concentrations under 30 mg/m^3 is associated with disturbances of the menstrual cycle and a slight increase in miscarriages. These findings were not confirmed by other studies.

Mutagenic and carcinogenic effects

No reports are available to indicate any carcinogenic or mutagenic effects of carbon disulfide.

Sensory effects

Carbon disulfide is an odorant that has a detection threshold of 200 μg/m^3 *(22–24)*. Its sweet and aromatic smell appears in its pure form at concentrations 3–4 times above this detection threshold. In its technical grade it has a very unpleasant smell *(22)*. In effluents from the viscose industry, it is accompanied by other odorous substances, such as hydrogen sulfide and carbonyl sulfide *(22)*. The odour quality of the emission changes according to the specific composition of these mixtures. The formation of carbonyl sulfide from carbon disulfide by photooxidation probably leads to variations in the qualities of the viscose odour in ambient air and to the characteristic smell of burnt rubber *(22,23)*.

It seems unlikely that carbon disulfide in emissions from the viscose industry is the major contributor to odour nuisance. Probably, the nuisance caused by viscose odours is mainly due to hydrogen sulfide and carbonyl sulfide; therefore, it occurs at concentrations below the carbon disulfide odour detection threshold of 200 μg/m^3.

Evaluation of Human Health Risks

Exposure

Inhalation represents the main route of entry of carbon disulfide into the human organism. Values in the vicinity of viscose rayon plants range from 0.01 to about 1.5 mg/m^3, depending mostly on the distance from the source.

Health risk evaluation

A summary of the most relevant concentration–response findings is given in Table 1.

In the light of numerous epidemiological studies, it is very difficult to establish the exact exposure–time relationship. During the approximate period 1955–1965 carbon disulfide concentrations in viscose rayon plants averaged about 250 mg/m^3; they were subsequently reduced to 50–150 mg/m^3 and more recently to 20–30 mg/m^3. Therefore, it is practically impossible to evaluate the long-term (five or more years) exposure level in a retrospective study. Moreover, most exposure data in occupational studies are not reliable,

Table 1. Some concentration–response relationships in occupational exposure to carbon disulfide

Carbon disulfide concentration (mg/m^3)	Duration of exposure (years)	Symptoms and signs	Reference
500-2500	0.5	Polyneuritis, myopathy, acute psychosis	*6*
450-1000	< 0.5	Polyneuritis, encephalopathy	*7*
200- 500	1-9	Increased ophthalmic pressure	*12*
60- 175	5	Eye burning, abnormal pupillary light reactions	*13*
31- 137	10	Psychomotor and psychological disturbances	*8*
29- 118	15	Polyneuropathy, abnormal EEG, conduction velocity slowed, psychological changes	*9,10*
29- 118	10	Increase in coronary mortality, angina pectoris, slightly higher systolic and diastolic blood pressure	*14-17*
40- 80	2	Asthenospermia, hypospermia, teratospermia	*19*
22- 44	>10	Arteriosclerotic changes and hypertension	*25*
30- 50	>10	Decreased immunological reactions	*26*
30	3	Increase in spontaneous abortions and premature births	*20*
20- 25	< 5	Functional disturbances of the CNS	*27,28*
10	10-15	Sensory polyneuritis, increased pain threshold	*11*
10	10-15	Depressed blood progesterone, increased estriol, irregular menstruation	*21*

owing to poor measurement methodology. It is necessary to keep this in mind also when studying Table 1.

At exposure levels of 30 mg/m^3 and above, observable adverse health effects have been well established. The CHD rate increases at levels of 30–120 mg/m^3 of carbon disulfide after an exposure of more than 10 years. Effects on the central and peripheral nervous systems and the vascular system have been established in the same range of concentrations after long-term exposure. Functional changes of the CNS have even been observed at lower concentrations (20–25 mg/m³).

Some authors claim to have observed adverse health effects in workers exposed to 10 mg/m³ of carbon disulfide for 10–15 years. However, because of the lack of reliable retrospective data on exposure levels, the dose–response relationship governing these findings is difficult to establish.

Guidelines

The lowest concentration of carbon disulfide at which an adverse effect was observed in occupational exposure was about 10 mg/m³, which may be equivalent to a concentration in the general environment of 1 mg/m³. In selecting the size of the protection (safety) factor, the expected variability in the susceptibility of the general population was taken into account, and a protection factor of 10 was considered appropriate. This leads to the recommendation of a guideline value of 100 μg/m³, with an averaging time of 24 hours. It is believed that below this value adverse health effects of environmental exposure to carbon disulfide (outdoor or indoor) are not likely to occur.

If carbon disulfide is used as the index substance for viscose emissions, odour perception is not to be expected when carbon disulfide peak concentration is kept below 1/10 of its odour threshold value, i.e. below 20 μg/m³. Based on the sensory effects of carbon disulfide, a guideline value of 20 μg/m³ (average time 30 minutes) is recommended.

References

1. **National Institute for Occupational Safety and Health.** *Criteria for a recommended standard occupational exposure to carbon disulfide.* Washington, DC, US Department of Health, Education, and Welfare, 1977.
2. **Izmerov, N.F., ed.** *Carbon disulfide.* Moscow, Centre of International Projects (GKNT), 1983 (Scientific reviews of Soviet literature on toxicity and hazards of chemicals, No. 41).
3. **Struwe, W. & Sprinzl, G.** *Transmission von H_2S und CS_2* [Transmission of H_2S and CS_2]. Vienna, Österreichisches Bundesinstitut für Gesundheitswesen, 1984.
4. *Carbon disulfide.* Geneva, World Health Organization, 1979 (Environmental Health Criteria, No. 10).
5. **Beauchamp, R.O., Jr. et al.** A critical review of the literature on carbon disulfide toxicity. *CRC critical reviews in toxicology,* **11**: 169–278 (1983).
6. **Vigliani, E.C.** Chronic carbon disulfide poisoning: a report on 100 cases. *Medicina del lavoro,* **37**: 165–193 (1946).
7. **Vigliani, E.C.** Carbon disulphide poisoning in viscose rayon factories. *British journal of industrial medicine,* **11**: 235–244 (1954).
8. **Hänninen, H.** Psychological picture of manifest and latent carbon disulphide poisoning. *British journal of industrial medicine,* **28**: 374–381 (1971).

9. **Seppäläinen, A.M. & Tolonen, M.** Neurotoxicity of long-term exposure to carbon disulfide in the viscose rayon industry: a neurophysiological study. *Work, environment, health,* **11**: 145–153 (1974).
10. **Tolonen, M.** Chronic subclinical carbon disulfide poisoning. *Work, environment, health,* **11**: 154–161 (1974).
11. **Martynova, A.P. et al.** [Clinical, hygienic and experimental investigations of the action on the body of small concentrations of carbon disulfide]. *Gigiena i sanitarija,* (5): 25–28 (1976) (in Russian).
12. **Maugeri, U. et al.** La oftalmodinamografia nella intossicazione solofocarbonica professionalle [Ophthalmodynamography in occupational carbon disulfide poisoning]. *Medicina del lavoro,* **57**: 730–740 (1966).
13. **Savić, S.** Influence of carbon disulfide on the eye. *Archives of environmental health,* **14**: 325–326 (1967).
14. **Tolonen, M. et al.** A follow-up study of coronary heart disease in viscose rayon workers exposed to carbon disulphide. *British journal of industrial medicine,* **32**: 1–10 (1975).
15. **Hernberg, S. et al.** Excess mortality from coronary heart disease in viscose rayon workers exposed to carbon disulfide. *Work, environment, health,* **10**: 93–99 (1973).
16. **Nurminen, M.** Survival experience of a cohort of carbon disulphide exposed workers from an eight-year prospective follow-up period. *International journal of epidemiology,* **5**: 179–185 (1976).
17. **Hernberg, S. et al.** Coronary heart disease among workers exposed to carbon disulphide. *British journal of industrial medicine,* **27**: 313–325 (1970).
18. **Cavalleri, A. et al.** L'excrétion urinaire de testosterone et de gonadotrophine stimulant les cellules interstitielles (ICSH) chez les sujets exposés au sulfure de carbone [Urinary excretion of testosterone and gonadotrophine stimulating the interstitial cells (ICSH) in subjects exposed to carbon disulfide]. *Archives des maladies professionnelles, de médecine du travail et de sécurité sociale,* **31**: 23–30 (1970).
19. **Lancranjan, I. et al.** Changes of the gonadic function in chronic carbon disulfide poisoning. *Medicina del lavoro,* **60**: 556–571 (1969).
20. **Petrov, M.V.** [Course and termination of pregnancy in women working in the viscose industry]. *Pediatrija, akušerstvo i ginekologija,* **3**: 50–52 (1969) (in Russian).
21. **Vasiljeva, I.A.** [Effect of low concentrations of carbon disulfide and hydrogen sulfide on the menstrual function in women and on the estrous cycle under experimental conditions]. *Gigiena i sanitarija,* **38**: 24–27 (1973) (in Russian).
22. **National Research Council.** *Odors from stationary and mobile sources.* Washington, DC, National Academy of Sciences, 1979.
23. **Brunekreef, B. & Harssema, H.** Viscose odors in ambient air: a study of the relationship between the detectability of viscose odors and concentrations of H_2S and CS_2 in ambient air. *Water, air and soil pollution,* **13**: 439–446 (1980).

24. **van Gemert, L.J. & Nettenbreijer, A.H., ed.** *Compilation of odour threshold values in air and water.* Zeist, Central Institute for Nutrition and Food Research, 1977 and Supplement V, 1984.
25. **Gavrilescu, N. & Lilis, R.** Cardiovascular effects of long-extended carbon disulphide exposure. *In:* Brieger H. & Teisinger J., ed. *Toxicology of carbon disulphide.* Amsterdam, Excerpta Medica, 1967, pp. 165–167.
26. **Kašin, L.M.** [Overall immunological reactivity and morbidity of workers exposed to carbon disulfide]. *Gigiena i sanitarija,* **30**: 331–335 (1965) (in Russian).
27. **Gilioli, R. et al.** Study of neurological and neurophysiological impairment in carbon disulphide workers. *Medicina del lavoro,* **69**: 130–143 (1978).
28. **Cassito, M.G. et al.** Subjective and objective behavioural alterations in carbon disulphide workers. *Medicina del lavoro,* **69**: 144–150 (1978).

8

1,2-Dichloroethane

General Description

1,2-Dichloroethane ($C_2H_4Cl_2$) (DCE, ethylene dichloride) is a flammable, colourless liquid with a sweet taste. The compound has a boiling-point of 83 °C and a freezing-point of −35 °C. The solubility in water at 20 °C is 8.7 g/litre. The vapour pressure is 64 mmHg at 20 °C.

Sources

1,2-Dichloroethane is not reported to occur as a natural product *(1)*. World production in 1979 was estimated to be 23 130 kilotonnes *(2)*. In the countries of the European Community an estimated 5290 kilotonnes was produced in 1977 *(3)*. In the United States production increased from 4160 kilotonnes in 1974 to 4750 kilotonnes in 1977 *(4)*. Actual production figures may be higher, since a part of it (used as intermediate) is not separated and therefore is not always reported by some producers *(1)*. In the manufacturing process, DCE is derived from ethene via (catalyzed) reaction with chlorine or with oxygen and hydrogen chloride *(1)*.

The major industrial use of DCE is in the synthesis of other chemicals, among which vinyl chloride is the most important (80–90% of the DCE produced is used for production of this compound). A less important use is the production of ethylene diamines. 1,2-Dichloroethane is also used as a lead scavenger in petrol. A minor use is as a solvent and a fumigant *(2,5,6)*.

Major sources of DCE emission to the environment include industrial use, manufacture of the compound, and inappropriate disposal of "EDC-tars", the heavy ends in vinyl chloride production. Emissions occur directly to the atmosphere both during the production process and during handling and storage. In addition, fugitive emissions contribute significantly. Emission of DCE also occurs when the compound is used for extraction purposes in the pharmaceutical industry or for crystallization in the chemical industry *(5)*.

Besemer et al. *(5)* give, as a first approximation, an emission factor of 10 kg DCE per tonne production for western Europe. This value applies for controlled emissions. For uncontrolled emissions the estimated emission factor is 40 kg/tonne. Thus, total emissions of 61 000 and 110 000 tonnes DCE have been estimated for western and eastern Europe respectively *(5)*.

Occurrence in air

Rural or "background" levels of DCE in the USA have been reported to be up to 0.20 μg/m³ *(7)*. Similar values were estimated for the Netherlands *(5)*.

In the air of seven cities in the USA average DCE levels were between 0.5 and 6.1 μg/m³ (the median level was 1.0 and the maximum concentration 30.0 μg/m³) *(7)*. 1,2-Dichloroethane concentrations near sites of production and in dispersive use are higher: at 12 locations near production facilities in each of three areas in the USA average levels gradually decreased from 61 μg/m³ at about 1 km distance to 2 μg/m³ at 3–4 km distance *(8,9)*. Near petrol stations concentrations may also be elevated *(10)*. In Sweden an average of 4.0 μg/m³ was measured in the air of petrol stations. In parking garages and car repair shops average concentrations were between 2.0 and 6.5 μg/m³. Inside cars, averages of 0.4–1.2 μg/m³ were found. It is noteworthy that, besides being directly volatilized from petrol, DCE is also released via engine exhaust *(11)*. The few indoor measurement data available indicate that indoor levels of DCE are not higher than outdoor levels *(12)*.

Degradation of DCE in the atmosphere proceeds mainly by reaction with hydroxyl radicals, a reaction that presumably will ultimately lead to carbon monoxide and hydrogen chloride. The degradation process is relatively slow (half-time = 29 days), as a consequence of which the compound will be ubiquitous. Dry deposition (half-time = 1 year) and wet deposition (half-time = 390 years), when compared with degradation via hydroxyl radicals (half-time = 29 days), are negligible *(5)*.

Conversion factors

1 ppm = 4.12 mg/m³ (20 °C, 101.3 kPa)

1 mg/m³ = 0.242 ppm

Routes of Exposure

Air

The intake from urban air in the USA was estimated to be between 8 and 80 μg per day *(13)*, with an average of about 20 μg per day.

On the basis of model calculations an estimate of DCE exposure among the Dutch population was made. According to this, out of a total population of 14 million, 1000 are exposed to levels of about 1.5 μg/m³, 35 000 to 1 μg/m³, 300 000 to 0.5 μg/m³, and the remainder to below 0.2 μg/m³ *(5)*.

No data were available concerning occupational exposure in DCE- and vinyl chloride-synthesizing industries *(1)*. During use of the compound as a solvent in several industries, concentrations of between 11 and 800 mg/m³ were measured (data from before 1960) *(14,15)*. More recently, time-weighted averages of 0.1 and 1.0 mg/m³ were reported for two different occupational categories in an anti-knock blending plant in the USA. The maximum exposure measured was 8.9 mg/m³ *(16)*.

Drinking-water

Average levels found in drinking-water were generally below 1 μg/litre *(17–19)*.

Food

Reports on DCE residues in food are scarce. Bauer *(17)* found that levels were generally low in foods in the Federal Republic of Germany and reported an average of 0.8 μg/kg for milk products with added fruits. Significant residues of DCE in food (spice, grains) are possible when the compound has been used as an extractant or fumigant *(1,5)*.

Relative significance of different routes of exposure

The scarcity of data on concentrations of DCE in foodstuffs does not allow a meaningful estimate of the total daily intake via this route. If a conservative upper level of 1 μg/litre is taken for the concentration of DCE in drinking-water, the daily intake from water would be 2 μg per person (assuming a water intake of 2 litres/day).

The average daily intake of DCE from urban air in the USA is estimated at about 20 μg *(13)*. The intake via ambient air in the Netherlands is reckoned to be 30 μg/day for the 1000 subjects estimated to be exposed to 1.5 μg/m^3, 20 μg/day for 35 000 exposed to 1 μg/m^3, 10 μg/day for the 300 000 exposed to 0.5 μg/m^3 and 4 μg/day for the 13 million exposed to concentrations below 0.2 μg/m^3 *(5)*.

Kinetics and Metabolism

Absorption

In guinea pigs, mice and rats it was shown that the absorption of DCE is a rapid process: almost immediately after exposure the compound is found in the blood *(20–23)*. During inhalation, steady-state levels in the blood were reached within 2–3 hours. A comparison of blood concentrations in rats indicates that at doses of 113–150 mg/kg body weight, peak levels after oral dosing are markedly higher than those observed after inhalation, a finding that indicates a difference in absorption dynamics between the two routes *(21,22,24)*. A similar rapid rate of absorption would be expected in people exposed to DCE.

Distribution

After oral and inhalatory dosing of rats, the amount of DCE found in adipose tissue exceeded the levels present in liver, brain, spleen, kidneys and lungs *(24)*. In one rat study in which radiolabelled DCE was used, residual radioactivity was highest in liver and kidneys *(21)*.

Vosovaja *(25)* found accumulation of DCE in placental and fetal tissues of rats after inhalation of 1000 mg/m^3 for 7 days. It has also been found in breast-milk and cows' milk *(26,27)*.

Elimination and biotransformation

The excretion of DCE from rodents was rapid. In mice and rats, after oral or inhalatory treatment with radiolabelled compound, 90% or more was excreted within 48 hours *(21,28)*. Excretion occurred mainly via the lungs and urine. About 90% of the body-burden was excreted via the lungs and urine within 24 hours in intraperitoneally injected mice and within 48 hours in orally dosed mice. Little radioactivity was found in the faeces. The elimination rate from blood and tissues was inversely related to exposure levels: the higher the dose, the longer the half-time for clearance from the blood. After oral dosing in rats the half-time ranged from 20 to 90 minutes and after inhalation the values ranged from 13 to 22 minutes *(21,24)*. The rates of elimination from tissues after oral dosing were comparable to that for blood, with the exception of the liver, where the elimination rate was higher. After inhalation of DCE, elimination was most rapid via the lungs and slowest in adipose tissue *(24)*.

At increased exposure levels the concentrations of unmetabolized DCE in the blood increased hyperproportionally, a phenomenon that can be assumed to be due to the saturation of metabolic pathways. At high levels the exhalation of unchanged DCE markedly increased *(21,24,28)*. In the biotransformation of DCE two major pathways are oxidation to chloroethanol and chloroacetic acid, and conjugation with glutathione, directly or after microsomal oxidation. This has been shown to lead to the formation of carboxymethylcysteine and thiodiacetic acid (and its sulfoxide), compounds that have been identified in the urine of rats. Other urinary metabolites found in rodents *in vivo* could result from direct glutathione conjugation *(21,28–30)*. Biotransformation to urinary metabolites ranges from 55% to 90% in rats and mice after oral, parenteral or inhalation exposure or dosing *(1)*.

Health Effects

Effects on experimental animals and *in vitro* test systems

Toxicological effects

LC_{50} values (6-hour exposure) were reported to be 5100 and 6666 mg/m^3 in rats and 1060 mg/m^3 in mice *(31–35)*. Oral LD_{50} values of 680 and 450 mg/kg were found in rats and mice respectively.

Several inhalation studies involving exposure of rats, mice and guinea pigs to DCE at various exposure levels ranging from 400 to 3900 mg/m^3, for several weeks to 36 weeks, have been conducted *(35–37)*. At levels above 1420 mg/m^3 high mortality rates were observed in all the studies. However, in one study mortality also occurred in all three species at levels of 730 mg/m^3. All studies demonstrated histological changes (fatty degeneration, cloudy swelling and necrosis) in the liver at the highest levels. The lowest level at which some of these effects were reported was 730 mg/m^3. Similar effects, of varying degrees of severity, were found at levels of 730 mg/m^3 and above in the kidneys. Furthermore, histological changes were observed in the myocardium, adrenals and lungs (oedema) of moribund and dead animals.

In each of the above-mentioned studies, a few animals of other species (cats, rabbits, dogs and monkeys) were also exposed to similar levels.

Although only a few animals were exposed, the results, when evaluated collectively, are not inconsistent with the observations in rats and guinea pigs, in that no adverse histological changes were observed at an exposure level of about 400 mg/m^3. Of all the species studied, mice and rats appear to be more sensitive than other species. Signs of CNS depression observed in these studies *(35–37)* included apathy in guinea pigs at 1980 mg/m^3 *(37)* and 3900 mg/m^3 *(36)* and coma in dogs and monkeys at 3900 mg/m^3 *(36)*.

Spreafico et al. *(24)* exposed rats in a long-term inhalation study (3–18 months) to 0, 5, 40, 202 or 1012 mg/m^3 for 7 hours per day, 5 days per week. Animals of each sex were exposed, starting at 3 months of age, for 3, 6 or 18 months. In addition, animals that were 14 months old when the study started were exposed for 12 months. The highest exposure level was reduced to 607 mg/m^3 after a few weeks because of high mortality. Slight changes in SGOT, SGPT and γ-glutamyltranspeptidase activities were observed in the older animals exposed for 12 months, but not in those exposed for 18 months at levels of 202 and 607 mg/m^3. Serum uric acid and blood urea nitrogen increased in the females exposed for 12 months, but not in animals exposed for 18 months. In addition, the animals exposed for 12 months, but not those exposed for 18 months, displayed decreases in serum cholesterol. The authors suggested that DCE in this exposure regimen lacks significant toxicity in spite of the changes in the biochemical parameters observed. The effects of DCE at the same exposure levels on older animals, compared with the absence of effects on younger animals exposed for a longer duration, make the results of this study inconclusive.

Fatty degeneration in the liver was also observed when DCE was administered orally, in oil. This effect was prominent at a dose of 300 mg/kg when administered daily for 5 days, but not at 150 mg/kg administered for 2 weeks *(21,38)*. In a 90-day study, increases in the weight of kidneys, liver and brain were observed at 90 mg/kg, but not at 30 and 10 mg/kg. These weight increases were not accompanied by consistent haematological, clinicochemical or histopathological abnormalities *(38)*. After oral administration of doses up to 35 mg/kg body weight per day for 2 years there were no effects on mortality rates, growth or serum biochemistry *(39)*.

The effects of DCE on reproduction, teratogenicity and related endpoints have been investigated in several published studies. In a one-generation inhalation study, male and female rats were exposed for 60 days (6 hours per day, 5 days per week) to DCE up to a level of 615 mg/m^3 and then bred to produce two litters. No effects on reproduction were reported and there was no evidence of fetal or maternal toxicity *(40)*. In contrast, Vosovaja *(25,41)* reported a variety of adverse effects (e.g. fetal toxicity and estrous cycle prolongation) in a two-generation study of rats exposed to 15 mg/m^3 and 57 mg/m^3. However, the lack of data concerning the protocol, test results and statistical tests employed makes it difficult to evaluate these findings.

In a two-generation reproduction study of mice given oral doses of 5, 15 and 50 mg/kg, no adverse effects were reported *(42)*. Maternal toxicity was not observed.

Mutagenic and carcinogenic effects

1,2-Dichloroethane mutagenicity has been studied in several test species and in *in vitro* tests using mammalian cells. It was weakly mutagenic in *Salmonella typhimurium* in the presence and absence of microsomal activation systems *(1)*. In the presence of cytosolic glutathione S-transferase, a stronger positive response was obtained. Glutathione conjugates of DCE consistently showed mutagenic activity in *S. typhimurium* strains TA 100 and TA 1535. Forward mutation tests in fungi were positive. In *Drosophila melanogaster* DCE induced both sex-linked recessive lethal mutations and somatic mutations. Mutagenicity tests in mammalian cells *in vitro* also showed a positive result, particularly in a cell line with high levels of glutathione S-transferase. *In vivo* mutagenicity tests (dominant lethal assay, micronucleus test), however, were mostly negative. A weak effect was reported in a spot test for somatic mutations in mice. DNA damage has been observed in bacteria, in mammalian cells *in vitro,* and in mammals *in vivo (1,5)*.

On the basis of the above evidence DCE must be considered a mutagenic agent.

Inhalation carcinogenicity studies were performed in Swiss mice and Sprague-Dawley rats at exposure levels of 20, 40, 202 and 1012 mg/m^3 for 7 hours per day, 5 days per week over a period of 78 weeks. Thereafter the animals were maintained until spontaneous death. The 1012 mg/m^3 dose level was reduced to 607 mg/m^3 because of the high mortality in the first weeks of the test. The only effect was a nonstatistically significant increased incidence of fibromas and fibroadenomas in the mammary gland of female rats at 20, 202 and 607 mg/m^3. A dose–response relationship was not apparent *(43)*.

Oral tests, performed in mice and rats, involved the dosing of DCE in corn oil by intragastric intubation. B6C3F1 mice were dosed with time-weighted average doses of 97 and 195 mg/kg body weight (males) and 149 and 299 mg/kg body weight (females) over a period of 78 weeks. After this the animals were observed for a further 12–13 weeks. Survival rates were decreased in a dose-related manner in females. There were dose-related increases in the incidence of alveolar or bronchiolar adenomas (both sexes), hepatocellular carcinomas (males), adenocarcinomas of the mammary gland (females), endometrial stromal polyps or stromal sarcomas (females) and squamous-cell carcinomas of the forestomach (females; in males hyperplastic changes at this site). Seven mice developed metastatic tumours *(44,45)*. A similar test was conducted in Osborne-Mendel rats. The average dose levels were 47 and 95 mg/kg body weight for both males and females. The test period was 78 weeks and the post-dosing observation period lasted 15–32 weeks. A dose-related increase in mortality occurred. The incidence of squamous-cell carcinoma in the forestomach was increased in males only (incidences 0/60, 3/50 and 9/50 in control, 47 mg/kg and 95 mg/kg groups respectively); the incidence of subcutaneous fibroma was also elevated in males only (control: 0/60; 47 mg/kg: 5/50; 95 mg/kg: 6/50). The incidence of adenocarcinoma in the mammary gland was increased in females (control: 1/59; 47 mg/kg: 1/50; 95 mg/kg: 18/50). The haemangiosarcoma incidence was elevated both in males (control: 1/60; 47 mg/kg: 9/50;

95 mg/kg: 7/59) and in females (control: 0/59; 47 mg/kg: 4/50; 95 mg/kg: 4/50). In addition, treated females showed hyperplastic lesions in the forestomach. Nine rats developed metastatic tumours *(44,45)*.

A dermal carcinogenicity test was performed in Ha:ICR mice. The animals were treated 3 times per week with 42 or 146 mg DCE (dissolved in acetone) for 440–594 days. At the highest dose level, DCE induced lung papillomas. The study also included a dermal test for tumour-initiating properties; phorbol myristate acetate was used as a tumour promoter. No initiating effect was observed *(46)*.

Effects on humans

Toxicological effects

Inhalation of DCE adversely effects the CNS. Symptoms of intoxication include headache, dizziness, weakness, spasms, muscular hypotonia, vomiting and unconsciousness. Death often follows. Autopsy reports frequently mention damage to lungs, liver and kidneys. Heart rhythm disturbances have also been observed. A number of case studies related to accidental oral ingestion of DCE and resulting in fatalities have been reported. After acute oral ingestion, symptoms include CNS depression, gastroenteritis, functional disorders of liver and kidneys and cardiovascular insufficiency *(1,47)*. Accidental ingestion of 10–250 g of DCE resulted in death in all instances *(1)*.

Only two limited studies on long-term occupational exposure, both dating back to the 1950s, are available. In one study workers exposed to DCE concentrations of 40–800 mg/m^3 for 2–8 months were examined. In particular, workers exposed to the higher concentrations complained of a burning sensation of the eyes, lacrimation, dizziness, nausea, vomiting and constipation; abnormalities in liver, CNS, gastrointestinal tract and haematological parameters were also found *(14)*. The second study involved workers who were exposed to a time-weighted average DCE concentration of 114 mg/m^3 during 20% of their worktime. Morbidity among these workers in comparison with nonexposed workers was elevated for all disease categories. Upon closer examination of a group of 83 of the exposed workers, neurotic conditions, autonomic dystonia, hyperthyroidism, goitre, visual-motor reaction impairment, neuromyalgia and/or tenovaginitis were found *(15)*. These latter conditions were stated by the authors to be unlikely to be related to DCE exposure.

No epidemiological studies are available regarding carcinogenicity.

Evaluation of Human Health Risks

Exposure

Rural or background atmospheric concentrations in western Europe and North America are approximately 0.2 μg/m^3, and the limited data available on indoor concentrations show that they are about the same. Average levels in cities vary from 0.4 μg/m^3 to 1.0 μg/m^3, increasing to 6.1 μg/m^3 near petrol stations, parking garages and production facilities.

Health risk evaluation

Human studies point to effects on the CNS and the liver, but the limited data do not allow a definitive conclusion regarding a lowest-observed-adverse-effect level or no-observed-effect level. In animals, long-term inhalation exposure (>6 months) to DCE levels of approximately 700 mg/m^3 and above has been shown to result in histological changes in the liver *(35–37)*. The same animal studies reported no adverse histological changes in the liver and kidneys of guinea pigs and rats at levels of about 400 mg/m^3. Findings concerning effects on reproduction are contradictory.

Animal data suggest a no-observed-effect level in laboratory animals of 400 mg/m^3 and a lowest-observed-adverse-effect level of 700 mg/m^3.

With regard to mutagenicity as an endpoint and to the causal connections between DNA damage and the initiation of carcinogenicity, DCE has been shown to be weakly mutagenic in *Salmonella typhimurium,* both in the absence and in the presence of microsomal activation systems. It has also been demonstrated to be mutagenic in other test species and in *in vitro* tests using mammalian cells.

In a lifetime study in rats and mice in which DCE was administered by gavage, DCE caused tumours at multiple sites in both species. In the only inhalation study performed *(24)*, DCE exposure did not result in an increased tumour incidence. The negative results obtained in this study, however, do not detract from the positive findings of the oral study *(44,45)* when differences in total dose, exposure time and pharmacokinetics are considered.

1,2-Dichloroethane was evaluated in 1979 by IARC as a chemical for which there is sufficient evidence of carcinogenicity in experimental animals and inadequate evidence in humans *(6)*. To date there are two publications giving quantitative carcinogenic risk estimates based on animal data. One, developed by the National Institute of Public Health in the Netherlands on the basis of oral exposure of rats by gavage *(45)*, indicates a lifetime risk of one in a million from exposure to 0.48 μg/m^3 *(5)*, which corresponds to a unit risk of about 2×10^{-6}. The US Environmental Protection Agency *(48)* has estimated an incremental unit risk of 2.6×10^{-5} on the basis of data from gavage studies and of 1×10^{-6} on the basis of a negative inhalation study.

Guidelines

Evidence of carcinogenicity in animals is sufficient on the basis of oral ingestion data. However, animal inhalation data do not at present provide positive evidence. Because of deficiencies in extrapolation from oral data to inhalation, the two risk estimates available are not used in the guidelines.

For noncarcinogenic endpoints, data from animal studies imply a no-observed-adverse-effect level of about 400 mg/m^3 and suggest a lowest-observed-adverse-effect level of about 700 mg/m^3. A protection (safety) factor of 1000 is considered appropriate in extrapolation of animal data to the general population. In selecting such a large protection factor, variations in exposure time, the limitation of the data base and the fact that a no-effect level in man cannot be established are of decisive importance. The resulting value of 0.7 mg/m^3 for continuous exposure (averaging time 24 hours) is recommended as a guideline value. Since this value is above current environmental levels and present exposures are not of concern to health, this guideline relates only to accidental release episodes or specific indoor pollution problems.

References

1. *1,2-Dichloroethane.* Geneva, World Health Organization (in press) (Environmental Health Criteria, No. 62).
2. **Gold, L.S.** Human exposures to ethylene dichloride. *Banbury reports,* **5**: 209–225 (1980).
3. **Atri, F.R.** *Studie zur Bewertung der Toxizität wassergefährdender Stoffe. Band III. 1,2-Dichlorethan* [Studies on evaluating the toxicity of water pollutants. Vol. III. 1,2-Dichloroethane]. Berlin, Umweltbundesamtes, 1984 (Forschungsbericht 106 07 047).
4. **Drury, J.S. & Hammons, A.S.** *Investigations of selected environmental pollutants: 1,2-dichloroethane.* Washington, DC, US Environmental Protection Agency, Office of Toxic Substances, 1979 (Report No. EPA-560/2-78-006, PB 295865).
5. **Besemer, A.C. et al.** *Criteriadocument over 1,2-dichloorethaan* [1,2-Dichloroethane criteria document]. The Hague, Ministerie van Volkshuisvesting, Ruimtelijke Ordening en Milieubeheer, 1984 (Publikatiereeks Lucht, No. 30).
6. 1,2-Dichloroethane. *In: Some halogenated hydrocarbons.* Lyon, International Agency for Research on Cancer, 1979, pp. 429–448 (IARC Monographs on the Evaluation of the Carcinogenic Risk of Chemicals to Humans, Vol. 20).
7. **Singh, H. B. et al.** Distribution of selected gaseous organic mutagens and suspect carcinogens in ambient air. *Environmental science and technology,* **16**: 872–880 (1982).
8. **Elfers, L.A.** *Monitoring of ambient levels of ethylene dichloride (EDC) in the vicinity of EDC production and user facilities.* Research Triangle Park, NC, US Environmental Protection Agency, 1979 (Report No. EPA-600/4-79-029, PB 298303).
9. **Kellam, R.G. & Dusetzina, M.G.** Human exposure to ethylene dichloride: potential for regulation via EPA's proposed airborne carcinogen policy. *Banbury reports,* **5**: 265–276 (1980).
10. **Tsani-Bazaca, E. et al.** Ambient concentrations and correlations of hydrocarbons and halocarbons in the vicinity of an airport. *Chemosphere,* **11**: 11–23 (1982).
11. **Jonsson, A. & Berg, S.** Determination of 1,2-dibromoethane, 1,2-dichloroethane and benzene in ambient air using porous polymer traps and gas chromatographic-mass spectrometric analysis with selected ion monitoring. *Journal of chromatography,* **190**: 97–106 (1980).
12. **Hartwell, T.D. et al.** Comparative statistical analysis for volatile halocarbons in indoor and outdoor air. *In:* Berglund, B. et al., ed. *Indoor air: chemical characterization and personal exposure.* Stockholm, Swedish Council for Building Research, 1984, Vol. 4, pp. 57–61.

13. **Singh, H.B. et al.** Measurements of some potentially hazardous organic chemicals in urban environments. *Atmospheric environment,* **15**: 601–612 (1981).
14. **Cetnarowicz, J.** Badania eksperymentalne i Kliniczne nad Azialaniem dwuchloroetanu [Experimental and clinical investigations on the action of dichloroethane]. *Folia medica cracoviensia,* **1**: 169–192 (1959).
15. **Kozik, I.V.** [Some problems of occupational hygiene in the use of dichloroethane in the aircraft industry]. *Gigiena truda i professional'nye zabolevanija,* **1**: 31–38 (1957) (in Russian).
16. **Jacobs, E.S.** Use and air quality impact of ethylene dichloride and ethylene dibromide scavengers in leaded gasoline. *Banbury reports,* **5**: 239–255 (1980).
17. **Bauer, U.** Belastung des Menschen durch Schadstoffe in der Umwelt. Untersuchungen über leicht flüchtige organische Halogen-Verbindungen in Wasser, Luft, Lebensmitteln und im menschlichen Gewebe. I & IV Mitteilung [Human exposure to environmental chemicals — investigations on volatile organic halogenated compounds in studies on water, air, food and human tissues. I & IV]. *Zentralblatt für Bakteriologie, Mikrobiologie und Hygiene. I Abt. Orig. B,* **174**: 15–56, 556–583 (1981).
18. **Fujii, T.** Direct aqueous injection gas chromatography-mass spectrometry for analysis of organohalides in water at concentrations below the parts per billion level. *Journal of chromatography,* **139**: 297–302 (1977).
19. **Symmons, J.M. et al.** National organics reconnaissance survey for halogenated organics. *Journal of the American Water Works Association,* **67**: 634–648 (1975).
20. **Jakobson, I. et al.** Uptake via blood and elimination of 10 organic solvents following epicutaneous exposures of anaesthetized guinea pigs. *Toxicology and applied pharmacology,* **63**: 181–187 (1982).
21. **Reitz, R.H. et al.** Pharmacokinetics and macromolecular interactions of ethylene dichloride in rats after inhalation or gavage. *Toxicology and applied pharmacology,* **62**: 190–204 (1982).
22. **Sopikov, N.F. & Goršunova, A.I.** [Investigation of the intake, distribution and excretion of ethylene dichloride in rats]. *Gigiena truda i professional'nye zabolevanija,* **4**: 36–40 (1979) (in Russian).
23. **Tsuruta, H.** Percutaneous absorption of organic solvents I. Comparative study of the *in vivo* percutaneous absorption of chlorinated solvents in mice. *Industrial health,* **13**: 217–236 (1975).
24. **Spreafico, F. et al.** Pharmacokinetics of ethylene dichloride in rats treated by different routes and its long-term inhalatory toxicity. *Banbury reports,* **5**: 107–129 (1980).
25. **Vozovaja, M.A.** [Effect of dichloroethane on the reproductive cycle and embryogenesis in experimental animals]. *Akušerstvo i ginekologija,* **2**: 57–59 (1977) (in Russian).
26. **Urusova, T.P.** [The possibility of dichloroethane contaminating the milk of breastfeeding women following industrial exposure]. *Gigiena i sanitarija,* **3**: 36–37 (1953) (in Russian).

27. **Sykes, J.F. & Klein, A.K.** Chloro-organic residues in milk of cows orally administered ethylene dichloride. *Journal of the Association of Official Agricultural Chemists,* **40**: 203–209 (1957).

28. **Yllner, S.** Metabolism of 1,2-dichloroethane-146 in the mouse. *Acta pharmacologica et toxicologica,* **30**: 257–265 (1971).

29. **Johnson, M.K.** Metabolism of chloroethanol in the rat. *Biochemical pharmacology,* **16**: 185–199 (1967).

30. **Kokarovceva, M.G. & Kiseleva, N.** [Chloroethanol, a toxic metabolite of 1,2-dichloroethane]. *Farmakologija i toksikologija,* **41**: 118–120 (1978) (in Russian).

31. **Bonnet, P.** Détermination de la concentration léthale 50 des principaux hydrocarbures aliphatiques chlorés chez le rat [Determination of the lethal dose 50 of the principal chlorinated aliphatic hydrocarbons in the rat]. *Archives des maladies professionnelles, de médecine du travail et de sécurité sociale,* **41**: 317–321 (1980).

32. **Gradiski, D. et al.** Toxicité aiguë comparée par inhalation des principaux solvants aliphatiques chlorés [Comparative acute toxicity by inhalation of the principal chlorinated aliphatic solvents]. *Archives des maladies professionnelles, de médecine du travail et de sécurité sociale,* **39**: 249–257 (1978).

33. **McCollister, D.D. et al.** Comparative inhalation toxicity of fumigant mixtures. Individual and joint effects of ethylene dichloride, carbon tetrachloride and ethylene dibromide. *Archives of industrial health,* **13**: 1–7 (1956).

34. **Munson, A.E. et al.** *In vivo* assessment of immunotoxicity. *Environmental health perspectives,* **43**: 41–52 (1982).

35. **Spencer, H.C. et al.** Vapour toxicity of ethylene dichloride determined by experiments on laboratory animals. *Archives of industrial hygiene and occupational medicine,* **4**: 482–493 (1951).

36. **Heppel, L.A. et al.** The toxicology of 1,2-dichloroethane (ethylene dichloride). V. The effect of daily inhalations. *Journal of industrial hygiene and toxicology,* **28**: 113–120 (1946).

37. **Hofmann, H.T. et al.** Zur Inhalationstoxizität von 1,1- und 1,2-Dichloräthan [On the inhalation toxicity of 1,1- and 1,2-dichloroethane]. *Archiv für Toxikologie,* **27**: 248–265 (1971).

38. **van Esch, G.J. et al.** *Ninety-day toxicity study with 1,2-dichloroethane (DCE) in rats.* Bilthoven, National Institute of Public Health and Environmental Hygiene, 1977 (Report 195/77 Alg.Tox).

39. **Alumot, E. et al.** Tolerance and acceptable daily intake of chlorinated fumigants in the rat diet. *Food and cosmetics toxicology,* **14**: 105–110 (1976).

40. **Rao, K.S. et al.** Teratogenicity and reproduction studies in animals inhaling ethylene dichloride. *Banbury reports,* **5**: 149–161 (1980).

41. **Vozovaja, M.A.** [Development of progeny of two generations obtained from females subjected to the action of dichloroethane]. *Gigiena i sanitarija,* (7): 25–28 (1974) (in Russian).

42. **Lane, R.W. et al.** Effects of 1,2-dichloroethane and 1.1.1-trichloroethane in drinking water on reproduction and development in mice. *Toxicology and applied pharmacology,* **63**: 409–421 (1982).
43. **Maltoni, C. et al.** Long-term carcinogenic bioassays on ethylene dichloride administered by inhalation to rats and mice. *Banbury reports,* **5**: 3–29 (1980).
44. **US National Cancer Institute.** *Bioassay of 1,2-dichloroethane for possible carcinogenicity.* Bethesda, MD, US Department of Health, Education and Welfare, 1978 (DHEW Publication No. (NIH) 78-1305).
45. **Ward, J.M.** The carcinogenicity of ethylene dichloride in Osborne-Mendel rats and B6C3F1 mice. *Banbury reports,* **5**: 35–53 (1980).
46. **Van Duuren, B.L. et al.** Carcinogenicity of halogenated olefinic and aliphatic hydrocarbons in mice. *Journal of the National Cancer Institute,* **63**: 1433–1439 (1979).
47. **National Institute for Occupational Safety and Health.** *Criteria for a recommended standard occupational exposure to ethylene dichloride (1,2-dichloroethane).* Washington, DC, US Government Printing Office, 1976 (DHEW Publication No. (NIOSH) 76-139).
48. *Health assessment document for 1,2-dichloroethane (ethylene dichloride).* Washington, DC, US Environmental Protection Agency, 1985 (Report No. EPA-600/8-84-0067).

9

Dichloromethane

General Description

Dichloromethane (CH_2Cl_2, methylene chloride, DCM), a clear liquid, is nonflammable and nonexplosive when mixed with air. It is relatively soluble in water (20 g/litre). In the atmosphere, it is removed principally via reaction with hydroxyl radicals. In the presence of moisture it hydrolyses very slowly. No appreciable decomposition occurs at room temperature when the dry compound comes into contact with common metals. Phosgene and hydrochloric acid are formed by contact with hot surfaces or flames.

Sources

Worldwide production of DCM in 1980 amounted to 570 000 tonnes, of which 270 000 tonnes were produced in western Europe *(1)*. In the United States, production amounted to 276 000 tonnes in 1983 *(2)*. In Japan, 35 000 tonnes were produced in 1980 *(3)*.

Approximately 80% of all DCM produced is emitted into the atmosphere as a result of use. Minor losses occur during production and shipping. Volatilization also appears to be the major process by which DCM is lost from water *(4)*.

Dichloromethane is used as a paint remover, a polyurethane foam-blowing agent, and a solvent. It is increasingly used as a replacement for fluorocarbon propellants in aerosols (e.g. insecticides, hair sprays, shampoos and paints). Dichloromethane is used as a solvent in pharmaceutical applications, in the manufacture of photographic film and synthetic fibres, and in the extraction of naturally occurring, heat-sensitive substances such as edible fats, cocoa, butter, caffeine and hops. It is also used as a component in fire-extinguishing products, as an insecticidal fumigant for grains, and as a coolant and refrigerant *(5,6)*.

Occurrence in air

The global average concentration of DCM is reported to be 0.10 $\mu g/m^3$ *(7)*. In surveys involving urban areas in the United States, average levels of DCM varied from about 1 to 13 $\mu g/m^3$ *(8,9)*. The same concentration range has been measured at three locations with different levels of air pollution in the Netherlands *(10)*. In cases involving exposure of people living near waste disposal sites, levels have been greatly in excess of average ambient concentrations or those predicted by modelling *(11)*. In nonoccupational indoor environments, during the use of paint removers containing DCM, the time-weighted averages in a room without ventilation varied between 460 and 2980 mg/m^3 *(12)*.

Dispersion models have been used to predict population exposure to ambient DCM. The predicted maximum annual average to which people may be exposed if living near DCM production facilities is 0.050 mg/m^3 *(6)*. People living near other DCM sources, such as organic solvent cleaning and paint and varnish removal operations, are expected to be exposed to concentrations that do not exceed an average of 0.025–0.050 mg/m^3 over a period of one year *(6)*.

In occupational environments, exceedingly high levels of DCM have been determined. For example, in a chemical plant in the United States, an upper level of 19 150 mg/m^3 was measured *(13)*. A comparable level (17 000 mg/m^3) was measured in a European manufacturing facility *(13,14)*. It is not likely that these high concentrations are representative of current occupational environments. In a more recent study in the United States, an 8-hour time-weighted average concentration in a manufacturing plant ranged from 486 mg/m^3 in areas of low DCM use to 1648 mg/m^3 in areas of high use *(15)*.

Conversion factors

1 ppm (in air, 25 °C) = 3.47 mg/m^3
1 mg/m^3 = 0.28 ppm

Routes of Exposure

Air
Inhalation represents the principal route of exposure. Indoor exposures to consumer products containing DCM may represent significant sources.

Drinking-water
The occurrence of DCM in drinking-water has not been well characterized. Measurable levels have been recorded in drinking-water supplies in the United States *(1)*. Dichloromethane can also be formed during chlorination treatment.

Food
Very little data are available. In one study, 0.3–0.4 mg DCM per kg decaffeinated coffee beans were found *(16)*. Because DCM is highly volatile, nearly all DCM applied to foods would be expected to volatilize during processing operations.

Kinetics and Metabolism

Absorption
Inhalation represents the most important route of absorption of DCM. Uptake and retention vary considerably in sedentary individuals. In exposures to 347 mg/m^3 for more than 2 hours, assuming a resting ventilation rate of

6 litres/minute, uptake varies between 50 and 75 mg/hour *(17,18)*. Physical exercise increases absolute uptake via an increase in the ventilation rate *(19,20)*. Absorption via drinking-water is essentially complete. Absorption through skin is slow.

Distribution
Available evidence suggests that DCM is probably distributed in all tissues, including passage across the placental barrier *(6)*.

Biotransformation
In man, the principal metabolite of health concern is carbon monoxide. Another metabolite of DCM is carbon dioxide (possibly as much as 65%). During metabolism, carbon monoxide is converted to carboxyhaemoglobin (COHb). The COHb produced is additional to that formed from exogenous carbon monoxide. A functional relationship exists between DCM exposure levels, duration of exposure, the time course and peak blood COHb levels *(6)*. It has been determined in humans and animals that the conversion of DCM to COHb is saturable *(21)*. In man, the kinetics of COHb formation are consistent with Michaelis-Menten kinetics and thus indicate that the DCM–carbon monoxide pathway becomes saturated at an exposure level of about 1400 mg/m^3 or greater ($2 \times K_M$). Because conversion of DCM to carbon monoxide is saturable, zero-order kinetics constrain blood levels of COHb to an upper range of between 12% and 15% *(6)*.

In a controlled study in which nonsmoking sedentary volunteers were exposed for 7.5 hours to 173.5, 347, 520.5 or 694 mg/m^3 of DCM, peak COHb blood levels of 1.9%, 3.4%, 5.0% and 6.8% respectively were formed, indicating a linear relationship between the exposure level to DCM and the COHb blood levels *(18)*. The results of other controlled studies are in general agreement *(22,23)*.

Since normal values of COHb in nonsmokers are generally in the range 0.5–2%, the amount of COHb formed as a result of exposure to the very low levels of DCM in the ambient air can be considered to be of no practical significance.

Elimination
Pulmonary elimination accounts for virtually all unchanged DCM. Since carbon monoxide and carbon dioxide are the principal metabolites, pulmonary elimination is the most significant means by which DCM and its metabolites are eliminated from the body.

Health Effects

Effects on experimental animals and *in vitro* test systems

Toxicological effects
Dichloromethane acts as a CNS depressant at acute levels above 1737 mg/m^3 *(6)*. Effects upon the cardiovascular system (cardiac arrhythmias) are associated with exposure levels exceeding 17 000 mg/m^3.

The available evidence indicates that effects on the liver occur only at exposure levels greatly in excess of those found in the general environment. Such changes include vacuolization, decrease in glycogen and cytochrome P-450 and increased SGPT *(24,25)*. In a study conducted by Burek et al. *(26)* on female rats, slight morphological effects on the liver after 18 months of exposure to levels up to 1735 mg/m^3 were observed. At levels of 347 mg/m^3, slight cytoplasmic vacuolization, increased fat staining, a decrease in cytochrome P-450, and fat infiltration were observed in rats and/or mice *(27–29)*. Haun et al. *(27)* found no observable liver effects after exposure of mice, rats, dogs and monkeys to 87 mg/m^3. Nonspecific renal tubular degeneration and regeneration were observed in rats upon exposure to 347 mg/m^3 for 100 days. No such effects were observed after exposure to 87 mg/m^3. This is the longest continuous exposure study reported.

There is no evidence from studies on laboratory animals to suggest that DCM has any teratogenic potential at levels that do not produce maternal toxicity *(1,6)*. It is probable that the minimal embryonic effects in animals noted in those studies are due to maternal toxicity induced by DCM, as well as to reduced levels of oxygen attributable to elevated COHb levels.

No studies have been found that address the relationship between inhalation exposure to DCM and reproductive effects. No reproductive impairment was found when rats were allowed to mate after receiving drinking-water containing 125 mg/litre for 13 weeks *(30)*.

Mutagenic and carcinogenic effects

Commercial samples of DCM have demonstrated weak positive mutagenic effects in *Salmonella*, yeast and *Drosophila*. Negative results have been reported for gene mutation tests in fungi and mammalian cells in culture. DCM induced chromosomal aberrations in cultured mammalian cells, but not in bone marrow cells *in vivo*. It caused an increase in sister chromatid exchange, but no unscheduled DNA synthesis or inhibition of DNA synthesis *(1,6)*.

In 1982 an IARC working group concluded that DCM could not be classified as to its carcinogenic potential for man *(31)*. The experimental evidence in laboratory animals available at that time was considered insufficient for drawing other conclusions. This evidence, plus additional data, was reviewed subsequently by a Task Group of the International Programme for Chemical Safety *(1)*. It was concluded that the expanded data base was still insufficient for drawing conclusions regarding carcinogenic potential. The additional data included the studies of Burek et al. *(26)*, and those of the National Coffee Association *(32,33)*. At present, the data base has been expanded further with the publication of the US National Toxicology Program report *(34)*. In this lifetime inhalation bioassay, Fischer 344/N rats were exposed to 3470, 6940 and 13 880 mg DCM per m^3, and B6C3F1 mice were exposed to 6940 and 13 880 mg DCM per m^3. It should be noted that the inhalation concentration saturating overall metabolism in rats is about 3000 mg/m^3 *(21)*. Some evidence of a positive carcinogenic response was found in female and male rats (mammary gland and subcutaneous neoplasms). In mice, the reported incidence of lung and liver tumours was statistically significant in both males and females at the two high exposure levels (6940 and 13 880 mg/m^3). At these exposure concentrations, however, toxic effects were apparent in the liver, kidneys and reproductive organs of the animals.

In February 1986 an IARC Working Group met to re-evaluate the carcinogenicity of DCM. This group concluded, on the basis of the above-mentioned recent data, that there was sufficient evidence of carcinogenicity of DCM in experimental animals. Evidence of carcinogenicity in humans was considered to be inadequate *(35)*.

Effects on humans

Toxicological effects

The predominant effects of DCM are nervous system disturbances and elevated COHb levels in the blood.

The lowest concentration reported to affect behaviour during short-term exposure of men and women was 696 mg/m^3 *(36)*. Blood levels of COHb were elevated, suggesting that carbon monoxide may account for performance decrements. Such effects are reversible.

Four studies have been conducted concerning long-term exposure of industrial workers to DCM at levels of about 347 mg/m^3 *(15,37–39)*. In each study, no effect of exposure was observed on the incidence of heart disease. In the study by Cherry et al. *(39)* no neurobehavioural impairments or motor conduction velocity changes attributable to DCM levels of between 260 and 347 mg/m^3 were found.

The available evidence, both epidemiological and experimental, shows a linear relationship between the level of exposure to DCM and COHb levels in the blood. It can be calculated from the data of the controlled study on volunteers by Di Vincenzo & Kaplan *(18)* that a 1% COHb increase would be associated with a 7.5-hour exposure level of about 90 mg DCM per m^3, corresponding to a 24-hour exposure level up to 28 mg/m^3. Because current ambient air concentrations of DCM, as well as maximum levels predicted from dispersion modelling, are orders of magnitude lower than 28 mg/m^3, the contribution of ambient DCM to the formation of COHb is negligible.

Mutagenic and carcinogenic effects

No significant differences in cancer deaths have been observed in the few studies undertaken *(15,37,38)*.

Sensory effects

Dichloromethane has an ethereal odour. The detection threshold level reported by Leonardos et al. *(40)* is 743 mg/m^3.

Evaluation of Human Health Risks

Exposure

Dichloromethane is widely used indoors in paint-stripping and manufacturing. These sources may predominate over outdoor exposures for many individuals. Ambient air levels in urban areas are likely to be below 15 μg/m³. The global average background is well below 1 μg/m³.

Health risk evaluation

Because air levels of DCM in the general environment are orders of magnitude lower than levels associated with adverse effects upon the CNS, it is unlikely that exposure represents a health concern as far as this endpoint is concerned. When the long-term effects of occupational exposure to DCM on the functioning of the CNS are considered, the apparent no-observed-adverse-effect level is about 347 mg/m³ (100 ppm). A similar no-observed-adverse-effect level can be identified in epidemiological studies on workers in whom exposure in relation to heart disease was evaluated. Similarly, the contribution of ambient DCM to the formation of COHb is negligible. Thus, it is unlikely that individuals with cardiovascular problems will be compromised.

Another biological endpoint of interest is the formation of COHb. The recommended maximum acceptable level of COHb in the blood is 3% *(41)*. Because carbon monoxide from other sources may contribute to the formation of COHb at levels approaching this, it is desirable to minimize the additive amounts of COHb derived from exposure to DCM. It is suggested that no more than 0.1% additional COHb be formed from ambient DCM. If 0.1% COHb is regarded as the maximum allowable increase as a consequence of endogenous carbon monoxide formation from exposure to DCM, the corresponding 24-hour exposure concentration of DCM would be about 3 mg/m³ if extrapolation were made from the results of a controlled human study *(18)*.

There are no data relating to effects on the liver or kidneys in humans. Available animal data suggest that 347 mg/m³ represent a lowest-observed-adverse-effect level, and no effect was observed upon exposure to 87 mg/m³. Both levels are considerably in excess of the ambient concentrations found or expected.

Dichloromethane is a weak mutagen in *Salmonella,* yeast and *Drosophila.* It induces chromosomal aberrations in mammalian cells *in vitro* but not *in vivo.* Ambient air levels of DCM are unlikely to represent a hazard to intrauterine development or to have adverse effects on reproduction.

According to recent re-evaluation by IARC, DCM shows sufficient evidence of carcinogenicity in experimental animals but inadequate evidence of carcinogenicity in humans. The same conclusion was recently drawn by the US Environmental Protection Agency *(42)*. The upper-bound incremental unit risk estimate for inhaling 1 μg DCM per m³ over a lifetime was calculated to be 4.1×10^{-6} *(42)*. No specific pharmacokinetic or metabolic data were used for this calculation.

Guidelines

There is sufficient evidence of carcinogenicity of DCM in experimental animals. Evidence of carcinogenicity in humans is inadequate. No risk estimates taking metabolic data into consideration are available; consequently, a cancer risk estimate is not presented in this section.

Using data from available controlled human studies, 24-hour exposure to DCM at a concentration of 3 mg/m^3 has been calculated to increase the blood COHb level by 0.1%, which is considered the maximum allowable increase due to endogenous carbon monoxide formation from exposure to DCM. Since an increase of 0.1% COHb does not add significantly to endogenously formed COHb, the corresponding DCM exposure concentration of 3 mg/m^3 provides sufficient protection. Also, on the basis of animal data showing a lowest-observed-adverse-effect level of 347 mg/m^3 and a no-observed-adverse-effect level of 87 mg/m^3, a value of 3 mg/m^3 is considered appropriate.

On the basis of the above considerations, an air quality guideline for DCM of 3 mg/m^3 (24-hour average) is recommended.

References

1. *Methylene chloride.* Geneva, World Health Organization, 1984 (Environmental Health Criteria, No. 32).
2. *Synthetic organic compounds, US production and sales.* Washington DC, US International Trade Commission, 1984.
3. **International Register of Potentially Toxic Chemicals.** *Data profile on methylene chloride.* Geneva, United Nations Environment Programme, 1984.
4. **Dilling, W.L. et al.** Evaporation rates and reactivities of methylene chloride, chloroform, 1,1,1-trichloroethane, trichloroethylene, tetrachloroethylene, and other chlorinated compounds in dilute aqueous solutions. *Environmental science and technology,* **9**: 833–838 (1975).
5. Dichloromethane. *In: Some halogenated hydrocarbons.* Lyon, International Agency for Research on Cancer, 1979, pp. 449–465 (IARC Monographs on the Evaluation of the Carcinogenic Risk of Chemicals to Humans, Vol. 20).
6. *Health assessment document for dichloromethane (methylene chloride).* Washington, DC, US Environmental Protection Agency, 1985 (Final report No. EPA-600/8-82-004F).
7. **Singh, H.B. et al.** Selected man-made halogenated chemicals in the air and oceanic environment. *Journal of geophysical research,* **88**: 3675–3683 (1983).
8. **Singh, H.B. et al.** Measurements of some potentially hazardous organic chemicals in urban environments. *Atmospheric environment,* **15**: 601–612 (1981).
9. **Singh, H.B. et al.** Distribution of selected gaseous organic mutagens and suspect carcinogens in ambient air. *Environmental science and technology,* **16**: 872–880 (1982).
10. **Guicherit, R. & Schulting, F.L.** The occurrence of organic chemicals in the atmosphere of the Netherlands. *Science of the total environment,* **43**: 193–219 (1985).

11. **Pellizzari, E.D. & Bunch, J.E.** *Ambient air carcinogenic vapours: improved sampling and analytical techniques and field studies.* Washington, DC, US Environmental Protection Agency, 1979 (Report No. EPA-600/2-79-081).
12. **Otson, R. et al.** Dichloromethane levels in air after application of paint removers. *American Industrial Hygiene Association journal,* **42**: 56–60 (1981).
13. **National Institute for Occupational Safety and Health.** *Criteria for a recommended standard: occupational exposure to methylene chloride.* Washington, DC, US Department of Health, Education and Welfare, 1976 (DHEW Publication No. (NIOSH) 76-138).
14. **Kuželova, M. & Vlasak, R.** Vliv metylenchloridu na zdraví pracujících při výrobě filmové fólie a sledování kyseliny mravenčí jako metabolitu metylenchloridu [The effect of methylene dichloride on the health of workers in production of film foils and investigation of formic acid as a methylene dichloride metabolite]. *Pracovni lekarstvi,* **18**: 167–170 (1966).
15. **Ott, M.G. et al.** Health evaluation of employees occupationally exposed to methylene chloride. *Scandinavian journal of work, environment & health,* **9** (Suppl.1): 1–38 (1983).
16. **National Institute for Occupational Safety and Health.** *Technical report: extent-of-exposure survey of methylene chloride.* Washington, DC, US Department of Health and Human Services, 1980 (DHHS Publication No. (NIOSH) 80-181).
17. **Di Vincenzo, G.D. et al.** Human and canine exposure to methylene chloride vapor. *American Industrial Hygiene Association journal,* **33**: 125–135 (1972).
18. **Di Vincenzo, G.D. & Kaplan, C.J.** Uptake, metabolism, and elimination of methylene chloride vapor by humans. *Toxicology and applied pharmacology,* **59**: 130–140 (1981).
19. **Di Vincenzo, G.D. & Kaplan, C.J.** Effect of exercise or smoking on the uptake, metabolism, and excretion of methylene chloride vapor. *Toxicology and applied pharmacology,* **59**: 141–148 (1981).
20. **Åstrand, I. et al.** Exposure to methylene chloride. I. Its concentration in alveolar air and blood during rest and exercise and its metabolism. *Scandinavian journal of work, environment & health,* **1**: 78–94 (1975).
21. **McKenna, M.J. et al.** The pharmacokinetics of inhaled methylene chloride in rats. *Toxicology and applied pharmacology,* **65**: 1–10 (1982).
22. **Stewart, R.D. et al.** Experimental human exposure to methylene chloride. *Archives of environmental health,* **25**: 342–348 (1972).
23. **Ratney, R.S. et al.** *In vivo* conversion of methylene chloride to carbon monoxide. *Archives of environmental health,* **28**: 223–226 (1974).
24. **Balmer, M.F. et al.** Effects in the liver of methylene chloride inhaled alone and with ethyl alcohol. *American Industrial Hygiene Association journal,* **37**: 345–352 (1976).
25. **Morris, J.B. et al.** Studies on methylene chloride-induced fatty liver. *Experimental and molecular pathology,* **30**: 386–393 (1979).

26. **Burek, J.D. et al.** Methylene chloride: a two-year inhalation toxicity and oncogenicity study in rats and hamsters. *Fundamental and applied toxicology,* **4**: 30–47 (1984).
27. **Haun, C.C. et al.** Continuous animal exposure to low levels of dichloromethane. *In: Proceedings of the 3rd Annual Conference on Environmental Toxicology.* Wright-Patterson Air Force Base, OH, Aerospace Medical Research Laboratory, 1972, pp. 199–208 (AMRL-TR-72-130, Paper No. 12).
28. **Weinstein, R.S. & Diamond, S.S.** Hepatotoxicity of dichloromethane with continuous inhalation exposure at a low dose level. *In: Proceedings of the 3rd Annual Conference on Environmental Toxicology.* Wright-Patterson Air Force Base, OH, Aerospace Medical Research Laboratory, 1972, pp. 209–222 (AMRL-TR-72-130, Paper No. 13).
29. **Weinstein, R.S. et al.** Effects of continuous inhalation of dichloromethane in the mouse: morphologic and functional observations. *Toxicology and applied pharmacology,* **23**: 660–679 (1972).
30. **Bornmann, G. & Loeser, A.** Zur Frage einer chronisch toxischen Wirkung von Dichlormethan [The question of the chronic toxic action of dichloromethane]. *Zeitschrift für Lebensmittel-Untersuchung und Forschung,* **136**: 14–18 (1967).
31. Dichloromethane. *In: Chemicals, industrial processes and industries associated with cancer in humans. IARC Monographs, Volumes 1 to 29.* Lyon, International Agency for Research on Cancer, 1982, pp. 111–112 (IARC Monographs on the Evaluation of the Carcinogenic Risk of Chemicals to Humans, Supplement 4).
32. *Methylene chloride, final report, 24-month chronic toxicity and oncogenicity study in rats.* Washington, DC, Hazelton Laboratories America Inc., National Coffee Association, 1982 (Project No. 2112-101).
33. *24-month oncogenicity study of methylene chloride in mice. Final report.* Washington, DC, Hazelton Laboratories America Inc., National Coffee Association, 1983 (Project No. 2112-101).
34. *Toxicology and carcinogenesis studies of dichloromethane (methylene chloride) (CAS No. 75-09-2) in F344/N rats and B6C3F1 mice (inhalation studies).* Research Triangle Park, NC, National Toxicology Program, 1986 (NIH Publication No. 86-2562).
35. *Some halogenated hydrocarbons and pesticide exposures.* Lyon, International Agency for Research on Cancer, 1986 (IARC Monographs on the Evaluation of the Carcinogenic Risk of Chemicals to Humans, Vol. 41).
36. **Putz, V.R. et al.** A comparative study of the effects of carbon monoxide and methylene chloride on human performance. *Journal of environmental pathology and toxicology,* **2**: 97–112 (1976).
37. **Friedlander, B.R. et al.** Epidemiologic investigation of employees chronically exposed to methylene chloride. *Journal of occupational medicine,* **20**: 657–666 (1978).
38. **Hearne, T. & Friedlander, B.R.** Follow-up of methylene chloride study. *Journal of occupational medicine,* **23**: 660 (1981).

39. **Cherry, N. et al.** Some observations on workers exposed to methylene chloride. *British journal of industrial medicine,* **38**: 351–355 (1981).
40. **Leonardos, G. et al.** Odor threshold determinations of 53 odorant chemicals. *Journal of the Air Pollution Control Association,* **19**: 91–95 (1969).
41. *Carbon monoxide.* Geneva, World Health Organization, 1979 (Environmental Health Criteria, No. 13).
42. *Addendum to health assessment document for dichloromethane (methylene chloride).* Washington, DC, US Environmental Protection Agency, 1985 (Final Report No. EPA-600/-8-82-004F).

10

Formaldehyde

General Description

Formaldehyde (HCHO) is the simplest and most common aldehyde found in the environment. At normal room temperatures it is a colourless gas with a pungent odour. Formaldehyde is soluble in water and is used in solution or in polymerized form (paraformaldehyde).

Sources

In the natural environment, formaldehyde is an intermediary in the methane cycle, with low background concentrations. Anthropogenic sources include direct emissions, especially from the production and use of formaldehyde, and secondary reactions of oxidized hydrocarbons from stationary and mobile sources. The major anthropogenic sources affecting humans are in the indoor environment. Products containing formaldehyde, such as resins, are common. Primary sources include insulating materials, chipboard and plywood, and fabrics; other sources are cigarette smoke, heating and cooking.

Occurrence in air

The natural background concentration is a few $\mu g/m^3$. In urban air the annual average is approximately 0.005–0.01 mg/m^3; it is higher in the vicinity of industrial processes. Short-term peaks (e.g. at peak traffic times in built-up urban areas or in conditions of photochemical smog) are about one order of magnitude greater.

Indoor concentrations can be very high in prefabricated buildings where chipboard is the main construction material (0.1–5.0 mg/m^3); however, the highest values in this range occur only very rarely. Present averages in these buildings tend to be lower (0.1–1.0 mg/m^3), due to better product control in the past few years. Recent surveys indicate concentrations in conventional buildings averaging between 0.05 and 0.1 mg/m^3, depending on the age of the building and the products used; rarely do these concentrations exceed 0.2 mg/m^3. Concentrations in buildings whose residents have reported symptoms are generally somewhat higher; they are influenced by temperature, humidity and ventilation.

Conversion factors

1 ppm (20–25 °C) = 1.2 mg/m³

1 mg/m³ = 0.833 ppm

Routes of Exposure

The possible routes of exposure to formaldehyde are ingestion, inhalation, dermal absorption and, rarely, blood exchange (as in dialysis).

Air

Assuming a breathing rate of 20 m³ per day for an average adult, given the exposures mentioned above and making different assumptions of the time spent in various environments, one can calculate inhalation exposure per day. Average time estimates lead to the conclusion that people spend 60–70% of their time in the home, 25% at work and 10% outdoors. If one assumes that normal work exposures are similar to home exposures and the data given on the occurrence of formaldehyde in air are used, the daily exposure resulting from breathing is about 1 mg per day, with a few exposures at >2 mg per day. This compares favourably with the estimated range of 0.3–2.1 mg per day, based on the work of Kalinic et al. *(1)*, with estimated weighted average exposures of 0.02–0.14 mg/m³.

Occupational exposure

Occupational exposure contributes to total exposure. For example, a high occupational exposure (e.g. in formaldehyde or resin production, or during disinfection procedures) of 1 mg/m³ for a 25% time-weighted period would give a daily intake of about 5 mg per day.

Smoking

Concentrations of 60–130 mg/m³ were measured in mainstream cigarette smoke *(2)*. For a person smoking 20 cigarettes per day this would lead to an exposure of 1 mg per day. Exposure to sidestream smoke (or environmental tobacco smoke: ETS) can be estimated from chamber measurements. When 6 cigarettes were smoked in a 50 m³ test chamber with one air change per hour (ACH), formaldehyde levels were over 0.12 mg/m³ *(3)* within 15 minutes. Weber-Tschopp and co-workers *(4)* measured the yield of 5–10 cigarettes in a 30 m³ chamber with low ACH (0.2–0.3) as 0.21–0.35 mg/m³, which would be about 0.05–0.07 at 1 ACH. This concentration is in the same range as that likely to be found in rooms of most conventional buildings without cigarette smoke (see the section on the occurrence of formaldehyde in air, p. 91). If, however, ETS is a source, and there are other sources, then the equilibrium concentration is not the sum of these concentrations. When actually measured, ETS contributes only 10–25% of total indoor exposure.

Drinking-water

Except for accidental contamination of water with formaldehyde, concentrations in drinking-water can be expected to be less than 0.1 mg/litre; intake from this source can therefore be considered negligible (below 0.2 mg/day).

Food

Some formaldehyde occurs naturally in raw food (e.g. meat, certain kinds of fruit and vegetable) and some accidental contamination results from fumigation processes (e.g. of grain). Sources include the cooking or smoking of food. The daily intake is difficult to evaluate, but a rough estimate from the available data is in the range of 1.5–14 mg per day for an average adult, most of it in a bound and unavailable form.

Other routes of exposure

Skin absorption

Dermal contact and absorption may occur during the use of cosmetics, household products, disinfectants, textiles (especially synthetics) and orthopaedic casts. Skin exposure from most of these sources is localized, although some formaldehyde is available for inhalation. Systemic absorption through the entire epidermal layer and beyond, including across the circulatory layer, is estimated to be negligible. Contact with liquid barriers, as in the eyes, does not appear to lead to absorption.

Blood exchanges

In certain rare events, formaldehyde in aqueous solution enters the bloodstream directly. These events are most likely to occur in dialysis or in surgery with assisted circulation, in which the dialysis machine and tubes are disinfected with formaldehyde. Formaldehyde from adsorption or backwashes can then enter the patient's bloodstream.

Relative significance of different routes of exposure

As the kinetics of formaldehyde uptake and metabolism are a function of the mode of entry, biological effects may vary with the range of exposure.

It is assumed that inhalation is the primary route by which formaldehyde affects the organism. Absorption by the skin predominantly affects the first three layers, and little if any reaches the bloodstream. There is a relatively large exposure to formaldehyde from ingestion, but almost all of it is excreted as harmless metabolites. Blood exchange is a critical form of exposure but is very rare, even in the very small segment of the population at risk. It is only in inhalation that the irritant and other properties of formaldehyde directly affect the organism. An accurate estimate of its relative contribution would require measurement of the proportionate weights of formaldehyde, its metabolites, excretion rates, etc.

Kinetics and Metabolism

Formaldehyde gas is readily absorbed in the respiratory and gastrointestinal tract and metabolized; its metabolites circulate to other organ systems; the half-time in the blood is short. Most metabolites are excreted quickly, as is bound formaldehyde.

Absorption
Owing to its solubility in water, formaldehyde is rapidly absorbed in the upper respiratory tract. Formaldehyde adducts can cross the membranes. Percutaneous absorption (i.e. localized to the point of contact) does occur.

Distribution
Studies of exposed humans have shown no differences in blood levels compared with those of controls. Orally administered formaldehyde resulted in a rapid increase in formate levels (but not formaldehyde levels) in the blood of dogs. These increases are considered transient.

Biotransformation
Formaldehyde reacts virtually instantaneously with primary and secondary amines, thiols, hydroxyls and amides to form methylol derivatives. Formaldehyde is oxidized to formate along three pathways; hydrogen peroxide constitutes one of these. In the body formaldehyde is produced in small quantities as a normal metabolite and also in the oxidative demethylation of drugs and other foreign substances; it may therefore be found in the liver *(5)*.

Formaldehyde, like many other chemical carcinogens, acts as an electrophile and can react with macromolecules such as DNA, RNA and protein to form reversibly bound adducts or irreversible crosslinks.

Elimination
Formaldehyde disappears from the plasma with a half-time of about 1–1.5 minutes, most of it being converted to carbon dioxide and exhaled via the lungs. Smaller amounts are excreted via the urine as formate salts and several other metabolites.

Health Effects

Effects on experimental animals and *in vitro* test systems

Toxicological effects
Acute exposures of mice and rats to formaldehyde concentrations of 120 mg/m^3 produced salivation, dyspnoea, vomiting, spasms and death *(6)*. At concentrations of 12–50 mg/m^3 rats, mice and guinea pigs showed decreased respiratory rate, increased airway resistance and decreased compliance *(7–9)*, as well as eye irritation *(10)*. Acute exposure to low concentrations (1.2 mg/m^3) resulted in decreases in the respiratory rate of rats and mice *(8)*, and increased airway resistance and decreased compliance in guinea pigs *(11)*. Marked differences are found between species in the irritancy response to formaldehyde. Mice are more responsive than rats and are measured by the reduction in their respiratory rate. Exposure of rats to 20 mg/m^3 for several days resulted in dose-related mucostasis and ciliastasis. Cell degeneration, inflammation and necrosis, as well as early squamous-cell metaplasia, were found in the nasal epithelium at 20 mg/m^3 *(12,13)*. Increased cell proliferation was noted in the nasal cavity of rats at between 7 and 25 mg/m^3, but not at concentrations below this range in rats

and mice *(13)*. Effects were more clearly related to concentration than to concentration × time *(13)*.

Rats exposed to 5.7 mg/m^3 showed a slower weight gain during a 3-month exposure; at 10 mg/m^3 there was a decrease in liver weight and in macrophage activity and food intake *(14)*. F/344 rats exposed for 2 years to concentrations of 2.5–18 mg/m^3 developed dose-related nasal lesions *(15)*. The lesions included dysplasia and squamous metaplasia of the respiratory epithelium, the olfactory epithelium and the respiratory epithelium in the trachea. Some regression was observed after exposure ceased. Golden hamsters remained free of metaplasia at concentrations of up to 3.7 mg/m^3 for 6 months, with minimal occurrence at 12 mg/m^3 for a lifetime; cynomolgus monkeys developed an increased incidence of squamous metaplasia after 6 months at a concentration of 3.7 mg/m^3 *(16)*. Animal studies on the effect of formaldehyde on the immune system failed to show a clear effect in the hamster, except at lethal concentrations *(17)*; guinea pigs showed sensitization from dermal exposure to formaldehyde *(18)*.

Primary irritant effects often include inflammation, oedema, and increased permeability of the cell wall. Formaldehyde inhibits ciliary movement and reduces mucociliary clearance.

Although a number of animal studies have been conducted, the data do not indicate that formaldehyde is embryotoxic or teratogenic. The total evidence is still somewhat limited.

Mutagenic and carcinogenic effects

Formaldehyde has been shown to be mutagenic in several *in vitro* systems, and would have to be classified as a weak mutagen. *In vivo,* formaldehyde has been shown to be mutagenic only in male *Drosophila* larvae, but most studies have reported negative results.

Formaldehyde has been found to produce a high incidence of squamous-cell carcinomas in the nose of two strains of rats of both sexes, but only at the highest dose (18.7 mg/m^3). The dose–response relationship was clearly nonlinear, with a disproportionate increase in tumours at higher concentrations. In mice no statistically significant increase in nasal tumours, even at 18.7 mg/m^3, was found *(15)*. Hamsters also show no carcinogenic effect from formaldehyde *(17)*.

Effects on humans

Toxicological effects

Data on acute toxicity are available mainly from epidemiological studies of occupationally exposed populations and residents of buildings constructed of materials containing formaldehyde, and from controlled human exposure studies.

The symptoms displayed by humans after short-term exposure to formaldehyde are similar to those observed in animals: irritation of the eyes, nose and throat, together with exposure-dependent discomfort, lachrymation, sneezing, coughing, nausea and dyspnoea *(19)*. Symptoms are often more severe at the start of exposure and after minutes or hours they diminish. Children have been reported to be more sensitive *(20)*, their condition improving after removal of the sources. Table 1 shows that human responses start with an odour threshold at about 0.1 mg/m^3, progressing to eye and throat irritation at 0.5 mg/m^3; irritation and discomfort sharply increase between 1 and 20 mg/m^3, with danger to life at concentrations of 30 mg/m^3 and higher.

Table 1. Effects of formaldehyde in humans after short-term exposure

Effect	Formaldehyde concentration (in mg/m^3)	
	Estimated median	Reported range
Odour detection threshold (including repeated exposures)	0.1	0.06-1.2
Eye irritation threshold	0.5	0.01-1.9
Throat irritation threshold	0.6	0.1-3.1
Biting sensation in nose, eye	3.1	2.5-3.7
Tolerable for 30 minutes (lachrymation)	5.6	5-6.2
Strong lachrymation, lasting for 1 hour	17.8	12-25
Danger to life, oedema, inflammation, pneumonia	37.5	37-60
Death	125	60-125

Numerous reports show that exposure to formaldehyde vapour causes direct nonimmunological irritation of the skin *(18)*. A single application of 1% formalin in water on human skin will produce an irritant response in about 5% of the population *(21)*. Allergic contact dermatitis develops in man, but the threshold for induction is uncertain *(18)*.

A number of reports also show that formaldehyde gas exposure causes direct irritation of the respiratory tract (reviewed in *(18)*). Because of absorption in the upper respiratory tract, higher concentrations of formaldehyde are required to stimulate bronchial receptors than those needed to cause sensory irritation. If formaldehyde is absorbed by particles, peripheral lung tissue receptors might be stimulated by the hydrolytic release of formaldehyde *(22)*. No precise thresholds exist for the irritant effects of inhaled formaldehyde; most people experience irritation of the throat within the range 0.12–3.7 mg/m^3 *(18)*. Between 12 and 25 mg/m^3 symptoms are severe and it becomes difficult to breathe normally *(23)*. Effects in the pulmonary tissue and the lower airways are likely at concentrations of 6–40 mg/m^3, while pulmonary oedema, pneumonitis and pneumonia occur at concentrations of 60–120 mg/m^3 *(24)*. Some investigators report little or no adverse effect of formaldehyde on airway resistance in the range 1–3 mg/m^3 *(24)*, but above this range, with moderate to heavy exercise, some decrease in lung function can occur.

A number of studies point to formaldehyde as a potential factor predisposing certain groups, particularly children, to respiratory tract infections *(25)*.

Sensitization arising from the release of formaldehyde into the circulation of chronic haemodialysis patients shows evidence of formaldehyde-dependent immunization *(18)*.

No sufficiently well controlled scientific studies have been made that establish definitely that formaldehyde gas is able *per se* to cause a respiratory tract allergy. Clinical reports describe tests for respiratory sensitivity to formaldehyde gas in which allergic reactions were elicited, but their interpretation is uncertain. There may be susceptible groups or genetic differences in the population. Occupational studies indicate that 1–2% of the population exposed to high concentrations may develop asthma *(26)*.

Neurochemical and morphological studies indicated changes in the nervous system of experimental animals. In short-term exposures of humans to concentrations of 0.3–2.5 mg/m^3 for 5 hours no changes in the performance of mathematical tests were observed *(19)*. Two studies of long-term formaldehyde exposure were carried out in adult populations. Olsen & Dossing *(27)* observed no effects on the memory or concentration of subjects who underwent long-term exposure to 0.008–0.43 mg/m^3; similar results were obtained by Schenker et al. *(28)*.

Epidemiological studies in which neuropsychological symptoms due to occupational or environmental exposure to formaldehyde were evaluated have failed to overcome the problems commonly associated with such studies *(18)*.

Two occupational studies have been made on the irritant potential of formaldehyde in chronic disease. In a study at wood-processing plants, workers exposed to formaldehyde had a higher incidence of chronic upper respiratory disease *(29)*. Particle-borne formaldehyde was not measured in this study, and the air concentrations of formaldehyde gas in all locations were below 0.4 mg/m^3. Workers in an acrylic-wool filter department where the work environment had phenol levels of 7–10 mg/m^3 and formaldehyde levels of 0.5–1.0 mg/m^3 had lower lung function values, but they may also have been exposed to particles and fibres that were not monitored *(30)*.

A number of other studies are even more difficult to interpret. It is unlikely that chronic obstructive lung disease would occur in people exposed to less than 1.8 mg/m^3 and it is possible that some chronic disease might occur in those exposed to concentrations above 10 mg/m^3. Several studies have addressed the possible links between formaldehyde and pregnancy outcome, frequency of menstrual disorders, pregnancy complications and low-birth-weight babies, and reported increased effects. However, owing to the scarcity and limitation of the available data, it is not possible to draw definite conclusions *(18)*.

Mutagenic and carcinogenic effects

Data concerning the potential carcinogenicity of formaldehyde in humans are available only from studies of occupationally exposed populations. To date, five historical cohort studies *(31–35)*, three proportional mortality studies *(36–38)* and six case control studies *(39–44)* have been conducted. Those studied were professionals who used formaldehyde in the preservation of biological tissues (embalmers, anatomists and pathologists) and industrial workers involved in the production and use of formaldehyde. All these studies have limitations: confounding factors such as smoking or exposures to other chemicals could not be taken into account in the design or analysis of most of the investigations. In addition, quantitative information about exposure is usually not available. None of the studies provides conclusive evidence about the carcinogenicity of formaldehyde in humans. However, owing to the low statistical power of the best studies conducted to date, the possibility cannot be excluded that formaldehyde is a human carcinogen. One study indicates a possible slight excess risk of nasal cancer *(42)*, and several studies found a slight excess in cancer of other sites such as brain or skin. Again, it is not possible to attribute the risk clearly to formaldehyde.

No detectable differences in the frequency of chromosome aberrations and sister chromatid exchange were found in six pathology workers compared with unexposed controls *(45)*. A study of haemodialysis patients showed marked chromosome abnormalities. Such patients are at risk with regard to direct entry of formaldehyde via equipment, but the study did not investigate control subjects and is therefore difficult to interpret *(46)*.

Sensory effects

Formaldehyde has a pungent odour with an odour detection threshold of 0.06 mg/m^3 *(47)*; the odour recognition threshold is not known. It poses nuisance problems in indoor environments owing to its release from building materials or furnishings. It is not possible, on the basis of the scientific literature, to state a specific limit concentration of formaldehyde at which odour nuisance starts to appear. Indoor air usually contains other organic compounds which, in combination with formaldehyde or by themselves, may have odorous and irritating properties causing discomfort *(48)*. Some sensitive individuals can sense formaldehyde concentrations of 0.01 mg/m^3 and perhaps even lower by a "warm" feeling on the face *(48)*.

Evaluation of Human Health Risks

Exposure

Table 2 shows the contribution of various atmospheric environments to the average exposure of humans to formaldehyde. Ingestion contributes about 1–10 mg per day, almost completely from food intake. Because formaldehyde

Table 2. Contribution of various atmospheric environments to average exposure to formaldehyde[a]

Source	mg/day
Air	
Ambient air (10% of time)	0.02
Indoor air	
Home (65% of time)	
— conventional	0.5– 2.0
— prefabricated (chipboard)	1.0–10.0
Workplace (25% of time)	
— without occupational exposure[b]	0.2– 0.8
— with 1 mg/m^3 occupational exposure	5
— environmental tobacco smoke (ETS)	0.1– 1.0
Smoking (20 cigarettes/day)	1.0

[a] The contribution of food and water is ignored here, but was discussed earlier.

[b] Assuming the normal formaldehyde concentration in conventional buildings.

in food is not available in free form this type of exposure can be disregarded. The major route of exposure is thus inhalation via ambient air, indoor air, or from smoking. Table 2 shows that inhalation of indoor air is the major route of entry, with releases from chipboard and other building and furnishing materials constituting the bulk of the exposure. The Table also suggests that occupants of prefabricated buildings incorporating chipboard are likely to inhale 2–3 times as much formaldehyde as occupants of conventional buildings.

Health risk evaluation

Available clinical and epidemiological data indicate that substantial variations in individual responses to formaldehyde exist. Table 1 lists the ranges of the effects of short-term exposure to formaldehyde.

The threshold of irritation is reported to be as low as 0.1 mg/m^3, but significant increases in symptoms of irritation occur at levels between 0.3 and 1.0 mg/m^3 in healthy subjects. At concentrations above 1.2 mg/m^3 a progression of symptoms and effects occurs.

Estimating the human carcinogenic risk from formaldehyde on the basis of animal data is a matter of great uncertainty, since the dose–response curve is nonlinear at the higher exposure levels and the situation is even more complex if such nonlinearity continues throughout the entire dose-response curve.

The nonlinear relationship between administered and delivered formaldehyde doses has a substantial impact on quantitative estimates of risk associated with low-level exposures. This was illustrated *(49)* by comparing estimates of risk obtained using administered dose (airborne formaldehyde concentration) as the measure of exposure with corresponding estimates obtained using delivered dose (the concentration of covalently bound formaldehyde in target tissue DNA). The risk estimates based on delivered dose were approximately 53 times smaller when expressed as the maximum likelihood estimates (Table 3).

Table 3. Multistage model estimates of risk of squamous-cell carcinoma in rats, based on administered dose and delivered dose at selected airborne formaldehyde concentrations

Airborne concentration (ppm)	Dose measure	Maximum likelihood risk estimate	Upper 95% confidence limit
1.0	Administered	251×10^{-6}	16.0×10^{-4}
	Delivered	4.7×10^{-6}	6.2×10^{-4}
0.5	Administered	314×10^{-7}	8.1×10^{-4}
	Delivered	5.9×10^{-7}	3.1×10^{-4}
0.1	Administered	251×10^{-9}	15.6×10^{-5}
	Delivered	4.7×10^{-9}	6.2×10^{-5}

In spite of the relatively large amount of information gathered on formaldehyde in terms of its genotoxicity, pharmacokinetics and carcinogenicity, differences in assumptions can lead to risk estimate differences running into orders of magnitude. Therefore, no risk estimate calculation is indicated because available animal data do not allow a reasonable use of existing models.

Guidelines

There is sufficient evidence of the carcinogenicity of formaldehyde in experimental animals; the evidence of carcinogenicity in humans is inadequate. Formaldehyde was classified in Group 2B by IARC *(50)*. However, no risk estimate calculation is indicated because available animal data do not allow a reasonable use of existing models.

Formaldehyde is an odorant with an odour detection threshold of 0.06 mg/m^3, at which level there is little or no concern about human discomfort. The threshold of irritation is reported to be as low as 0.1 mg/m^3

after short-term exposure, but significant increases in symptoms of irritation start at levels above 0.3 mg/m^3 in healthy subjects. In order to avoid complaints of sensitive people about indoor air in nonindustrial buildings, the formaldehyde concentration should be below 0.1 mg/m^3 as a 30-minute average, and this is recommended as an air quality guideline value.

Building codes and production and processing regulations should take into account the numerous sources that may contribute to indoor formaldehyde levels.

References

1. **Kalinic, N. et al.** Formaldehyde levels in selected indoor microenvironments. *In:* Berglund, B. et al., ed. *Indoor air: sensory and hyperreactivity reactions to sick buildings.* Stockholm, Swedish Council for Building Research, 1984, Vol. 3.
2. **Weber-Tschopp, A. et al.** Reizwirkung von Formaldehyd auf dem Menschen [Irritant effects of formaldehyde on man]. *International archives of occupational and environmental health,* **39**: 207–218 (1977).
3. **Menzel, W. et al.** *Formaldehyd-Emission-Zigaretten/Tabak-Prüfraum* [Methods of formaldehyde measuring]. Braunschweig, Wilhelm Klauditz Institute, 1982 (WKI Kurzbericht, No. 13).
4. **Weber-Tschopp, A. et al.** Objektive und subjektive physiologische Wirkungen des Passivrauchens [Objective and subjective physiological effects of passive smoking]. *International archives of occupational and environmental health,* **37**: 277–288 (1976).
5. **Kleeberg, U. & Klinger, W.** Sensitive formaldehyde determination with Nash's reagent and tryptophan reaction. *Journal of pharmacology methods,* **8**: 19–31 (1982).
6. Formaldehyde. *In: Some industrial chemicals and dyestuffs.* Lyon, International Agency for Research on Cancer, 1982, pp. 345–389 (IARC Monographs on the Evaluation of the Carcinogenic Risk of Chemicals to Humans, Vol. 29).
7. **Chang, J.C.F. et al.** Effect of single or repeated formaldehyde exposure on minute volume of B6C3F1 mice and F-344 rats. *Toxicology and applied pharmacology,* **61**: 451–459 (1981).
8. **Barrow, C.S. & Steinhagen, W.S.** Sensory irritation by inhaled formaldehyde in B6C3F1 mice and F-344 rats following single or repeated exposure. *Toxicologist,* **1**: 5–6 (1981).
9. **Kane, L.E. & Alarie, Y.** Sensory irritation to formaldehyde and acrolein during single and repeated exposures in mice. *American Industrial Hygiene Association journal,* **38**: 509–522 (1977).
10. **Salem, H. & Cullumbine, H.** Inhalation toxicities of some aldehydes. *Toxicology and applied pharmacology,* **2**: 183–187 (1960).
11. **Amdur, M.O.** The response of guinea pigs to inhalation of formaldehyde and formic acid alone and with a sodium chloride aerosol. *International journal of air pollution,* **3**: 201–220 (1960).

12. **Morgan, K.T. et al.** Formaldehyde and the nasal mucociliary apparatus. *In:* Clary, J.J. et al., ed. *Formaldehyde, toxicology, epidemiology and mechanisms.* New York, Marcel Dekker, 1983, pp. 193–209.
13. **Swenberg, J.A. et al.** Mechanisms of formaldehyde toxicity. *In:* Gibson, J.E., ed. *Formaldehyde toxicity.* New York, Hemisphere Publishing Corporation, 1983.
14. **Dubreuil, A. et al.** Inhalation, en continu, de faibles doses de formaldehyde : étude experimentale chez le rat [Continuous inhalation of low doses of formaldehyde: experimental study on the rat]. *European journal of toxicology,* **9**: 245–250 (1976).
15. **Kerns, W.D. et al.** The chronic effects of formaldehyde inhalation in rats and mice: a preliminary report. *In:* Gibson, J.E., ed. *Formaldehyde toxicity.* New York, Hemisphere Publishing Corporation, 1983.
16. **Rusch, G.M. et al.** A 26-week inhalation toxicity study with formaldehyde in the monkey, rat and hamster. *Toxicology and applied pharmacology,* **68**: 329–343 (1983).
17. **Dalbey, W.E.** Formaldehyde and tumors in hamster respiratory tract. *Toxicology,* **24**: 9–14 (1982).
18. Report on the Consensus Workshop on Formaldehyde. *Environmental health perspectives,* **58**: 323–381 (1984).
19. **Andersen, I. & Molhave, L.** Controlled human studies with formaldehyde. *In:* Gibson, J.E., ed. *Formaldehyde toxicity.* New York, Hemisphere Publishing Corporation, 1983.
20. **Burdach, S. & Wechselberg, K.** Gesundheitsschäden in der Schule [Damage to health at school]. *Fortschritte der Medizin,* **98**: 379–384 (1980).
21. **Maibach, H.** Formaldehyde: effects on animal and human skin. *In:* Gibson, J.E., ed. *Formaldehyde toxicity.* New York, Hemisphere Publishing Corporation, 1983.
22. **Alarie, Y.** Toxicological evaluation of airborne chemical irritants and allergens using respiratory reflex reactions. *In:* Leong, B.K.J., ed. *Proceedings of a Symposium on Inhalation Toxicology and Technology.* Ann Arbor, MI, Ann Arbor Science, 1981.
23. **Fassett, D.W.** Aldehydes and acetals. *In:* Patty, F.E., ed. *Industrial hygiene and toxicology,* 2nd ed. New York, Interscience, 1963, Vol. 2.
24. **Schachter, E.N. et al.** Respiratory effects of exposure to 2.0 ppm formaldehyde in healthy subjects. *American review of respiratory disease,* **129**: A151 (1984).
25. **Helwig, H.** Wie ungefährlich ist Formaldehyd? [How safe is formaldehyde?]. *Deutsche medizinische Wochenschrift,* **102**: 1612–1613 (1977)
26. **Frazier, C.A.** *Occupational asthma.* New York, Van Nostrand Reinhold, 1980.
27. **Olsen, H. & Dossing, M.** Formaldehyde induced symptoms in day care centres. *American Industrial Hygiene Association journal,* **43**: 366–370 (1982).
28. **Schenker, M.B. et al.** Health effects of residence in homes with urea formaldehyde foam insulation: a pilot study. *Environment international,* **8**: 359–363 (1982).

29. **Efremov, G.G.** [The upper respiratory tract in formaldehyde production workers]. *Žurnal ušnyh nosovyh, i gorlovyh boleznej,* **30**: 11–15 (1970) (in Russian).
30. **Schoenberg, J.B. & Mitchell, C.A.** Airway disease caused by phenolic (phenol-formaldehyde) resin exposure. *Archives of environmental health,* **30**: 574–577 (1975).
31. **Acheson, E.D. et al.** Formaldehyde in the British chemical industry. *Lancet,* **I**: 611–616 (1984).
32. **Wong, O.** An epidemiologic mortality study of a cohort of chemical workers potentially exposed to formaldehyde, with a discussion on SMR and PMR. *In:* Gibson, J.E., ed. *Formaldehyde toxicity.* New York, Hemisphere Publishing Corporation, 1983.
33. **Harrington, J.M. & Shannon, H.S.** Mortality study of pathologists and medical laboratory technicians. *British medical journal,* **4**: 329-332 (1975).
34. **Levine, R.J. et al.** The mortality of Ontario undertakers and a review of formaldehyde-related mortality studies. *Journal of occupational medicine,* **26**: 740–746 (1984).
35. **Harrington, J.M. & Oakes, D.** Mortality study of British pathologists 1974–80. *British journal of industrial medicine,* **41**: 188–191 (1984).
36. **Liebling, T. et al.** Cancer mortality among workers exposed to formaldehyde. *American journal of industrial medicine,* **5**: 423–428 (1984).
37. **Marsh, G.M.** Proportional mortality patterns among chemical plant workers exposed to formaldehyde. *British journal of industrial medicine,* **39**: 313–322 (1982).
38. **Walrath, J. & Fraumeni, J.F.** Mortality patterns among embalmers. *International journal of cancer,* **31**: 407–411 (1983).
39. **Hernberg, S. et al.** Nasal and sinonasal cancer. *Scandinavian journal of work, environment & health,* **9**: 315–326 (1983).
40. **Krüger Anderson, S. et al.** Formaldehydeksponering og lungecancer blandt danske læger [Formaldehyde exposure and lung cancer among Danish physicians]. *Ugeskrift for læger,* **144**: 1571–1573 (1982).
41. **Olsen, J.H. et al.** Occupational formaldehyde exposure and increased nasal cancer risk in man. *International journal of cancer,* **34**: 639–644 (1984).
42. **Brinton, L.A. et al.** A case-control study of cancers of the nasal cavity and paranasal sinuses. *American journal of epidemiology,* **119**: 896–906 (1984).
43. **Fayerweather, W.E. et al.** Case-control study of cancer deaths in Du Pont workers with potential exposure to formaldehyde. *In:* Clary, J.J. et al., ed. *Formaldehyde, toxicology, epidemiology and mechanisms.* New York, Marcel Dekker, 1983.
44. **Hayes, R.B. et al.** *Tumors of the nose and nasal sinuses: a case-control study.* Rotterdam, Erasmus University, Department of Public Health and Social Medicine, 1984.
45. **Thomas, E.J. et al.** Chromosomal aberrations and sister-chromatid exchange frequency in pathology staff occupationally exposed to formaldehyde. *Mutation research,* **141**: 89–93 (1984).

46. **Goh, K. & Cestero, R.V.M.** Chromosomal abnormalities in maintenance haemodialysis patients. *Journal of medicine,* **10**: 167–174 (1979).
47. **Berglund, B. et al.** Measurement of formaldehyde odor indoors. *In:* Fanger, P.O. ed. *Clima 2000: indoor climate.* Copenhagen, VVS Kongress-VVS Messe, 1985, Vol. 4, pp. 251–257.
48. **Ahlström, R. et al.** Odor interaction between formaldehyde and the indoor air of a "sick building". *In:* Berglund, B. et al., ed. *Indoor air: sensory and hyperreactivity reactions to sick buildings.* Stockholm, Swedish Council for Building Research, 1984, Vol. 3, pp. 461–466.
49. **Swenberg, J.A. et al.** A scientific approach to formaldehyde risk assessment. *Banbury reports,* **19**: 255–267 (1985).
50. *Chemicals, industrial processes and industries associated with cancer in humans. IARC Monographs, Volumes 1 to 29.* Lyon, International Agency for Research on Cancer, 1982 (IARC Monographs on the Evaluation of the Carcinogenic Risk of Chemicals to Humans, Supplement 4).

11

Polynuclear aromatic hydrocarbons (PAH)

General Description

Polynuclear (or polycyclic) aromatic hydrocarbons (PAH) are a large group of organic compounds with two or more benzene rings. They have a relatively low solubility in water but are highly lipophilic. Almost the total amount of PAH with low vapour pressure in the air is adsorbed onto particles. Polynuclear aromatic hydrocarbons dissolved in water or adsorbed on particulate matter can undergo photodecomposition when exposed to ultraviolet light from solar radiation. Some microorganisms in soil can degrade PAH.

Sources

Polynuclear aromatic hydrocarbons are formed mainly as a result of pyrolytic processes, especially the incomplete combustion of organic materials, as well as in natural processes such as carbonization *(1,2)*. There are several hundred PAH; the best-known is benzo[*a*]pyrene (BaP). The names of several PAH are shown in Fig. 1. In addition, a number of heterocyclic aromatic compounds (e.g. carbazole, acridine), as well as PAH with a NO_2-group (nitro-PAH), can be generated by incomplete combustion. These polycyclic compounds are not discussed here because research findings on these substances are still inconclusive.

The emissions of BaP into the air from several sources in the Federal Republic of Germany in 1981 were estimated to amount to 18 tonnes *(4)*: about 30% was caused by coke production, 56% by heating with coal, 13% by motor vehicles and less than 0.5% by heating with oil and by coal-fired power generation. Other BaP sources were not taken into consideration. In previous years BaP emission resulting from the use of hard-coal briquettes for heating purposes was much higher because such briquettes were produced using as a binder 7% pitch, which contains about 1% BaP.

Fig. 1. Relative proportions of 135 PAH (taking the concentration of chrysene as a reference substance at 100) isolated from suspended particulate matter collected in five cities[a] of the Federal Republic of Germany

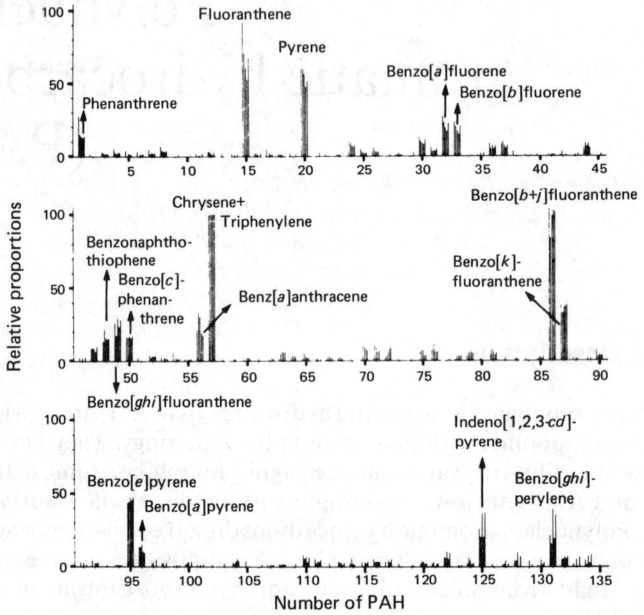

[a] The five peaks in each group relate to Bremen, Duisburg, Frankfurt, Karlsruhe and Munich in that order.

Source: adapted from König et al. *(3)*.

An analogous estimate for the United States has been made using data about 15 years older *(5)*. It shows high BaP emissions from open fires (550 tonnes/year) and lesser amounts from refuse-burning (35 tonnes/year) and cars (22 tonnes/year). In some countries biomass fuel combustion contributes substantially to PAH emissions *(6)*.

Occurrence in air

About 500 PAH have been detected in the air *(7)*, but most measurements have been made on BaP *(8)*. Data obtained prior to the mid-1970s may be comparable only to a limited extent with later data because of different sampling and analytical procedures.

The natural background level of BaP (excluding forest fires) may be nearly zero. In the United States in the 1970s the annual average value of BaP in urban areas without coke-ovens was less than 1 ng/m^3 and in other cities between 1 and 5 ng/m³ *(9)*. In several European cities in the 1960s the annual average concentration of BaP was higher than 100 ng/m³ *(8)*. However, in the Federal Republic of Germany BaP concentrations have decreased substantially in the last 15 years, generally to below 10 ng BaP per m³ *(10)*. There is also evidence that concentrations of BaP in large cities of the United Kingdom such as London have fallen to only a few per cent of former values over the past 30 years, as a result of controls on smoke emissions and the virtual disappearance of the open coal fire for domestic heating purposes *(11)*. This development is probably mainly due to modifications in heating systems and in the kind of heating fuel used. Coal burning in small domestic open fires or stoves produces amounts of BaP that are several magnitudes higher than those emitted by oil-fired central heating systems *(4)*. However, in some areas the emissions from special industrial sources are responsible for relatively high levels of BaP in the ambient air. Close to coke-oven plants levels of 40 ng BaP per m³ have been found *(12)*.

The relation between the amount of BaP and some other PAH is termed the "PAH profile" (Fig. 1). If PAH profiles are analysed by a routine method, only 6–15 of several hundred existing PAH are measured quantitatively. Although the PAH profiles of emissions can differ widely, they are relatively similar in the ambient air of several towns or cities. This is demonstrated by the findings of several scientific groups *(10)*. The different PAH profiles of emissions appear mixed, producing a relatively uniform PAH profile in the ambient air. Most important, these relations seem to be independent of the PAH concentration in the air. Differences in profiles seem to be more marked between individual investigators than when the same investigators analysed the air of different cities; this may be due to the different sampling methods used.

Under special conditions PAH can increase to very high concentrations indoors. Benzo[*a*]pyrene levels of 6 μg/m³ were found in houses without chimneys in southern China *(13)*. In India the BaP exposure averaged about 4 μg/m³ during cooking with biomass fuels *(6)*.

Very high concentrations of BaP can occur in workplaces. Recent measurements, using stationary samplers or personal samplers over an 8-hour average, showed average BaP concentrations of between 22 and 37 μg/m³ on the topside of older coke-oven batteries and between 1 and 5 μg/m³ at several other worksites in the same plants *(14,15)*. High values have also been reported in the retort-houses of coal-gas works in the United Kingdom, ranging from 3 μg/m³ in mask samples (representing the potential average exposure of workers) to more than 2 mg/m³ in peak emissions from the retorts. In the aluminium-smelting industry concentrations much higher than 10 μg/m³ were found at some workplaces *(16,17)*. The PAH profiles at different workplaces in some cases show marked differences.

Routes of Exposure

Air

The intake of BaP by inhalation of polluted ambient air can be estimated on the basis of the data given for the occurrence of PAH in air. Assuming, for example, an exposure to a relatively high concentration of 50 ng BaP per m^3 and a deposition rate of 40% from 20 m^3 air inhaled per day, the daily intake would be 400 ng of BaP. However, the BaP intake in clean rural areas may be no more than 1% of this amount, and even in major cities of Europe and North America, where adequate smoke control has been achieved, it may be only a few per cent of it.

The average total BaP content in the mainstream smoke of 1 cigarette was 35 ng before 1960 and 18 ng in 1978/1979; modern "low-tar" cigarettes deliver 10 ng BaP *(18)*. The concentration of BaP in a room extremely polluted with cigarette smoke was 22 ng/m^3 *(19)*. (Regarding the great difference between the carcinogenic potential of coke-oven emissions and cigarette smoke related to the BaP concentration, see the section on evaluation of human health risks, pp. 111–114.)

Drinking-water

Examination of a number of drinking-water supplies for six PAH (fluoranthene, benzo[*b*]fluoranthene, benzo[*k*]fluoranthene, benzo[*a*]pyrene, benzo[*ghi*]perylene, indeno[1,2,3-*cd*]pyrene) indicated that the collective concentrations generally did not exceed 0.1 μg/litre. The concentrations of these six PAH were between 0.001 and 0.01 μg/litre in 90% of the samples and higher than 0.11 μg/litre in 1%. Concentrations of benzo[*a*]pyrene were shown to range from 0.1 to 23.4 μg/litre in drinking-water *(1)*.

Food

Polynuclear aromatic hydrocarbons are found in substantial quantities in some foods, depending on the method of cooking, preservation and storage, and are detected in a wide range of meat, fish, vegetables and fruits. American sources indicate an intake of total PAH from food in the order of 1.6–16 g per day. The content of benzo[*a*]pyrene in various processed foods (refined, broiled, smoked) was repeatedly found to measure up to 50 μg/kg *(1)*. However, nearly all the cited data were published between 1965 and 1975 when the method of PAH measurement was less sophisticated.

Relative significance of different routes of exposure

There are data only on the local carcinogenic effect of PAH. Therefore, the ingestion and inhalation of PAH may have local effects independent of each other. Inhaled PAH are suspected to be a cause of some lung cancers. Little research has been done on the carcinogenic effect of ingested PAH, even though the oral intake of BaP may be much higher than the inhaled amount in the general population.

Kinetics and Metabolism

Polynuclear aromatic hydrocarbons are highly lipid-soluble and are absorbed by the lungs and gut of mammals. Inhaled PAH are predominantly adsorbed on soot particles. After deposition in the airways, the particles containing PAH can be eliminated by bronchial clearance. Polynuclear aromatic hydrocarbons might be partly removed from the particles during transport on the ciliated mucosa and may penetrate into the bronchial epithelium cells where metabolism takes place. The toxicological importance of the relatively small amount of nonparticulate PAH (i.e. in the vapour phase) is unknown *(20)*.

Polynuclear aromatic hydrocarbons are metabolized via the mixed-function oxidase system, with oxidation as the first step. The resultant epoxides or phenols may then go through detoxification reactions. After being metabolized to dihydrodiols, some of the epoxides can be further oxidized to diol-epoxides. These latter compounds are thought to be the ultimate carcinogens. The water-soluble metabolites of PAH are primarily eliminated in the urine and faeces *(1,21)*. Despite their high lipid solubility, PAH are rapidly and extensively metabolized and show little tendency towards bioaccumulation in the fatty tissues.

Benzo[*a*]pyrene is metabolized to approximately 20 primary and secondary oxidized metabolites and to a variety of conjugates. Several metabolites can induce mutations, transform cells and/or bind to cellular macromolecules. The 7,8-diol-9,10- epoxide is considered to be an ultimate carcinogenic metabolite *(1)*.

Polynuclear aromatic hydrocarbons can induce the synthesis of the enzymes responsible for its own metabolism. However, the effects of various exogenous and endogenous inducers and inhibitors of metabolism (genetic factors, age, sex and nutritional status) which have been shown to influence these enzyme systems make it difficult to predict whether the carcinogenicity of PAH will be increased or decreased.

Health Effects

Effects on experimental animals and *in vitro* test systems

Toxicological effects

Little information is available on the acute, subacute and chronic toxicity of PAH after ingestion, because overt signs of toxicity are not usually produced until the dose is sufficient to produce tumours *(1)*.

Limited data are available on the teratogenic effects of PAH, but one study of pregnant rats in which BaP was applied by subcutaneous injection indicates that it acts as a transplacental carcinogen only at comparatively high doses *(22)*.

Mutagenic and carcinogenic effects

A number of PAH have been shown to be mutagenic in bacterial systems and *in vitro* cell lines, and *in vivo* by sister chromatid exchange. Sufficient evidence for activity in short-term tests exists for six PAH: benz[*a*]anthracene, benzo[*a*]pyrene, cyclopenta[*cd*]pyrene, dibenz[*a,c*]anthracene, dibenz[*a,h*]anthracene, 1-methylphenanthrene *(1)*. The significance for human populations is not clear; however, there is a fairly good correlation between *in vitro* tests for mutagenesis and carcinogenic potency determined by skin-painting tests.

Many PAH are capable of producing tumours in numerous test species. Most experiments were carried out by skin-painting in mice, many by subcutaneous injection, and a substantial number by intratracheal instillation or by intrapulmonary injection in rodents. Oral intake has produced positive results, particularly with regard to mammary gland and forestomach tumours. After long-term inhalation of 10 mg BaP per m^3, cancer of the respiratory tract occurred in 35% of golden hamsters; additionally, a substantial number of tumours of the gastrointestinal tract were found *(23)*. The carcinogenic potencies per unit dose of tested PAH vary widely. Dibenz[*a,h*]anthracene and BaP are the strongest carcinogenic PAH repeatedly measured in the ambient air; concentrations of other, stronger PAH have not been reported, probably partly because of analytical problems, although they may be of importance for the total carcinogenic effect. Benz[*a*]anthracene and chrysene are relatively weak carcinogens, while the benzo[*a*]fluoranthenes are moderately carcinogenic. This ranking order is similar in the skin-painting test *(24)*, and the subcutaneous test in mice *(10)*.

The 4–7 ring PAH fraction of condensate from car exhaust and domestic coal stove emissions contains nearly all the carcinogenic potential of PAH. This was found after skin-painting, subcutaneous injection and intrapulmonary injection of fractions *(10,25,26)*. Polynuclear aromatic hydrocarbons with 2 or 3 rings are not yet proven to be carcinogenic.

It may be concluded from the skin-painting tests of different condensates that BaP represents about 5–15% of the carcinogenic potency of exhaust condensates from petrol-driven vehicles and coal-fired domestic stoves. In contrast to this calculation, the portion of BaP contributing to the carcinogenic effect of cigarette smoke condensate seems to represent only 1–2% *(27)*. However, the total carcinogenic potential of exhausts cannot be established from condensates alone, since the gases are not included.

In inhalation experiments with fumes from pyrolysed pitch (90 μg BaP per m^3, 80 hours/week), lung tumours were found in 18% of the exposed rats (no lung tumours in the control group) and in 86% of the exposed mice (3.5% in the control group) *(28)*. With regard to their composition, the emissions of heated pitch in an atmosphere of nitrogen are comparable with emissions from coke-ovens. Also, diesel-engine emissions with a low concentration of PAH (less than 100 ng BaP per m^3) induced lung tumours in rats after long-term inhalation *(29–31)*. These results indicate the presence in diesel-engine emissions of carcinogenic or cocarcinogenic substances other than PAH, capable of causing tumours at least in rats if inhaled in high doses.

Effects on humans

Toxicological effects
On the basis of experimental results, toxic effects other than carcinogenicity are not to be expected.

Carcinogenic effects
In the past, chimney sweeps and tar-workers were dermally exposed to substantial amounts of PAH and there is sufficient evidence that skin

cancer in many of these workers was caused by PAH *(32)*. Epidemiological studies in coke-oven workers, coal-gas workers and employees in aluminium production plants provide sufficient evidence of the role of inhaled PAH in the induction of lung cancer *(17, 33–37)*. In the extensive study by Redmond *(36)*, an excessively high rate of lung cancer mortality was found in coke-oven workers. Table 1 illustrates that increases in lung cancer cases correlate closely with the time spent working on the top of ovens where an average BaP concentration of about 30 $\mu g/m^3$ has been detected *(14,15)*.

Because several PAH are carcinogenic, a suitable index for the carcinogenic potential of the fraction of PAH in the ambient air has to be found. If all PAH profiles were identical, the concentration of a single PAH would be a good index of the carcinogenic potential of the total PAH fraction. The PAH profiles of different emissions are far from this ideal, but, as stated, the variations of PAH profiles in workplaces are not so wide and the deviation from the mean is relatively low in ambient air. Therefore, as 5–15% of the total carcinogenic effect from PAH fractions of different exhaust condensates is due to BaP according to skin-painting studies, BaP may be provisionally regarded as a sufficient index for the carcinogenic potential of the PAH fraction in ambient air. If more PAH were included in the index, only those errors resulting from insufficient detection of BaP could be diminished. As mentioned, the PAH profiles detected in different emissions and workplaces sometimes differ widely from each other and from PAH profiles in the ambient air. Moreover, it cannot be excluded that PAH profiles in the ambient air vary under special conditions. Considerably more data are necessary in order to develop a precise index for the carcinogenic potential of all PAH profiles which can occur under conditions relevant for lung cancer risk estimates. Furthermore, the carcinogenicity of PAH mixtures may be influenced by the synergistic and antagonistic effects of other components emitted together with PAH during incomplete combustion.

Evaluation of Human Health Risks

Exposure

Annual mean concentrations of BaP as measured in the 1950s and 1960s in European cities ranged from 1 to over 100 ng/m^3, the highest being found where coal was widely used for domestic heating. Smoke control policies or voluntary changes in the fuels used have since led to a downward trend in concentrations in areas that were formerly highly polluted, and available measurements indicate that annual means in major urban areas are now mainly in the range of 1–10 ng BaP per m^3.

Health risk evaluation

Risk estimates for coke-oven emissions have been carefully considered in the United States *(38)*, with particular reference to the studies in Allegheny County reported by Redmond *(36)*. Estimates of exposure to the benzene-soluble component of the emissions have been made for defined job categories. Individual exposure histories and mathematical models were developed to relate respiratory cancer mortality rates to exposure, allowing

Table 1. Deaths from cancer of the respiratory system, 1953–1970, among male coke-oven workers by length of employment (oven top, full-time workers only)

Years of exposure	Number employed	Observed deaths	Expected deaths	Relative risk	Percentage risk of lung cancer in controls in 17 years of observations
≧5	150	25	[2.72]	9.19	[1.81]
5–9	[78]	[9]	[1.36]	[6.60]	[1.74]
≧10	72	16	[1.36]	11.79	[1.89]
10–14	[43]	[8]	[0.85]	[9.43]	[1.98]
≧15	29	8	[0.51]	15.72	[1.76]

Note. The data were presented as a cumulative series of years of employment. Figures in brackets have been obtained by subtraction or multiplication/division from surrounding ones. The number of expected deaths estimated in this way can only be approximate, since the relative risks were originally calculated as weighted averages, and the basic data cannot be reconstructed precisely without knowing the weighting factors.

Sources: Redmond *(36)* and US Environmental Protection Agency *(38)*.

for trends both within and beyond the periods of observation reported. The conclusion was that a linearized multistage model would be the most conservative. When this model was used, the most plausible upper-bound individual lifetime unit risk estimate associated with a continuous exposure to $1 \mu g/m^3$ of benzene-soluble compounds of coke-oven emissions in ambient air was approximately 6.2×10^{-4} *(38)*.

For the present purpose, with BaP regarded as an index of general PAH mixtures from emissions of coke ovens and of similar combustion processes in urban air, information on the BaP content of coke-oven emissions is required in order to proceed further. A value of 0.71% BaP in benzene-soluble coke-oven emissions has been reported *(16)*. This would then yield a lifetime risk of respiratory cancer of 8.7×10^{-5} per ng BaP per m^3 or 8.7×10^{-2} per μg BaP per m^3.

This risk assessment would imply that about 9 per 100 000 exposed people may die from cancer of the respiratory tract as a result of spending a lifetime in ambient air containing an average level of 1 ng BaP per m^3 mixed with all the other PAH and related substances in coke-oven emissions.

An alternative link with a BaP index has been explored by applying the results of measurements above old coke-oven plants in the Federal Republic of Germany, believed to represent concentrations analogous to those in former years in Allegheny installations, to the data in Table 1 (these form a subset of the series examined by the US Environmental Protection Agency (EPA) *(38)*).

The BaP concentration measured in the Federal Republic of Germany averaged $30 \mu g/m^3$ on the top of the coke-oven, and $3 \mu g/m^3$ at the oven side *(15)*. Using these exposure data and interpolating according to a linear nonthreshold model, Pott has calculated a risk of death from respiratory tract cancer of 1% per μg BaP per m^3 at the workplace after exposure for 25 years and in 5 per 100 000 population exposed to 1 ng BaP per m^3 for 50 years *(10)*.

Some other risk estimates of respiratory tract cancer related to BaP in the ambient air have been summarized for the US EPA *(39)*. The estimated risk per year, as shown in these studies, ranges from 0.11×10^{-5} to 1.4×10^{-5} per ng BaP per m^3.

Rough extrapolation from the results of the first study of inhalation by rats of pyrolysed pitch emissions *(28)* gives a lung cancer risk that is not substantially lower than that stated in the epidemiological study by Redmond *(36)* (see above).

It can be assumed that the carcinogenic potential of coke-oven emissions and emissions of heated pitch — apart from the aromatic amines — is concentrated largely in the PAH group, because their PAH content is extremely high in comparison with other substances. In this case, BaP may be a suitable index for the carcinogenic potential of the total emission in relation to lung cancer, but, as mentioned above, this does not apply to emissions of other combustion processes (e.g. diesel engine exhaust or cigarette smoke). In diesel engine exhaust and cigarette smoke BaP represents only the carcinogenic potential of the PAH. These emissions obviously contain some other very effective carcinogenic or cocarcinogenic agents. According to epidemiological studies and interpolations, the inhalation of just a few (5, for example) milligrams of BaP in cigarette smoke as a lifetime total might have a carcinogenic effect 50 times higher than the inhalation of the same amount of BaP in coke-oven emissions *(10)*. These differences cannot be explained by differences in the PAH profiles measured.

Guidelines

Owing to its carcinogenicity, no safe level of PAH can be recommended. There is no known cancer threshold for BaP, the most thoroughly studied PAH, nor is there an ambient mixture of PAH that does not contain BaP and other substances for which there is sufficient evidence of carcinogenicity in animals.

A number of different risk estimates for PAH have been made, based primarily on using BaP as the index compound. The US EPA has offered an upper-bound lifetime cancer risk estimate of 62 per 100 000 exposed people per μg benzene-soluble coke-oven emission per m^3 ambient air. Assuming a 0.71% content of BaP in these emissions, it can be estimated that 9 out of 100 000 people exposed to 1 ng BaP per m^3 over a lifetime would be at risk of developing cancer.

References

1. *Polynuclear aromatic compounds. Part 1. Chemical, environmental and experimental data.* Lyon, International Agency for Research on Cancer, 1983 (IARC Monographs on the Evaluation of the Carcinogenic Risk of Chemicals to Humans, Vol. 32).
2. **Schmeltz, I. & Hoffmann, D.** Formation of polynuclear aromatic hydrocarbons from combustion of organic matter. *In:* Freudenthal, R. & Jones, P.W., ed. *Carcinogenesis — a comprehensive survey. Vol. 1. Polynuclear aromatic hydrocarbons: chemistry, metabolism, and carcinogenesis.* New York, Raven Press, 1976, pp. 225–239.
3. **König, J. et al.** Untersuchung von 135 polyzyklischen aromatischen Kohlenwasserstoffen in atmosphärischen Schwebstoffen aus 5 Städten der Bundesrepublik Deutschland [Comparing investigations on samples of airborne particulate matter from 5 cities of the Federal Republic of Germany on their content of 135 polycyclic aromatic hydrocarbons]. *Staub, Reinhaltung der Luft,* **41**: 73–78 (1981).
4. **Ahland, E. et al.** Luftverschmutzung und Krebs. Prüfung von Schadstoffen aus verschiedenen Emissionsquellen auf ihre krebserzeugende Wirkung [Air pollution and cancer. Investigation of hazardous constituents from various emission sources for their carcinogenic impact]. *Münchener medizinische Wochenschrift,* **127**: 218–221 (1985).
5. **Committee on Biologic Effects of Atmospheric Pollutants.** *Particulate polycyclic organic matter.* Washington, DC, National Academy of Sciences, 1972.

6. **Smith, K.R. et al.** Air pollution and rural biomass fuels in developing countries: a pilot village study in India and implications for research and policy. *Atmospheric environment,* **17**: 2343–2362 (1983).
7. **Herlan, A.** Kanzerogene polyzyklische Aromate und Metabolite als mögliche Bestandteile von Emissionen [Carcinogenic polycyclic aromatics and metabolites as possible components of emissions]. *Zentralblatt für Bakteriologie, Mikrobiologie und Hygiene, I Abt. Orig. B,* **165**: 174–191 (1977).
8. **Sawicki, E.** Analysis of atmospheric carcinogens and their co-factors. *In:* Rosenfeld, C. & Davis, E., ed. *Environmental pollution and carcinogenic risks.* Paris, INSERM, 1976, pp. 297–354 (INSERM Symposia Series, Vol. 52, IARC Scientific Publications No. 13).
9. **Faoro, R.B. & Manning, J.A.** Trends in benzo[*a*]pyrene, 1966–77. *Journal of the Air Pollution Control Association,* **31**: 62–64 (1981).
10. **Pott, F.** Pyrolyseabgase, Profile von polyzyklischen aromatischen Kohlenwasserstoffen und Lungenkrebsrisiko — Daten und Bewertung [Pyrolytic emissions, profiles of polycyclic aromatic hydrocarbons and lung cancer risk — data and evaluation]. *Staub, Reinhaltung der Luft,* **45**: 369–379 (1985).
11. **Waller, R.E.** Lungenkrebs und städtische Luftschadstoffe [Lung cancer and urban air pollutants]. *Atemwegs- und Lungenkrankheiten,* **8**: 150–153 (1982).
12. **Grimmer, G. et al.** Comparison of the profiles of polycyclic aromatic hydrocarbons in different areas of a city by glass-capillary-gas-chromatography in the nanogram-range. *International journal of environmental analytical chemistry,* **10**: 265–276 (1981).
13. **National Centre for Preventive Medicine.** [Determination of air pollutants in high incidence and low incidence areas of lung cancer in Xuanwei County]. *Journal of the Institute of Health (Peking),* **13**: 20–25 (1984) (in Chinese).
14. **Björseth, A.** Determination of polynuclear aromatic hydrocarbons in the working environment. *In:* Jones, P.W. & Leber, P., ed. *Polynuclear aromatic hydrocarbons. Third International Symposium on Chemistry and Biology — Carcinogenesis and Mutagenesis.* Ann Arbor, MI, Ann Arbor Science, 1979, pp. 371–381.
15. **Blome, H.** Messungen polyzyklischer aromatischer Kohlenwasserstoffe an Arbeitsplätzen — Beurteilung der Ergebnisse [Measurements of polycyclic aromatic hydrocarbons at places of work — assessment of results]. *Staub, Reinhaltung der Luft,* **41**: 225–229 (1981).
16. **Lindstedt, G. & Sollenberg, J.** Polycyclic aromatic hydrocarbons in the occupational environment, with special reference to benzo[*a*]pyrene measurements in Swedish industry. *Scandinavian journal of work, environment & health,* **8**: 1–19 (1982).
17. *Polynuclear aromatic compounds. Part 3. Industrial exposures in aluminium production, coal gasification, coke production, and iron and steel founding.* Lyon, International Agency for Research on Cancer, 1984 (IARC Monographs on the Evaluation of the Carcinogenic Risk of Chemicals to Humans, Vol. 34).

18. **Hoffmann, D., et al.** *In: Luftverunreinigung durch polyzyklische aromatische Kohlenwasserstoffe — Erfassung und Bewertung — Kolloquium Hannover 1979* [Air pollution by polycyclic aromatic hydrocarbons — determination and evaluation — Hannover Colloquium, 1979]. Düsseldorf, VDI-Verlag, 1980, pp. 335–350 (VDI-Berichte No. 358).
19. **Grimmer, G. et al.** Zum Problem des Passivrauchers: Aufnahme von polyzyklischen aromatischen Kohlenwasserstoffen durch Einatmen von zigarettenrauchhaltiger Luft [Passive smoking: intake of polycyclic aromatic hydrocarbons by breathing of air containing cigarette smoke]. *International archives of occupational and environmental health,* **40**: 93–99 (1977).
20. *Feasibility of assessment of health risks from vapor-phase organic chemicals in gasoline and diesel exhaust.* Washington, DC, National Research Council, 1983.
21. **Dipple, A. et al.** Polynuclear aromatic carcinogens. *In:* Searle, C.E., ed. *Chemical carcinogens.* Washington, DC, American Chemical Society, 1984, Vol. 1, pp. 41–163 (ACS Monograph No. 182).
22. **Bulay, O.M. & Wattenberg, L.W.** Carcinogenic effects of subcutaneous administration of benzo[*a*]pyrene during pregnancy on the progeny (34993). *Proceedings of the Society for Experimental and Biological Medicine,* **135**: 84–86 (1970).
23. **Thyssen, J. et al.** *In: Luftverunreinigung durch polyzyklische aromatische Kohlenwasserstoffe — Erfassung und Bewertung — Kolloquium Hannover 1979* [Air pollution by polycyclic aromatic hydrocarbons — determination and evaluation — Hannover Colloquium, 1979]. Düsseldorf, VDI-Verlag, 1980, pp. 329–333 (VDI-Berichte No. 358).
24. **Hoffmann, D. & Wynder, E.L.** Environmental respiratory carcinogenesis. *In:* Searle, C.E., ed. *Chemical carcinogens.* Washington, DC, American Chemical Society, 1976, pp. 324–365 (ACS Monograph No. 173).
25. **Grimmer, G. et al.** On the contribution of polycyclic aromatic hydrocarbons to the carcinogenic impact of automobile exhaust condensate evaluated by local application onto mouse skin. *Cancer letters,* **21**: 105–113 (1983).
26. **Grimmer, G. et al.** Contribution of polycyclic aromatic hydrocarbons to the carcinogenic impact of gasoline engine exhaust condensate evaluated by implantation into the lungs of rats. *Journal of the National Cancer Institute,* **72**: 733–739 (1984).
27. **Dontenwill, W. et al.** Experimentelle Untersuchungen über die tumorerzeugende Wirkung von Zigarettenrauch-Kondensaten an der Mäusehaut. VI Mitteilung: Untersuchungen zur Fraktionierung von Zigarettenrauch-Kondensat [Experimental studies on the tumorigenic activity of cigarette smoke condensate on mouse skin. VI. Fractionation of cigarette smoke condensate]. *Zeitschrift für Krebsforschung und klinische Onkologie,* **85**: 155–167 (1976).
28. **Heinrich, U. et al.** Comparison of chronic inhalation effects in rodents after long-term exposure to either coal oven flue gas mixed with pyrolized pitch or diesel engine exhaust. *In:* Ishinishi, N. et al., ed. *Carcinogenic and mutagenic effects of diesel engine exhaust.* Amsterdam, Elsevier, 1986.

29. **Brightwell, J. et al.** Neoplastic and functional changes in rodents after chronic inhalation of engine exhaust emissions. *In:* Ishinishi, N. et al., ed. *Carcinogenic and mutagenic effects of diesel engine exhaust.* Amsterdam, Elsevier, 1986, pp. 471–485.
30. **Heinrich, U. et al.** Chronic effects on the respiratory tract of hamsters, mice and rats after long-term inhalation of high concentrations of filtered and unfiltered diesel engine emissions. *Journal of applied toxicology,* **6**: 383–395 (1986).
31. **Mauderly, J. et al.** Carcinogenicity of diesel exhaust inhaled chronically by rats. *In:* Ishinishi, N. et al., ed. *Carcinogenic and mutagenic effects of diesel engine exhaust.* Amsterdam, Elsevier, 1986, pp. 397–409.
32. *Polynuclear aromatic compounds. Part 4. Bitumens, coal-tars and derived products, shale-oils and soots.* Lyon, International Agency for Research on Cancer, 1985 (IARC Monographs on the Evaluation of the Carcinogenic Risk of Chemicals to Humans, Vol. 35).
33. **Kawahata, K.** Über die gewerblich hervorgerufenen Lungenkrebse bei Generator-Gas-Arbeitern in den Stahlwerken [Occupational lung cancer in gas generator workers in steel works]. *GANN journal,* **32**: 367–387 (1938).
34. **Doll, R. et al.** Mortality of gasworkers — final report of a prospective study. *British journal of industrial medicine,* **29**: 394–406 (1972).
35. **Hammond, E.C. et al.** Inhalation of benzpyrene and cancer in man. *Annals of the New York Academy of Sciences,* **271**: 116–124 (1976).
36. **Redmond, C.K.** Epidemiological studies of cancer mortality in coke plant workers. *In: Seventh Conference on Environmental Toxicology 1976.* Washington, DC, US Environmental Protection Agency, 1976, pp. 93–107 (AMRL-TR-76-125, Paper No. 3).
37. **Manz, A. et al.** *Zur Frage des Berufskrebses bei Beschäftigten der Gasindustrie (Cohortenstudie)* [Occupational cancer in gas industry workers (cohort study)]. Dortmund, Bundesanstalt für Arbeitsschutz und Unfallforschung, 1982.
38. *Carcinogen assessment of coke oven emissions.* Washington, DC, US Environmental Protection Agency, 1984 (Final report No. EPA-600/6-82-003F).
39. **Nisbet, I.C.T. et al., ed.** *Review and evaluation of the evidence for cancer associated with air pollution.* Washington, DC, US Environmental Protection Agency, 1984 (Final report No. EPA-450/5-83-006R).

12

Styrene

General Description

Styrene ($C_6H_5 \cdot CH=CH_2$) is a volatile colourless liquid that has a tendency to polymerize.

Sources

Styrene is mainly used in the manufacture of polymers, copolymers and reinforced plastics, particularly polystyrene. Worldwide production is around 10 million tonnes per year *(1)*.

Styrene is not known to occur as a natural product. In the atmosphere, it has been detected in a wide variety of locations. Its presence is thought to be mainly due to emissions from industrial processes, particularly from the petrochemical industry. Styrene has been detected in emissions from sources such as motor vehicles and from other forms of combustion and incineration *(1)*. Some liberation of styrene may also take place from recently manufactured plastic goods. While this may contribute to indoor levels of styrene, the effect on total emissions to the environment is negligible.

The concentration of styrene in urban air is relatively low compared with that of aromatic hydrocarbons, such as toluene and xylene. This appears to be due to the ready reactivity of styrene with ozone to yield benzaldehyde and peroxides, all of which are irritants; one of the peroxides, peroxybenzyol nitrate, is a potent eye irritant. Styrene is an active generator of photochemical smog.

Occurrence in air

Data on atmospheric concentrations over Los Angeles, CA, in 1965 showed styrene to be present at levels of 8–63 $\mu g/m^3$, with an average of 21 $\mu g/m^3$ *(2)*. Styrene was also found in the atmosphere above Nagoya, Japan, in 1977, at levels of 0.84 $\mu g/m^3$ *(3)* and in Delft, Netherlands, at levels of 0.4 $\mu g/m^3$, with a maximum of 2.9 $\mu g/m^3$ *(4)*. It has also been detected in rural air, but the concentrations were not given *(4)*.

The median concentration of styrene inside residential buildings in New Jersey was 1.8 $\mu g/m^3$ (maximum concentration 54 $\mu g/m^3$), which was about three times higher than the outdoor concentrations *(5)*. In Czechoslovakia, concentrations of styrene in the mg/m^3 range were reported in newly finished buildings where floor-covering containing styrene was used.

In factories producing styrene there may be exposure to benzene and ethylbenzene while, in the manufacture of polystyrene, environmental contamination may be due entirely to styrene. In plastic applications where styrene is a solvent-reactant for copolymerization, styrene is the major air contaminant; however, there may be concomitant exposures to fibrous glass, catalysts, accelerators, cleaning agents, and other chemicals. In many applications, the operations involve potential contact of the skin with liquid styrene.

In polystyrene production plants, styrene concentrations were often less than 21 mg/m^3 (5 ppm) and rarely exceeded 210 mg/m^3 (50 ppm) *(1)*. Concentrations of styrene found during the production of reinforced plastics were generally much higher, with peak concentrations up to 6300 mg/m^3 (1500 ppm).

The main sources of styrene contamination near a butadiene-styrene rubber plant in Czechoslovakia *(6)* were aeration towers used to eliminate the pollutant content of wastewaters, the warehouse for liquid materials and the conduit between the warehouse and the polymerization area. Fifty metres downwind from the towers "several tenths of a mg/m^3" were determined. In the immediate downwind vicinity of the warehouse, 0.07 mg/m^3 styrene was found, compared with a concentration of 0.03 mg/m^3 at a distance of 800 m. In readings other than negative, the minimum concentration of styrene found outside the factory fluctuated around 0.01 mg/m^3. In the USA, measurements of styrene within 200 m of processing plants have given a wide range of values: 0.3–2900 $\mu g/m^3$ *(7)*.

Styrene enters the environment through combustion and pyrolysis processes in addition to industrial leakages. When different styrene-containing plastics were degraded in a laboratory at 210–260 °C, 330–7400 μg styrene, among other chemicals, was liberated per gram of plastic *(8)*. At higher temperatures even more styrene was released.

Other sources of styrene in the environment are vehicle exhaust emissions. Although little styrene was found in the exhaust of conventional engines (0.76%), somewhat more was found in that of rotary engines (2.67%) *(9)*. Other studies have shown that styrene can be found in small quantities in spark-ignition engines *(10)*, and in oxyacetylene and oxyethylene flames *(11)*.

Conversion factors

1 ppm = 4.2 mg/m^3

1 mg/m^3 = 0.24 ppm

Routes of Exposure

Air

Styrene is present in unpolluted rural air in small concentrations. The concentrations in urban atmosphere are around 0.3 $\mu g/m^3$, leading to a daily intake of about 6 μg/person. In polluted urban air and within 1 km of styrene polymerization units, the concentration can be 20–30 μg/m^3. Persons living in such areas would inhale 400–600 μg of styrene per day. Indoor sources may also contribute to the level of exposure.

Occupational exposure
The concentration of styrene in the reinforced plastics industry may be around 200 mg/m^3 (48 ppm) and in styrene polymerization operations 10 mg/m^3 (2.4 ppm); the exposure per working day can amount to 2 g and 100 mg respectively *(1)*.

Smoking
Styrene has also been identified in cigarette smoke condensates *(12)*. Levels ranging from 20 to 48 μg per cigarette were reported.

Drinking-water
Styrene has been detected in drinking-water as well as in industrial effluents, although the latter source is much more frequent *(13)*. It has been found at a concentration of 1 μg/litre (0.001 ppm) in the Scheldt river in the Netherlands, in the Kanawha river in West Virginia *(4)* and in effluent discharged from petroleum-refining (31 μg/litre), chemical (30 μg/litre), rubber-manufacturing (2.6–3 μg/litre) and textile-manufacturing plants in the United States *(4)*.

Styrene can migrate through soil, as was demonstrated in an incident described by Grossman *(14)*. After approximately one year, well-water in an area where two drums of styrene had been buried began to have an obnoxious odour, which was traced to the presence of 0.1–0.2 mg styrene per litre.

Valenta *(6)* found styrene in the water of wells surrounding the dump of a butadiene-styrene plant at levels of 0.01–0.02 mg/litre. After sanitation measures were begun, levels increased to 1.0–2.7 mg styrene per litre during the period of work (approximately 4 months) and then decreased.

Styrene has also been detected in finished drinking-water in the USA at concentrations of less than 1 μg/litre, and specifically in commercial, charcoal-filtered drinking-water in New Orleans, LA *(4)*.

Styrene evaporates readily from water to air. The evaporation half-time of styrene in water at a depth of 1 m is estimated to be about 5.9 hours *(15)*. Styrene is not thought to bioaccumulate or bioconcentrate in organisms and food chains to any measurable extent *(15)*.

Food
Polystyrene and its copolymers are widely used as food-packaging materials. The ability of styrene monomer to migrate from polystyrene packaging to food has been reported in a number of publications and probably accounts for the greatest contamination of foods by styrene monomer. In a study designed to detect styrene monomer at 0.05 mg/kg, no migration of the monomer was detected in milk samples stored in polystyrene containers for up to 8 days *(16)*. Styrene has been reported to convey disagreeable odours and taste to dairy products at 0.2–0.5 mg/kg *(17)*.

Styrene has been found at concentrations of 2.5–80 μg/kg in yogurt and other milk products packaged in polystyrene containers *(17)*. The styrene content in the products increases in the course of storage *(17–19)*.

Relative significance of different routes of exposure

The concentration of styrene in various exposure situations is shown in Table 1. Occupational exposures by far exceed any other type of exposure. In the general population, indoor and outdoor air account for the largest exposures. However, smokers inhale high amounts of styrene in cigarette smoke.

Kinetics and Metabolism

Absorption

In animal and human controlled studies, the uptake of styrene has been found to be rapid. The principal routes of exposure are pulmonary and, to a lesser extent, dermal *(1)*. Depending on exposure conditions, some 50–100% of styrene is absorbed in the airways *(1)*.

Distribution

Styrene is widely distributed throughout the body. The distribution and sequestration of styrene to lipid depots and its subsequent slow elimination would indicate a potential for accumulation in situations of repeated daily exposure.

Table 1. Relative significance of different routes of exposure

Exposure situation	Styrene concentration (μg/m^3)	Nominal daily intake[a]
Reinforced plastics industry	200 000	2 g
Styrene polymerization	10 000	100 mg
Within 1 km of the production unit	30	600 μg
Polluted urban atmosphere	20	400 μg
Urban atmosphere	0.3	6 μg
Indoor air	0.3–50	6–1 000 μg
Polluted drinking-water (2 litres per day)	1 μg/litre	2 μg
Cigarette smoke (20 cigarettes per day)	20–48 μg per cigarette	400– 960 μg

[a] Nominal daily intake is calculated assuming a daily breathing volume of 10 m^3 at work or 20 m^3 at home or in an urban atmosphere.

Biotransformation

Styrene is largely biotransformed via the 7,8-epoxide by the mixed-function oxidase system, principally to mandelic and phenylglyoxylic acids, which are excreted in the urine. Recent evidence suggests that other minor metabolic pathways may also be important in the toxicological assessment, and further study is needed in this area *(1)*.

Elimination

The elimination of styrene and its metabolites appears to involve, at least, a two-compartment pharmacokinetic model, and its dose-dependency would suggest the involvement of saturable metabolic pathways. The elimination half-times of styrene from the venous blood and alveolar air are about 1 hour and 10 hours for the fast and slow component, respectively *(1)*.

Biological monitoring to reflect previous exposure has involved the quantitative analysis of alveolar air and the urinary metabolites, mandelic, hippuric and phenylglyoxylic acid. The monitoring of urinary mandelic acid appears to give the best correlation with styrene in ambient air. Mandelic acid is excreted with half-times of 4 hours and 25 hours for the fast and slow component, respectively *(1)*.

Health Effects

Effects on experimental animals and *in vitro* test systems

Toxicological effects

One mammalian study has suggested that inhalation exposure to styrene is associated with embryotoxicity *(20)*. The ratio between the adult toxic dose and the developmental toxic dose is not known.

Mutagenic and carcinogenic effects

The mutagenicity of styrene, when metabolically activated, has been confirmed in various experimental systems *in vitro* and *in vivo*. Discrepant findings have, however, been obtained, presumably depending on the specific metabolic activation/inactivation pattern. Styrene-7,8-oxide, one of the reactive metabolites of styrene, shows chemical reactivity and mutagenicity in almost all the test systems studied *(1,21)*.

Studies in experimental animals have provided limited evidence that styrene may be carcinogenic *(1,4,22)*. In one study, oral administration of styrene to mice induced a significant increase in pulmonary tumours at a dose of 1350 mg/kg and a doubtful increase in another strain of mice at a dose of 300 mg/kg *(1)*. In two studies on rats, styrene-7,8-oxide, the primary metabolite of styrene, was found to be carcinogenic after oral administration *(21)*.

Effects on humans

Toxicological effects

A higher prevalence of respiratory tract irritation was reported in workers exposed to time-weighted average concentrations of 84 mg/m^3 *(23,24)*.

Exposure to 420 mg/m^3 and above caused acute irritation of mucous membranes in the eyes and the upper respiratory tract.

Several studies among workers revealed that styrene can cause depression of the CNS. Prenarcotic symptoms such as weakness, headache, fatigue, malaise, tension, nausea and dizziness were reported where styrene concentrations in the air exceeded 200 mg/m^3. Workers with these complaints were often employed in the production of glass fibre reinforced by polyester resin; in this industry, styrene concentrations are high. As exposure increased to above 840 mg/m^3, drowsiness, nausea and disturbances in equilibrium became evident *(1)*. Some studies reported the abovementioned CNS symptoms even at concentrations below 200 mg/m^3, e.g. at 70.7–164 mg/m^3 *(25)* and at 84 mg/m^3 *(23,24)*.

Slight disturbances of visual-motor accuracy and psychomotor performance were noted at styrene levels exceeding 210 mg/m^3 and an increased incidence of abnormalities in EEGs was detected at styrene concentrations below 420 mg/m^3 *(1)*. However, Cherry et al. *(26,27)* noted slower simple reaction times in workers exposed to average concentrations of 84 mg/m^3 (concentrations ranged from 25.2 to 130 mg/m^3).

Epidemiological studies among workers with long-term occupational exposure to styrene have shown an increased frequency of abnormal EEGs in correlation with the exposure level as indicated by the excretion of 700 mg or more of mandelic acid per litre of urine. The same workers showed a decline in psychomotor performance and visual-motor accuracy in psychological tests at mandelic acid levels of 1600 mg per litre of urine, this also indicating an exposure–response relationship. Definitive evidence on lesions in the peripheral nervous system is still lacking *(1)*.

Clear evidence of hepatotoxicity in workers exposed to styrene could not be demonstrated, although significantly elevated gamma-GT activity was found in workers exposed to time-weighted average concentrations of 84 mg/m^3 *(23,24)*.

Mutagenic and carcinogenic effects

Several studies indicate that structural chromosome aberrations are increased in the peripheral blood lymphocytes of workers exposed to styrene in the reinforced plastics industry. Negative results have also been reported, especially in the production of styrene monomer or polymer, where the range of exposure to styrene is low *(1,4)*.

Even though the exact role of styrene in inducing somatic chromosome aberrations cannot be convincingly shown because of the multiple exposures to various chemicals on the part of the workers studied, the simultaneous experimental clastogenicity of styrene in human cells firmly suggests that styrene induces chromosome damage in workers.

Several case reports and epidemiological studies have indicated an increased risk of cancer in the lymphatic and haematopoietic tissues of workers involved in the manufacture of styrene, polystyrene and butadiene-styrene rubber *(1,4,22)*. However, evidence is insufficient either to establish a direct cause–effect relationship or to determine the levels and duration of exposure to styrene required to produce such effects.

Sensory effects

Styrene in its pure form has an odour detection threshold of 70 μg/m^3 *(28)*. Its characteristic pungent odour is recognized at concentrations 3–4 times greater than this threshold value. When styrene is emitted into the air, its half-time is estimated to be 2 hours. In ambient air it is chemically transformed into benzaldehyde and formaldehyde, both of which are odorous air pollutants *(29)*. No data have been published on the odour nuisance of industrial emissions in which styrene is considered to be a main odour contributor.

Some individuals can perceive the odour of styrene at levels lower than 70 μg/m^3, but, in general, odour problems are not likely to occur if peak concentrations in the ambient air are kept below this threshold value.

Evaluation of Human Health Risks

Exposure

Styrene concentrations in rural or unpolluted urban areas and in unpolluted indoor environments are low (below 1 μg/m^3).

Levels in polluted urban areas may be up to 20 μg/m^3, but can be much higher in newly-built houses containing styrene-based materials. In the reinforced plastics industry, concentrations in air range from 10 mg/m^3 to 1000 mg/m^3; about 1 km away from styrene-producing factories, levels can be around 30 μg/m^3 or more.

Health risk evaluation

Data on mutagenicity are contradictory. Even though the biological significance of findings, especially at the individual level, is not at present understood, the induction of genetic damage in styrene-exposed populations should be considered as an indicator of possible adverse health effects. Evidence for carcinogenicity in experimental animals is limited, while that of carcinogenicity in humans is inadequate. The IARC categorization places styrene in Group 3.

Toxic effects in humans include CNS dysfunction and irritation of mucous membranes. Since an increased frequency of some effects, such as prenarcotic symptoms (lightheadedness, etc.), lower respiratory tract symptoms (cough, wheezing, tightness of chest) and elevated gamma-GT activities, has been reported at a concentration of 84 mg/m^3 *(23–27)*, this level is considered to be the lowest-observed-adverse-effect level.

Guidelines

The lowest-observed-adverse-effect level of styrene in workers was found to be 84 mg/m^3. Although the adverse nature of observed effects on the lower respiratory tract and the CNS was minimal at this concentration, a protection (safety) factor of 10 is considered appropriate, mainly because very limited data are available on the toxic effects of long-term inhalation exposure. If this factor was enlarged by an order of magnitude for the purpose of extrapolation to the general population, a value of 800 μg/m^3 would result. It is believed that below 800 μg/m^3 (averaging time 24 hours) adverse health effects of environmental exposure (outdoor or indoor) to styrene are not likely to occur and this level is recommended as a guideline value.

In view of the low toxicity of styrene, the air quality guideline value could then be based on its odour annoyance property. The peak concentration of styrene in the ambient air should be kept below the odour detection threshold level of 70 $\mu g/m^3$ as a 30-minute average.

References

1. *Styrene.* Geneva, World Health Organization, 1983 (Environmental Health Criteria, No. 26).
2. **Neligan, R.E. et al.** The gas chromatographic determination of aromatic hydrocarbons in the atmosphere. *American Chemical Society Division of Water Air Waste Chemistry Preprints,* **5**: 118–121 (1965).
3. **Hoshika, Y.** Gas chromatographic determination of styrene as its dibromide. *Journal of chromatography,* **136**: 95–103 (1977).
4. *Some monomers, plastics and synthetic elastomers, and acrolein.* Lyon, International Agency for Research on Cancer, 1979 (IARC Monographs on the Evaluation of the Carcinogenic Risk of Chemicals to Humans, Vol. 19).
5. **Hartwell, T.D. et al.** Comparison of indoor and outdoor levels for air volatiles in New Jersey. *In:* Berglund, B. et al., ed. *Indoor air: chemical characterization and personal exposure.* Stockholm, Swedish Council for Building Research, 1984, Vol. 4, pp. 81–85.
6. **Valenta, J.** [Air and water pollution by styrene from butadiene-styrene rubber production]. *Ceskoslovenska hygiena,* **11**: 349–352 (1966) (in Czech).
7. *Summary health assessment document for styrene.* Research Triangle Park, NC, US Environmental Protection Agency, 1985 (preliminary draft).
8. **Hoff, A. et al.** Degradation products of plastics. *Scandinavian journal of work, environment & health,* **8**: 1–60 (1982).
9. **Schofield, K.** Problems with flame ionization detectors in automotive exhaust hydrocarbon measurements. *Environmental science and technology,* **8**: 826–834 (1974).
10. **Fleming, R.D.** *Effect of fuel composition on exhaust emissions from a spark-ignition engine.* Washington, DC, US Department of the Interior, Bureau of Mines (Report of Investigations 7423).
11. **Crittenden, B.D. & Long, R.** The mechanisms of formation of polynuclear aromatic compounds in combustion systems. *In:* Freudenthal, R. & Jones, P.W., ed. *Carcinogenesis — a comprehensive survey. Vol. 1. Polynuclear aromatic hydrocarbons: chemistry, metabolism, and carcinogenesis.* New York, Raven Press, 1976, pp. 209–223.
12. **Johnstone, R.A.W. et al.** Composition of cigarette smoke: some low-boiling components. *Nature,* **195**: 1267–1269 (1962).
13. **Santodonato, J. et al.** *Investigation of selected potential environmental contaminants: styrene, ethylbenzene, and related compounds.* Washington, DC, US Environmental Protection Agency, 1980 (Report No. EPA-560/11-80-018).

14. **Grossman, I.G.** Waterborne styrene in a crystalline bedrock aquifer in the Gales Ferry Area, Fedyard, South Eastern Connecticut. *US Geological surveys, professional papers,* **700-B**: 203–209 (1970).
15. **Pervier, J.W. et al.** *Survey reports on atmospheric emissions from the petrochemical industry,* Vol. IV. Springfield, VA, National Technical Information Service, 1974, p. 287 (PB Rep. PB-245630).
16. **Jensen, F.** Determination of monomers from polystyrene in milk products. *Annali dell'Istituto Superiore di Sanità,* **8**: 443–448 (1972).
17. **Finley, J.W. & White, J.C.** Two methods to determine if styrene monomer is present in milk. *Bulletin of environmental contamination and toxicology,* **2**: 41–46 (1967).
18. **Withey, J.R.** Quantitative analysis of styrene monomer in polystyrene and foods including some preliminary studies of the uptake and pharmacodynamics of the monomer in rats. *Environmental health perspectives,* **17**: 125–133 (1976).
19. **Withey, J.R. & Collins, P.G.** Styrene monomer in foods: a limited Canadian survey. *Bulletin of environmental contamination and toxicology,* **19**: 86–94 (1978).
20. **Kankaanpää, J.T.J. et al.** The effect of maternally inhaled styrene on embryonal and foetal development in mice and chinese hamsters. *Acta pharmacologica et toxicologica,* **47**: 127–129 (1980).
21. *Allyl compounds, aldehydes, epoxides and peroxides.* Lyon, International Agency for Research on Cancer, 1985 (IARC Monographs on the Evaluation of the Carcinogenic Risk of Chemicals to Humans, Vol. 36).
22. *Chemicals, industrial processes and industries associated with cancer in humans. IARC Monographs, Volumes 1 to 29.* Lyon, International Agency for Research on Cancer, 1982 (IARC Monographs on the Evaluation of the Carcinogenic Risk of Chemicals to Humans, Supplement 4).
23. **Lorimer, W.V. et al.** Clinical studies of styrene workers: initial findings. *Environmental health perspectives,* **17**: 171–181 (1976).
24. **Lorimer, W.V. et al.** Health status of styrene-polystyrene polymerization workers. *Scandinavian journal of work, environment & health,* **4**(Suppl.2): 220–226 (1978).
25. **Dolara, P. et al.** Enzyme induction in humans exposed to styrene. *Annals of occupational hygiene,* **27**: 183–188 (1983).
26. **Cherry, N. et al.** An investigation of the acute behavioural effects of styrene on factory workers. *British journal of industrial medicine,* **37**: 234–240 (1980).
27. **Cherry, N. et al.** Acute behavioural effects of styrene exposure. A further analysis. *British journal of industrial medicine,* **38**: 346–350 (1981).
28. **Hellman, T.M. & Small, F.H.** Characterization of the odour properties of 101 petrochemicals using sensory methods. *Journal of the Air Pollution Control Association,* **24**: 979–982 (1974).
29. **van Gemert, L.J. & Nettenbreijer, A.H., ed.** *Compilation of odour threshold values in air and water.* Zeist, Central Institute for Nutrition and Food Research, 1977.

13

Tetrachloroethylene

General Description

Tetrachloroethylene (C_2Cl_4) is a nonflammable compound that is stable up to 500 °C in the absence of catalysts, moisture and oxygen. The boiling-point is 121 °C. It decomposes slowly in contact with moisture to yield trichloroacetic acid and hydrochloric acid. It is relatively insoluble in water (150 mg/litre).

Sources

Tetrachloroethylene is mainly used as a solvent in dry-cleaning and metal cleaning. It is also used in processing and finishing operations in the textile industry, as an extraction solvent, a heat-exchange fluid, in grain fumigation, in a variety of consumer products, and in the manufacture of fluorocarbons *(1,2)*.

The annual production is estimated to be 50 000–100 000 tonnes in eastern Europe, about 55 000 tonnes in Japan, 100 000–250 000 tonnes in western Europe *(3)* and about 265 000 tonnes in the USA *(4)*. Global emissions are estimated to be as high as 800 000 tonnes annually.

Occurrence in air

In the troposphere, tetrachloroethylene is photochemically degraded. Reaction with hydroxyl radicals is the principal mechanism by which it is removed. The estimated lifetime of tetrachloroethylene in the troposphere is less than one year. Levels in air fluctuate considerably on both a seasonal and a diurnal basis *(2)*.

Global background levels of tetrachloroethylene are around 0.2 $\mu g/m^3$ *(2)*. Average instantaneous concentrations in the ambient air of 62 out of 92 cities in the Federal Republic of Germany ranged between 1 and 10 $\mu g/m^3$ in autumn 1980 *(5)*. Von Düszeln & Thiemann *(6)* measured an average level of 4 $\mu g/m^3$ in the air of 14 cities in the Federal Republic of Germany during 1980–1981. Similar levels have been measured in the United States.

Some of the highest levels reported have been found near waste disposal sites and dry-cleaning establishments *(2)*. Levels in the latter are generally below 678 mg/m³ (100 ppm) *(1)*. Indoor levels up to 250 μg/m³ have been measured.

Conversion factors

1 ppm (in air, 25 °C)	=	6.78 mg/m³
1 mg/m³	=	0.14 ppm

Routes of Exposure

Air

Air pollution unquestionably represents a major source of exposure. Indoor pollution may be just as or more significant than exposure to ambient air.

Drinking-water

Exposure via drinking-water is minor, except in unusual contaminant cases *(7)*. Because of its relative insolubility and moderate vapour pressure (19 torr at 25 °C), tetrachloroethylene would volatilize rapidly from surface waters.

Food

Intake via food can be an important source of exposure and was at one time calculated to be as much as 160 μg per day *(8–10)*. However, improved food processing operations are currently believed to have resulted in a marked decrease of tetrachloroethylene in food.

Relative significance of different routes of exposure

Indoor exposure to tetrachloroethylene can often be higher than outdoor exposure *(11)*, the main sources being building materials and consumer products. In a study of 134 homes in the Netherlands, mean values ranged from 4 to 205 μg/m³ over a period of 5–7 days *(12)*. Fifteen Italian homes were found to have an exposure level of 3–47 μg/m³ over a period of 4–7 days *(13)*. In an American study, values up to 250 μg/m³ have been recorded *(14)*. High values have also been noted inside prefabricated housing *(15)*.

Estimated tetrachloroethylene exposure levels in different types of exposure are given in Table 1.

Kinetics and Metabolism

Absorption

Tetrachloroethylene is principally absorbed via the lungs. Data from human and laboratory animal studies suggest that uptake is approximately proportional to exposure concentration. Physical activity increases uptake *(1,2)*. More than 90% of the amount absorbed is eliminated unchanged via the lungs. Absorption through the skin is a minor source of exposure. Tetrachloroethylene is rapidly and almost completely absorbed via the gastrointestinal tract.

Table 1. Estimated tetrachloroethylene exposure levels in different types of exposure

Type of exposure	Observed range of concentration	Total volume inhaled or amount consumed per day	Inhalation or ingestion rate (mg/day)
General population			
Inhalation:			
urban	trace to 70 μg/m³	20 m³	trace to 1.4
rural	trace	20 m³	negligible
waste disposal sites	400 μg/m³	20 m³	≦8
Ingestion:			
drinking-water	≦2 μg/litre	2 litres	≦0.004
food			≦0.160
Occupational group			
Inhalation	≦340 mg/m³	10 m³	≦3400

Distribution

Studies on laboratory animals indicate that tetrachloroethylene is distributed to all the body tissues. However, no direct measurements of tissue levels of tetrachloroethylene in humans are available *(1)*.

Biotransformation

Tetrachloroethylene is chiefly metabolized in the liver. In man, trichloroacetic acid is the predominant metabolite and is excreted in the urine. Trichloroethanol has also been identified as a urinary metabolite *(16)*. There is no evidence that the pathways of metabolism in laboratory animals are qualitatively different from those in humans. From the studies of Ohtsuki et al. *(17)* it appears that saturation of metabolism in humans occurs at levels considerably in excess of 678 mg/m³ (100 ppm).

Elimination

In man, about 1–3% of the absorbed amount of tetrachloroethylene is excreted as urinary metabolites. A significant proportion of tetrachloroethylene (80–100%) is excreted in breath in the form of unchanged parent compound *(1,2)*. Tetrachloroethylene takes considerably longer to be eliminated from adipose tissue because of its high lipid solubility.

Health Effects

Effects on experimental animals and *in vitro* test systems

Toxicological effects

Acute toxicity studies have shown that the oral LD_{50} values in rodents range from 8400 to 13 000 mg/kg body weight, and LC_{50} values range from 20 to 35 mg/litre. Toxic effects are mainly related to the CNS and the liver. In liquid form, tetrachloroethylene has an irritant effect on the skin and eyes *(1)*.

Tetrachloroethylene has recently been evaluated in a 2-year chronic toxicology and carcinogenicity study by the US National Toxicology Program *(18)*. Rats were exposed to 1356 and 2712 mg/m^3, while mice were exposed to 678 and 1356 mg/m^3. Tetrachloroethylene caused pronounced renal tubular cell karyomegaly in both sexes of rats compared with control animals, and renal tubular cell hyperplasia in male rats. These effects have commonly been observed in a variety of rat strains exposed to chlorinated ethylenes. However, in the 13-week dose-finding study (680–10 800 mg/m^3), kidney lesions were not observed. Short-term exposure of rats to 1360–5400 mg/m^3 produced only minimal to mild liver congestion. In mice there was an increased incidence of liver degeneration in males at both levels, compared with controls. Necrosis of the liver was observed in males and high-dose females. In the 13-week subchronic study, only minimal to mild microscopic liver and kidney changes were observed in mice exposed to 1360–10 800 mg/m^3. There was an increased incidence of renal tubular cell karyomegaly in all exposed mice during the 2-year bioassay. Tubular epithelial cell hyperplasia was not observed.

In an earlier oral study by the US National Cancer Institute *(19)*, toxic nephropathy was also observed in B6C3F1 mice and Osborne-Mendel rats during the 78-week treatment period. In a 52-week inhalation exposure of Sprague-Dawley rats, Rampy et al. *(20)* observed chronic renal disease in animals exposed to both dose levels (2035 and 4070 mg/m^3 (300 ppm and 600 ppm)).

In one of the earliest studies, Carpenter *(21)* exposed albino rats to 3188 mg/m^3 for 150 days and found congested livers but no degeneration or necrosis. Rats also exhibited kidney damage and congested spleens. No evidence of damage in either kidneys or liver was observed at 475 mg/m^3. Microscopic examination of heart, brain, eye and nerve tissue revealed no damaging effects in any of the chronically exposed rats.

The relationship between dose, metabolism and indices of hepatotoxicity (e.g. SGOT, SGPT) has been investigated in an oral exposure study in mice. The hepatotoxicity of tetrachloroethylene appeared to be directly proportional to the extent to which it was metabolized *(22)*.

The mammalian tests to date (in rats, mice and rabbits) do not indicate any significant teratogenic potential of tetrachloroethylene. The anatomical effects observed in these inhalation experiments primarily reflect delayed development and generally can be considered reversible *(2)*.

Tetrachloroethylene has also been implicated in producing 20% abnormal sperm cell morphology in mice, but only at very high exposure levels *(23)*. No other reproductive effects have been reported.

Mutagenic and carcinogenic effects

Tetrachloroethylene has not been shown to be a mutagen. Highly purified tetrachloroethylene has been evaluated in the Ames (*Salmonella*) test, a yeast recombinogenic assay, a host-mediated assay using *Salmonella,* and DNA repair assays *(2)*. Responses were negative or weak and no dose-response relationships were established. The weak positive findings may be due to the presence of mutagenic contaminants and/or added stabilizers. The US National Toxicology Program has tested tetrachloroethylene (99.7%) in various *S. typhimurium* strains, both with and without metabolic activation from S9 fractions from rat and hamster *(18)*. The results were uniformly negative. It was also not shown to be mutagenic in L5178Y/$TK^{+/-}$ mouse lymphoma cells, with or without metabolic activation. It did not induce sex-linked recessive lethal mutations in *Drosophila.* Because the epoxide of tetrachloroethylene was mutagenic in bacterial studies, the possibility that tetrachloroethylene may pose a mutagenic hazard exists.

In a 2-year inhalation bioassay by the US National Toxicology Program *(18)*, a variety of neoplastic lesions were observed. Both levels of tetrachloroethylene (1356 mg/m^3 and 2712 mg/m^3) were associated with a statistically increased incidence of mononuclear cell leukaemia in F344/N rats, both male and female. The incidence of mononuclear cell leukaemia was already pronounced in the control population, but both concentrations decreased the time to diagnosis. Tetrachloroethylene also increased the incidence of renal tubular cell adenomas or adenocarcinomas (combined) in male rats. In B6C3F1 mice exposed to 680 and 1360 mg/m^3, the predominant findings were dose-related increases in the incidence of hepatocellular neoplasms.

In the earlier US National Cancer Institute lifetime gavage study *(19)*, a significant increase in the incidence of hepatocellular carcinomas in mice was observed at dose levels of 500 and 1000 mg/kg body weight. In a similar study with rats, no conclusions could be drawn because of the poor survival rate.

In a 52-week inhalation study *(20)*, no clear difference was found between exposed and control rats in the incidence of different tumour types. Rats were exposed to 2035 and 4070 mg/m^3 (300 ppm and 600 ppm) and then kept for lifetime.

Effects on humans

The known effects of tetrachloroethylene on man have been established primarily from cases in which individuals have been accidentally or occupationally exposed to very high concentrations. Effects on the CNS are generally most noticeable following acute or excessive occupational exposure.

Toxicological effects

The principal effect in single or short-term repeated exposure to levels above 678 mg/m^3 is alteration of the CNS function *(1)*. Effects upon the liver and kidneys *(1,2)* have usually been observed clinically during the period following exposure to concentrations in excess of 678 mg/m^3. There are no data to suggest that low-level exposures, such as those typical in the ambient environment, result in liver dysfunction. A no-observed-effect level of 136 mg/m^3 has been identified for man, based on subjective tests of the CNS function *(24)*. Effects upon the kidney are not well documented; cause–effect relationships with kidney function cannot be established *(2)*.

Useful information about the effects of long-term exposure to tetrachloroethylene is limited *(2)*. Extended exposure produces many of the symptoms also elicited during short-term exposure, but they are of a more continuous nature. Reports of frequent dizziness, headache, nausea, fatigue and disorientation are common, even for extended periods of time after exposure ceases. Subjects exposed for long periods are reported to have short-term memory deficits, ataxia, irritability, disorientation, sleep disturbances, and decreased alcohol tolerance. Such symptoms are sometimes reported to be irreversible. Disturbances of the autonomic nervous system, in addition to those of the CNS, have been reported in workers exposed to levels above 678 mg/m^3 *(25)*. More sensitive tests would have to be performed to determine whether tetrachloroethylene affects the CNS at even lower concentrations.

There have been many reports of alterations in the liver function of individuals exposed to unknown levels over extended periods *(2)*.

Carcinogenic effects

Two studies have been conducted to determine if there is a relationship between exposure of laundry and dry-cleaning workers to tetrachloroethylene and mortality due to cancer *(26,27)*. While an increased incidence of lung, cervical and skin cancer was observed in one study *(26)*, a cause–effect relationship with tetrachloroethylene could not be determined: workers were exposed to several other organic chemicals as well (e.g. carbon tetrachloride, trichloroethylene and benzene). In the second study *(27)*, an elevated risk of genital and kidney cancer, together with a smaller excess of bladder and skin cancer and lymphosarcoma, was found among female workers. Because of confounding factors and a lack of data on exposure, no cause–effect relationship with tetrachloroethylene can be ascribed.

Sensory effects

Tetrachloroethylene in its pure form has an odour detection threshold of 8 mg/m^3 *(28)*. It is recognized at concentrations 3–4 times higher than the odour detection threshold *(28)*. Its ethereal odour is that characteristically emitted by dry-cleaning establishments. It is also one of the many odorous substances contained in emissions from landfills.

Some individuals can perceive the odour of tetrachloroethylene at levels lower than 8 mg/m^3, but, in general, odour problems from dry-cleaning establishments are unlikely to occur if peak concentrations in the ambient air are kept below this threshold value.

Evaluation of Human Health Risks

Exposure

Maximum short-term concentrations in urban atmosphere are about 70 $\mu g/m^3$, averages ranging from 1 $\mu g/m^3$ to 10 $\mu g/m^3$. Global background levels are about 0.2 $\mu g/m^3$. The highest levels are likely to occur near dry-cleaning establishments and waste disposal sites. Indoor concentrations ranging from 3 $\mu g/m^3$ to 250 $\mu g/m^3$ have been measured.

Table 1 shows the amount of tetrachloroethylene inhaled by individuals in different types of exposure. The exposure of occupational groups far exceeds that of any other group.

Health risk evaluation

Because air levels of tetrachloroethylene in the general environment are orders of magnitude less than levels associated with adverse effects on the CNS or the autonomic nervous system, it is unlikely that exposure to tetrachloroethylene represents a health concern as far as these endpoints are concerned.

Available data on man, together with supportive evidence from laboratory animal studies, suggest that short-term exposure to 678 mg/m^3 (100 ppm) represents a likely threshold for nongenotoxic adverse effects (e.g. CNS disturbances). A no-observed-effect level of 136 mg/m^3 has been identified for man in short-term repeated exposure. There are no data to indicate what such a level would be in long-term exposures: more sensitive tests would have to be instituted.

While tetrachloroethylene is generally considered to be hepatotoxic and nephrotoxic in laboratory animals, the evidence regarding humans suggests that environmental levels pose no serious health problem.

Indices of hepatotoxicity have been linearly correlated with metabolism in animals *(22)*. In humans, transient liver damage is associated with short-term exposures in excess of 678 mg/m^3 (100 ppm). However, overexposure to tetrachloroethylene has not been associated with irreversible liver or kidney damage.

Tetrachloroethylene has been placed by IARC in Group 3, i.e. that in which chemicals cannot be classified as to their carcinogenicity in humans.

Guidelines

The lowest-observed-adverse-effect level of tetrachloroethylene in short-term repeated human exposure was found to be 678 mg/m^3. Although the adverse nature of observed effects on the CNS was minimal at this concentration, a protection (safety) factor of 100 is considered appropriate, mainly because of the very limited data available on the toxic effects of long-term inhalation exposure. A guideline value of 5 mg/m^3 (averaging time 24 hours) is recommended. This level is far above the current level of environmental exposure.

The odour detection threshold level is 8 mg/m^3. Odour problems should not occur when concentrations are kept below this value (averaging time 30 minutes).

References

1. *Tetrachloroethylene.* Geneva, World Health Organization, 1984 (Environmental Health Criteria, No. 31).
2. *Health assessment document for tetrachloroethylene (perchloroethylene).* Washington, DC, US Environmental Protection Agency, 1985 (Final report No. EPA-600/8-82-005F).
3. Tetrachloroethylene. *In: Some halogenated hydrocarbons.* Lyon, International Agency for Research on Cancer, 1979, pp. 491–514 (IARC Monographs on the Evaluation of the Carcinogenic Risk of Chemicals to Humans, Vol. 20).
4. *Synthetic organic chemicals, US production and sales (1982).* Washington, DC, US International Trade Commission, 1983 (USITC Publication 1183).
5. **Bauer, U. & Selenka, F.** Belastung der Bevölkerung durch Haloforme und chlorierte Lösemittel im Trinkwasser im Vergleich zu Luft und Lebensmitteln [Exposure of the population to haloforms and chlorinated solvents in drinking-water compared to air and food]. *Vom Wasser,* **59**: 7–16 (1982).
6. **von Düszeln, J. & Thiemann, W.** Volatile chlorinated hydrocarbons in a coastal urban atmosphere. *Science of the total environment,* **41**: 187–194 (1985).
7. **Zoeteman, B.C.J. et al.** Persistent organic pollutants in river water and ground water of the Netherlands. *Chemosphere,* **9**: 231–249 (1980).
8. **Bauer, U.** Belastung des Menschen durch Schadstoffe in der Umwelt. Untersuchungen über leicht flüchtige organische Halogen-Verbindungen in Wasser, Luft, Lebensmitteln und im menschlichen Gewebe. I, II, III & IV Mitteilung. [Human exposure to environmental chemicals — investigations on volatile organic halogenated compounds in water, air, food, and human tissues. I, II, III, IV]. *Zentralblatt für Bakteriologie, Mikrobiologie und Hygiene. I Abt. Orig. B,* **174**: 15–56, 200–237, 556–583 (1981).
9. **Zimmerli, B. et al.** Perchlorethylen in Lebensmitteln [Perchloroethylene in foodstuffs]. *Mitteilungen aus dem Gebiete der Lebensmitteluntersuchung und Hygiene,* **73**: 71–81 (1982).
10. **von Düszeln, J. et al.** Flüchtige Halogenkohlenwasserstoffe in Luft, Wasser und Nahrungsmitteln in der Bundesrepublik Deutschland [Volatile halogenated hydrocarbons in air, water, and food in the Federal Republic of Germany]. *Deutsche Lebensmittel-Rundschau,* **78**: 352–356 (1982).
11. **Wallace, L.A. et al.** Personal exposures, indoor–outdoor relationships, and breath levels of toxic air pollutants measured for 355 persons in New Jersey. *Atmospheric environment,* **19**: 1651–1661 (1985).
12. **Lebret, E. et al.** Volatile hydrocarbons in Dutch homes. *In:* Berglund, B. et al. ed. *Indoor air: chemical characterization and personal exposure.* Stockholm, Swedish Council for Building Research, 1984, Vol. 4, pp. 169–173.

13. **de Bortoli, M. et al.** *Measurements of indoor air quality and comparison with ambient air: a study of 15 homes in Northern Italy.* Luxembourg, Commission of the European Communities, 1985.
14. **Hartwell, T.D. et al.** Comparison of indoor and outdoor levels for air volatiles in New Jersey. *In:* Berglund, B. et al., ed. *Indoor air: chemical characterization and personal exposure.* Stockholm, Swedish Council for Building Research, 1984, Vol. 4, pp. 81–85.
15. **Monteith, D.K. et al.** Sources and characterization of organic air contaminants inside manufactured housing. *In:* Berglund, B. et al., ed. *Indoor air: chemical characterization and personal exposure.* Stockholm, Swedish Council for Building Research, 1984, Vol. 4, pp. 285–290.
16. **Ikeda, M. et al.** Urinary excretion of total trichloro-compounds, trichloroethanol and trichloroacetic acid as a measure of exposure to trichloroethylene and tetrachloroethylene. *British journal of industrial medicine,* **29**: 328–333 (1972).
17. **Ohtsuji, T. et al.** Limited capacity of humans to metabolize tetrachloroethylene. *International archives of occupational and environmental health,* **51**: 381–390 (1983).
18. *Toxicology and carcinogenesis studies of tetrachloroethylene (perchloroethylene) (CAS No. 127-18-4) in F344/N rats and B6C3F1 mice (inhalation studies).* Research Triangle Park, NC, National Toxicology Program, 1986 (NIH Publication No. 86-2567).
19. **US National Cancer Institute.** *Bioassay of tetrachloroethylene for possible carcinogenicity.* Washington, DC, US Government Printing Office, 1977 (DHEW Publication No. (NIH) 77-813).
20. **Rampy, L.W. et al.** *Results of a long-term inhalation toxicity study on rats of a perchloroethylene (tetrachloroethylene) formulation.* Midland, MI, Dow Chemical Company, Toxicology Research Laboratory, Health and Environmental Research, 1978.
21. **Carpenter, C.P.** The chronic toxicity of tetrachloroethylene. *Journal of industrial hygiene and toxicology,* **19**: 323–336 (1937).
22. **Buben, J.R. & O'Flaherty, E.J.** Delineation of the role of metabolism in the hepatotoxicity of trichloroethylene and perchloroethylene: a dose–effect study. *Toxicology and applied pharmacology,* **78**: 105–122 (1985).
23. **Beliles, R.P. et al.** *Teratogenic–mutagenic risk of workplace contaminants: trichloroethylene, perchloroethylene, and carbon disulfide.* Washington, DC, US Department of Health, Education, and Welfare, 1980 (Contract No. 210-77-0047).
24. **Hake, C.L. & Stewart, R.D.** Human exposure to tetrachloroethylene: inhalation and skin contact. *Environmental health perspectives,* **21**: 231–238 (1977).
25. **von Münzer, M. & Heder, K.** Ergebnisse der arbeitsmedizinischen und technischen Uberprüfung chemischer Reinigungsbetriebe [Results of the occupational medicinal and technical inspection of dry-cleaning establishments]. *Zentralblatt für Arbeitsmedizin, Arbeitsschutz und Prophylaxe,* **22**: 133–138 (1972).
26. **Blair, A. et al.** Causes of death among laundry and dry cleaning workers. *American journal of public health,* **69**: 508–511 (1979).

27. **Katz, R.M. & Jowett, D.** Female laundry and dry cleaning workers in Wisconsin: a mortality analysis. *American journal of public health,* **71**: 305–307 (1981).
28. [*Reports of studies on the measurement of offensive odors 1972–1980*]. Tokyo, Japan Environment Agency, 1980 (in Japanese).

14

Toluene

General Description

Sources

Toluene is a noncorrosive, volatile liquid with low solubility in water. Principal production sources include petroleum refining operations, coke-oven operations and the production of other chemicals (e.g. styrene). The bulk of production is in the form of a benzene-toluene-xylene mixture that is used in the backblending of petrol to enhance octane ratings. Purified toluene contains a small amount of benzene (<0.01%). Toluene is extensively used as a carrier in paints, inks, thinners and adhesives, as a component of cosmetic products, and in the production of other chemicals. It is estimated that the worldwide production of toluene amounts to some 10 million tonnes *(1)*.

Emissions to the atmosphere result from point sources (e.g. production) and area sources (e.g. marketing and use of petrol). Nonoccupational uses of paints and thinners, together with tobacco smoke, represent the principal sources of toluene in indoor environments. Emissions also occur via petroleum, coal and vegetation. It is difficult to estimate annual emissions because total use and source distribution vary widely from country to country.

Toluene is the most prevalent hydrocarbon in the troposphere. Its dispersion in the troposphere is largely dependent upon meteorological conditions and its atmospheric reactivity. Reaction with hydroxy radicals in the troposphere represents the principal mechanism by which toluene is removed *(2)*. The lifetime of toluene in the troposphere is dependent on hydroxy radical concentrations. In winter, the lifetime can be several months; in summer, several days *(2)*. In such meteorological conditions and together with other emitted pollutants associated with smog production, toluene may contribute significantly to the causation of smog. The contribution of toluene to the formation of ozone and formaldehyde is comparable to that of ethylene or propylene. Toluene can also be a precursor of peroxybenzoyl nitrate, a very potent eye irritant. Because of toluene's high volatility and low solubility in water, most toluene occurring in natural waters may be expected to be eventually released to the atmosphere.

Occurrence in air

Because most of the readily available data relate to the United States and a few European sites, it is difficult to make meaningful extrapolations to other regions of the world for which no data are available. An assessment of the quality and extent of available air data for toluene and other volatile organics is given in the summary report of Brodzinsky & Singh *(3)*.

Air monitoring data suggest that $0.75 \mu g/m^3$ could be regarded as an upper-bound background level to which all populations are exposed *(2)*. It has been suggested that urban residents worldwide are likely to be exposed to considerably higher levels *(3)*. The average concentrations found have varied widely, from 0.0005 to 1.31 mg/m^3 *(1)*. Instantaneous values could be as high as 20 mg/m^3 (5 ppm), which is approximately the upper range of the odour threshold *(3)*.

Conversion factors

1 ppm (in air, 25 °C) = 3.75 mg/m^3

1 mg/m^3 = 0.266 ppm

Routes of Exposure

Air

Air pollution is unquestionably the major source of exposure. Toluene levels in indoor environments are expected to be considerably higher than outdoor levels in those situations involving the nonoccupational use of paints and thinners, and also where tobacco smoke is present. Levels of toluene measured in homes have been as high as $610 \mu g/m^3$ *(4)*.

Occupational exposure

Occupational subpopulations are likely to be exposed to considerably higher levels of toluene than are general populations. In addition, air levels of toluene in the vicinity of industrial sources are likely to represent an additional burden to both workers and local residents *(5)*.

Smoking

Toluene is a major component of tobacco smoke *(6)*. The concentration of toluene in inhaled mainstream cigarette smoke is approximately 0.1 mg per cigarette; sidestream smoke may contain a higher amount.

Drinking-water

Exposure via drinking-water is minor, except in cases of unusually heavy contamination *(1)*.

Food

Exposure via food is also considered to be insignificant.

Relative significance of different routes of exposure

Inhalation is the predominant route of exposure. Worst-case exposure values relevant to a discussion of health effects can be constructed from the information presented in Table 1.

Kinetics and Metabolism

Absorption

The percentage of toluene retained by the human body after inhalation is about 40–60%. Uptake increases as the duration of exposure and the level of physical activity increase *(1,2)*. There appears to be sufficient evidence that, in the exposure range of 375–750 mg/m^3 (100–200 ppm), arterial blood levels in humans can be estimated from alveolar air concentrations, during and shortly after exposure *(1,2)*. The time required for uptake to reach an asymptotic blood level in humans could be as long as 2 hours. Absorption also occurs through the skin. However, dermal absorption is not likely to contribute significantly to the body-burden. Absorption of toluene from the gastrointestinal tract is considered to be complete.

Distribution

In inhalation experiments with laboratory animals, considerable amounts of toluene have been shown to partition in white adipose tissue, adrenals, kidneys, liver and brain *(8)*. Oral administration of tritium-labelled toluene produced a distribution pattern similar to that via inhalation *(9)*.

Biotransformation

Toluene is rapidly metabolized by the liver to benzoic acid. The latter is then conjugated with glycine to form hippuric acid, which is then excreted in the urine. Less than 1% of toluene undergoes ring hydroxylation to form o- and p-cresol. The pathway in humans and laboratory animals is believed to be very similar *(1,2)*.

While cresol production may be a source of concern with respect to potential mutagenicity and carcinogenicity, binding of toluene metabolites to protein and nucleic acids does not appear to occur to any significant extent. The level at which exposure may exceed conjugation capacity (assuming 50% retention) has been estimated at 3000 mg/m^3 (780 ppm) during light work or 1000 mg/m^3 (270 ppm) during heavy work *(10)*. Conjugation capacity appears to be limited by the availability of glycine. The overall metabolism of toluene may be as high as 70% of the absorbed dose *(11)*.

The ability of toluene to interfere with the biotransformation of other substances (e.g. styrene, acrylonitrile, benzene, trichloroethylene, carbon tetrachloride, and methylethylketone) has been reported by numerous authors *(1,2)*.

Elimination

About 20% of the absorbed dose is excreted as unchanged toluene in expired air. The elimination of toluene and metabolites from adipose tissue and

Table 1. Estimated toluene exposure levels in different types of exposure

Type of exposure	Observed range of concentration	Frequency of exposure	Total volume inhaled[a] or amount consumed per week	Inhalation or ingestion rate (mg/week)
General population				
Inhalation:				
urban areas	0.1–204 μg/m³	168 hours/week	140 m³	0.01–28
rural and remote areas	trace to 3.8 μg/m³	168 hours/week	140 m³	trace to 0.5
areas near manufacturing and user sites	0.1–20 mg/m³	168 hours/week	140 m³	14–2800[b]
indoor (nonindustrial)	17–700 μg/m³	168 hours/week	140 m³	2–98
Ingestion:				
drinking-water	0–19 μg/litre	2 litres/day	14 litres	0–0.3
food (fish only)	0–1 mg/kg	6.5 g/day	45.5 g	0–0.045
Occupational group				
Inhalation	375 mg/m³[c]	40 hours/week	32 m³	12 000
Dermal	0–170 μg/litre[d]	0–30 minutes/week		0–1.0
Cigarette smokers				
Inhalation	0.1 mg/cigarette[e]	20 cigarettes/day	140 cigarettes	14

[a] Based on a breathing rate of 20 m³/24 hours.

[b] The value of 2800 mg/week is based on the highest recorded instantaneous level assumed to be present for one week.

[c] This value is similar to permissible standards in various countries and represents the worst-case estimate.

[d] This value represents exposure to blood due to dermal contact and relates to absorbed levels *(7)*.

[e] From: Dalhamn et al. *(6)*; toluene content may be higher depending on tobacco type.

bone marrow is prolonged *(1)*. Hippuric acid is excreted in the urine, as are cresols (as sulfate or glucuronide conjugates). Neither hippuric acid nor cresols can provide a reliable measure of individual toluene uptake at low exposure levels in relation to varying workloads or food intakes *(1,2)*.

Health Effects

Effects on experimental animals and *in vitro* test systems

Toxicological effects

Experiments on laboratory animals indicate that the most pronounced effect, as in humans, is on the CNS. Levels below 3750 mg/m^3 (1000 ppm) have little or no effect on gross manifestations of animal behaviour *(1)*. Observations that lower levels caused changes in cognition and in brain neuromodulator levels *(12,13)* suggest that the potential effects of toluene on the CNS at levels below 375 mg/m^3 (100 ppm) cannot be ignored.

Exposure to toluene has not been shown to adversely affect the liver, kidneys or haematopoietic tissues; either there is an absence of effect or there are adaptive responses to exposure *(14,15)*.

Toluene has been shown to be fetotoxic at inhalation exposure levels up to 6000 mg/m^3 (rats), 3750 mg/m^3 (mice) and 847 mg/m^3 (rabbits) *(1,2)*. In rabbits exposed to 847 mg/m^3 during organogenesis, spontaneous abortions occurred, but no teratogenic effects were noted *(16)*.

The work of Andersson et al. *(17)* with male rats exposed to 1875 mg/m^3 and 3750 mg/m^3 for 6 hours per day for 3 and 5 days respectively, may indicate that toluene has potential gonadal effects. In addition to finding increases in brain neurotransmitters, the authors found an increase in corticosterone, prolactin and, on administration of a tyrosine hydroxylase inhibitor, a significant increase in follicle-stimulating hormone, suggesting that changes in brain neurotransmitters may be associated with menstrual cycle disturbances.

In male Donyru rats exposed to toluene levels of 375 and 750 mg/m^3 for one year, degeneration of the germinal cells of the testes was found in 4 of 12 animals exposed to the higher level, but not in controls *(18)*.

Mutagenic and carcinogenic effects

Only one chronic bioassay has been completed and published *(14)*. No increased incidence of neoplastic, proliferative, inflammatory or degenerative lesions was found in male and female rats exposed to 112, 375, or 1125 mg/m^3 for 24 months. Maltoni et al. *(19)* have recently reported the preliminary partial results of a long-term study on Sprague-Dawley rats (141 weeks by gavage; 500 mg/kg). Information was presented in tabular form only. The total number of malignant tumours was reported to be elevated compared with that found in controls. Statistical analyses were not provided in the report and the degree of purity of the test mixture was not stated.

While there are insufficient data to make a final evaluation of the effect of toluene on genetic activity, the *in vitro* studies performed on bacteria, yeast and mammalian cells are uniformly negative *(1,2)*. *In vivo* injections of toluene *(20,21)* and inhalation studies *(22)* have been associated with chromosomal aberration in rat bone marrow cells. However, the degree of purity of the test mixture was not stated. More recently, Gad-El-Karim et al. *(23)* found no evidence of a clastogenic effect when toluene was administered orally to CD-1 mice at a level of 1720 mg/kg in two doses 24 hours apart. Pretreatment with 3-methylcholanthrene did not alter these observations.

Effects on humans

Toxicity studies of humans have primarily involved evaluation of individuals exposed to toluene via inhalation in experimental or occupational settings or during episodes of intentional abuse. In many studies other chemicals were present in the exposure atmosphere as well. This makes it difficult to ascribe cause–effect relationships.

Toxicological effects

The health effect of primary concern is dysfunction of the CNS. Acute experimental and repeated occupational exposure at levels above 375 mg/m^3 have elicited dose-related CNS alterations such as fatigue, confusion and lack of coordination, as well as impairment of reaction time and perceptual speed *(1,2)*. The level of 375 mg/m^3 has been described as a no-effect level for reaction time tests *(24)*.

Von Oettingen et al. *(25,26)* have provided the most complete description of the effects of acute exposure of humans to toluene. The benzene level in the toluene was below 0.01%. In controlled 8-hour exposures of three individuals, effects ranged from moderate fatigue and sleepiness (3/3) at 375 mg/m^3 to rapid onset of severe fatigue, pronounced nausea, confusion, lack of self-control, considerable incoordination and staggering gait (3/3) at 3000 mg/m^3. One subject was reported to develop drowsiness and a mild headache during exposure to 187 mg/m^3, a level well above the odour threshold. Andersen et al. *(27)* reported statistically significant borderline impairment ($0.05 < P < 0.10$) in some tests of cognitive function at 375 mg/m^3, but no effect at 150 mg/m^3 or 37.5 mg/m^3, following 6 hours of exposure.

Gusev *(28)* examined the effects of acute, low-level (1 mg/m^3) toluene exposure for 6 minutes on the EEG activity of four subjects trained to develop synchronous and well marked alpha rhythms when stimulated by light. A statistically distinct change in EEG activity from the left temporal-occipital region of all subjects was observed. The toxicological significance of this observation is not clear, and there is no further information to confirm this report.

Toluene is an eye irritant at acute levels as low as 375 mg/m^3 *(27)*.

The effects associated with repeated occupational exposures to toluene over a period of weeks, described by Wilson *(29)*, are consistent with those reported by von Oettingen et al. *(25,26)*. Wilson also described myelotoxic effects. These may have been due to concurrent toluene-benzene exposure, because more recent evidence has not demonstrated any such effects. Takeuchi et al. *(30)* considered toluene exposure in a paint factory to be the likely cause of a diencephalon syndrome in two workers. Other solvents had been present in the exposure atmosphere, however. There are no reports of peripheral neuropathy associated with exposure to pure toluene, but psychophysiological disturbances in workers exposed to solvents have been found. For a variety of reasons, none of these changes can be clearly ascribed to toluene.

Data relating to individuals who intentionally abuse toluene indicate that chronic inhalation does not lead to chronic liver disease. Similarly, no significant cardiovascular effects are expected as a result of environmental exposures, since no such effects have been documented in intentional abuse situations at extremely high exposure levels. In the light of intentional abuse data, exposure to environmental levels of toluene is thought unlikely to have an adverse effect on the liver or kidneys *(1,2)*.

There are no adequate data on the teratogenic potential or reproductive effects of toluene exposure in humans. Hersh et al. *(31)* recently reported that three children exhibiting nonspecific teratogenic phenotypes were born to women who intentionally inhaled large quantities of pure toluene throughout pregnancy. The significance of this finding remains to be established. Two studies *(32,33)* reported menstruation disturbances in occupational settings. Because of confounding factors, no cause–effect relationship with toluene can be ascribed. If toluene exposure does alter key hormonal levels in women, this would have important implications.

Carcinogenic effects

There are no data relating to the incidence of cancer in humans exposed to toluene.

A number of studies have evaluated chromosome damage in the peripheral lymphocytes of workers exposed to toluene *(1)*. Studies by Bauchinger et al. *(34)* associated long-term exposure of rotogravure workers to toluene, with an increased incidence of sister chromatid exchanges and chromosomal aberrations. In the most recent extension of this work, Schmid et al. *(35)* reported a higher incidence of chromatid-type aberrations up to two years after cessation of exposure compared with that found in controls. After longer post-exposure periods, the number of chromosome aberrations became indistinguishable from that in the controls. Studies showing no chromosome aberrations in workers exposed to somewhat lower toluene levels were conducted by Mäki-Paakkanen et al. *(36)* and Forni et al. *(37)*.

Sensory effects

Toluene in pure form has an odour detection threshold of 1 mg/m^3 *(38,39)*. It has a solvent-like odour. The recognition threshold for toluene is about 10 times higher than the odour detection threshold *(38–40)*.

Although some individuals can perceive the odour of toluene at concentrations lower than 1 mg/m^3, odour problems are generally not likely to occur if peak concentrations in ambient air are kept below this threshold value.

Antti-Poika et al. *(41)* evaluated workers exposed to toluene concentrations of 255–694 mg/m^3 for 10 years and found no abnormalities with respect to autonomic nervous function, electroencephalography, psychological tests, or tomography of the brain. Cherry et al. *(42)* reported significant differences in a cohort of toluene-exposed workers (up to 1875 mg/m^3 for 9 years) given a reading test, but a cause–effect relationship could not be inferred from the data. Juntunen et al. *(43)* observed no adverse effects upon the nervous system when toluene-exposed rotogravure workers were compared with control workers. The duration of exposure was 22 years at an estimated long-term level of 439 mg/m^3. Münchinger et al. *(44)* reported CNS dysfunction in workers exposed to toluene for 12–18 years.

Evaluation of Human Health Risks

Exposure

Table 1 shows the amount of toluene inspired per week by individuals in certain types of exposure. The levels of exposure of occupational groups far exceed those of any other group.

If one assumes a worst-case situation based on Table 1, an individual would have the following profile: lives in an urban area near a manufacturing site, works in a toluene-containing work environment, is exposed to toluene in food (fish only) and in drinking-water, and smokes 40 cigarettes per day. The combined exposure is still predominantly that in the occupational setting. If one assumes that 50% of the toluene inspired (15 000 mg/week) is released by the lungs in unchanged form, the effective exposure dose is about 7500 mg/week on a continuous, long-term basis. This would correspond to an air exposure of about 100 mg/m^3 (30 ppm).

Health risk evaluation

There is a continuum in the effects of toluene on the CNS, ranging from clearly adverse effects, such as confusion and muscle weakness, occurring at 750 mg/m^3 and higher levels for very short periods, to slight changes in psychomotor function and mild symptoms detectable at 375 mg/m^3 after exposure for periods of hours. Eye and nose irritation is detectable at 375 mg/m^3 during a 6-hour exposure. Studies of occupationally exposed groups provide data reasonably consistent with those obtained from controlled acute human exposure experiments.

The available data for animals and humans are inadequate for evaluating the carcinogenicity of toluene.

Guidelines

The lowest-observed-effect level appears to be about 375 mg/m^3 (100 ppm), based on observations of the CNS and mucosal irritation.

For the CNS and mucosal irritant effects a protection (safety) factor of 50 (24-hour averaging time) appears adequate in view of the data on negative effects in humans, the lack of evidence of a chronic effect, and the minimally adverse nature of the observed effects at the lowest-observed-effect level. This results in a guideline level of 7.5 mg/m^3 for a 24-hour averaging time.

It may also be appropriate to use the odour detection threshold of approximately 1 mg/m^3 as an air quality guideline. No protection factor for a guideline based on the odour threshold would be necessary. However, the averaging time should be 30 minutes to avoid complaints regarding odour.

References

1. *Toluene.* Geneva, World Health Organization, 1985 (Environmental Health Criteria, No. 52).
2. *Health assessment document for toluene.* Washington, DC, US Environmental Protection Agency, 1983 (Report No. EPA-600/8-82-008F).
3. **Brodzinsky, R. & Singh, H.B.** *Volatile organic chemicals in the atmosphere: an assessment of available data.* Menlo Park, CA, Stanford Research Institute, 1982.
4. **Mølhave, L. et al.** *Afgasning fra byggematerialer — forekomst og hygiejnisk vurdering* [Emissions from building materials — occurrence and health evaluation]. Hørsholm, Statens Byggeforskningsinstitut, 1982 (SBI — Rapport 137).
5. **Sexton, K. & Westberg, H.** Ambient hydrocarbon and ozone measurements downwind of a large automotive painting plant. *Environmental science and technology,* **14**: 329–332 (1980).
6. **Dalhamn, T. et al.** Mouth absorption of various compounds in cigarette smoke. *Archives of environmental health,* **16**: 831–835 (1968).
7. **Sato, A. & Nakajima, T.** Differences following skin or inhalation exposure in the absorption and excretion kinetics of trichloroethylene and toluene. *British journal of industrial medicine,* **35**: 43–49 (1978).
8. **Carlsson, D. & Lindqvist, T.** Exposure of animals and man to toluene. *Scandinavian journal of work, environment & health,* **3**: 135–143 (1977).
9. **Pyykko, K. et al.** Toluene concentrations in various tissues of rats after inhalation and oral administration. *Archives of toxicology,* **38**: 169–176 (1977).
10. **Riihimäki, V.** Conjugation and urinary excretion of toluene and m-xylene metabolites in a man. *Scandinavian journal of work, environment & health,* **5**: 135–142 (1979).
11. **Veulemans, H. & Masschelein, R.** Experimental human exposure to toluene. III. Urinary hippuric acid excretion as a measure of individual solvent uptake. *International archives of occupational and environmental health,* **43**: 53–62 (1979).
12. **Horiguchi, S. & Inoue, K.** Effects of toluene on the wheel-turning activity and peripheral blood findings in mice: an approach to the maximum allowable concentration of toluene. *Journal of toxicological sciences,* **2**: 363–372 (1977).
13. **Gusev, I.S.** [Comparative toxicity of benzene, toluene and xylene]. *Biologičeskoe dejstvie i gigieničeskoe značenie atmosfernyh zagrjaznenij,* **10**: 96–108 (1967) (in Russian).
14. **Gibson, J.E. & Hardisty, J.F.** Chronic toxicity and oncogenicity bioassay of inhaled toluene in Fischer-344 rats. *Fundamental and applied toxicology,* **3**: 315–319 (1983).

15. **Ungváry, G. et al.** Effect of toluene inhalation on the liver of rats — dependence on sex, dose, and exposure time. *Journal of hygiene, epidemiology, microbiology and immunology,* **24**: 242–252 (1980).
16. **Ungváry, G. & Tátrai, E.** On the embryotoxic effects of benzene and its alkyl derivatives in mice, rats and rabbits. Receptors and other targets for toxic substances. *Archives of toxicology,* Suppl.8: 425–430 (1985).
17. **Andersson, K. et al.** Toluene-induced activation of certain hypothalamic and medial-eminence catecholamine nerve-terminal systems of the male rat and its effects on anterior pituitary hormone secretion. *Toxicology letters,* **5**: 393–398 (1980).
18. **Matsumoto, T. et al.** [Experimental studies on chronic toluene poisoning. III. Effects of toluene exposure on blood and organs in the rat]. *Japanese journal of industrial health,* **13**: 501–506 (1971) (in Japanese).
19. **Maltoni, C. et al.** Experimental studies on benzene carcinogenicity at the Bologna Institute of Oncology: current results and ongoing research. *American journal of industrial medicine,* **7**: 415–446 (1985).
20. **Dobrohotov, V.B.** [The mutagenic influence of benzene and toluene under experimental conditions]. *Gigiena i sanitarija,* **37**: 36–39 (1972) (in Russian).
21. **Lypkalo, A.A.** [Genetic activity of benzene and toluene]. *Gigiena truda i professional'nye zabolevanija,* **17**: 24–28 (1973) (in Russian).
22. **Dobrokhotov, V.B.** [Mutagenic effect of benzene, toluene, and a mixture of these hydrocarbons in a chronic experiment]. *Gigiena i sanitarija,* **41**: 36–39 (1976) (in Russian).
23. **Gad-El-Karim, M.M. et al.** Modifications in the myeloclastogenic effect of benzene in mice with toluene, phenobarbital, 3-methylcholanthrene, Aroclor 1254 and SKF-525A. *Mutation research,* **135**: 225–243 (1984).
24. **Gamberale, F. & Hultengren, M.** Toluene exposure. II. Psychophysiological functions. *Work-environment-health,* **9**: 131–139 (1972).
25. **von Oettingen, W.F. et al.** The toxicity and potential dangers of toluene: preliminary report. *Journal of the American Medical Association,* **118**: 579–584 (1942).
26. **von Oettingen, W.F. et al.** *The toxicity and potential dangers of toluene with special reference to its maximal permissible concentration.* Washington, DC, US Public Health Service, 1942 (Public Health Bulletin, No. 279).
27. **Andersen, I. et al.** Human response to controlled levels of toluene in six-hour exposures. *Scandinavian journal of work, environment & health,* **9**: 405–418 (1983).
28. **Gusev, I.S.** [Reflective effects of microconcentrations of benzene, toluene, xylene and their comparative assessment]. *Gigiena i sanitarija,* **30**: 6–10 (1965) (in Russian).
29. **Wilson, R.H.** Toluene poisoning. *Journal of the American Medical Association,* **123**: 1106–1108 (1943).
30. **Takeuchi, Y. et al.** [Diencephalic syndrome in two workers exposed mainly to toluene vapor]. *Japanese journal of industrial health,* **14**: 563–571 (1972) (in Japanese).

31. **Hersh, J.H. et al.** Toluene embryopathy. *Journal of pediatrics,* **106**: 922–927 (1985).
32. **Michon, S.** Zaburzenia w miesiaczkowaniu u kobiet pracujacych w "atmosferze" weglowodorow aromatycznych [Disturbance of menstruation in women working in an atmosphere polluted with aromatic hydrocarbons]. *Polski tygodnik lekarski,* **20**: 1648–1649 (1965).
33. **Syrovadko, O.N.** [Working conditions and health status of women handling organosiliceous varnishes containing toluene]. *Gigiena truda i professional'nye zabolevanija,* **12**: 15–19 (1977) (in Russian).
34. **Bauchinger, M. et al.** Chromosome changes in lymphocytes after occupational exposure to toluene. *Mutation research,* **102**: 439–445 (1982).
35. **Schmid, E. et al.** Chromosome changes with time in lymphocytes after occupational exposure to toluene. *Mutation research,* **142**: 37–39 (1985).
36. **Mäki-Paakkanen, J. et al.** Toluene-exposed workers and chromosome aberrations. *Journal of toxicology and environmental health,* **6**: 775–781 (1980).
37. **Forni, A. et al.** Chromosome studies in workers exposed to benzene or toluene or both. *Archives of environmental health,* **22**: 373–378 (1971).
38. **Hellman, T.M. & Small, F.H.** Characterization of petrochemical odors. *Chemical engineering progress,* **69**: 75–77 (1973).
39. **Nauš, A.** Čichové prahy nekterých prumyslových látek [Olfactory thresholds of industrial substances]. *Pracovni lekarstvi,* **34**: 217–218 (1982).
40. **Hellman, T.M. & Small, F.H.** Characterization of the odour properties of 101 petrochemicals using sensory methods. *Journal of the Air Pollution Control Association,* **24**: 979–982 (1974).
41. **Antti-Poika, M. et al.** Occupational exposure to toluene: neurotoxic effects with special emphasis on drinking habits. *International archives of occupational and environmental health,* **56**: 31–40 (1985).
42. **Cherry, N. et al.** British studies on the neuropsychological effects of solvent exposure. *Scandinavian journal of work, environment & health,* **10**(Suppl.1): 10–12 (1984).
43. **Juntunen, J. et al.** Nervous system effects of long-term occupational exposure to toluene. *Acta neurologica scandinavica,* **72**: 512–517 (1985).
44. **Münchinger, R.** Der Nachweis zentralnervöser Störungen bei Lösungsmittel-exponierten Arbeitern [Detection of dysfunctions of the central nervous system in workers exposed to solvents]. *In: Proceedings of the XIV International Congress of Occupational Health, Madrid, 16–21 September 1963.* Amsterdam, Excerpta Medica, 1964, Vol. 2, pp. 687–689.

15

Trichloroethylene

General Description

Trichloroethylene (C_2HCl_3, TCE), a colourless liquid of moderate volatility, is a powerful solvent. Its solubility at 20 °C is 0.1 g/100 ml water *(1)*.

Sources

There are no known natural sources of TCE. The estimated total production in western Europe was 270 000 tonnes in 1981. Production in the USA amounted to 133 000 tonnes in 1980. Most of the TCE produced commercially is derived from ethene or dichloroethane *(1–3)*.

Trichloroethylene is mainly used in the degreasing of fabricated metal parts. Other applications include industrial dry-cleaning, printing, the production of printing-ink, extraction processes, paint production and textile printing *(1,3)*. Trichloroethylene can be present in household products such as adhesives, spot removers and carpet cleaners *(1)*. All commercial grades of TCE contain stabilizers, added to avoid decomposition.

Most TCE (99%) is emitted into the environment as a result of use. Thus, the annual emissions in western Europe are virtually identical to the annual consumption, namely 219 000 tonnes in 1981. An estimate of the global emission, derived from production figures and usage patterns, is 435 000 tonnes ± 50% for the year 1977 *(1–3)*. Practically all TCE enters the atmosphere unchanged; some also enters water and wastewater.

Occurrence in air

Background levels, defined as levels measured in nonpolluted rural areas, are estimated to be in the nanogram/m^3 range *(1)*. It has been calculated that background levels in the northern hemisphere may be around 75 ng/m^3 *(2)*.

In European cities (Brussels, Grenoble, Moscow) levels of 5–30 μg/m^3 were found. In cities in the USA typical concentrations were between 0.5 and 3 μg/m^3. In industrial areas the levels are somewhat higher, the average level being 6 μg/m^3 (highest concentration 50 μg/m^3) *(2–4)*.

Data on indoor air concentrations are limited. Values tend to be in the same range as those of outdoor concentrations, but they may occasionally be substantially higher. Levels of TCE in 15 dwellings in northern Italy ranged from 1 to 86 μg/m^3, while corresponding outdoor concentrations were 1–24 μg/m^3 (average of 4–7 days) *(5)*.

Degradation of TCE in the atmosphere occurs by reaction with hydroxyl radicals and with ozone; the half-time was estimated to be about 70 hours at yearly average concentrations of 5×10^{-8} ppm for OH-radicals and of 0.040 ppm for ozone *(2)*. Due to degradation processes, TCE forms the highly toxic compound phosgene. The removal of TCE from the air by dry and wet deposition is negligible compared with the part played by chemical processes *(1)*.

Conversion factors

1 ppm = 5.4 mg/m^3
1 mg/m^3 = 0.18 ppm

Routes of Exposure

Air

In the Netherlands it was estimated that, of the total population of 14 million, 14 000 are exposed to an average concentration of 10 μg/m^3, resulting in a daily intake of 200 μg; an estimated 350 000 are exposed to 4 μg/m^3 (daily intake: 80 μg) and 13.6 million persons are exposed to 0.8 μg/m^3 (daily intake: 16 μg) *(2)*.

Evaporation from water used in homes (particularly in bathrooms) can contribute to indoor exposure, particularly if levels in tap water are relatively high *(6)*.

In the United States, metal degreasers were found to be exposed to TCE where 60% of the measured concentrations were below 270 mg/m^3 and 93% below 540 mg/m^3. For western Europe, data from 1980 show that, of the measured exposure levels for degreasers, 50–60% were between 0 and 135 mg/m^3, 72–79% between 0 and 270 mg/m^3, and 90–95% between 0 and 540 mg/m^3 *(2,3)*.

Drinking-water

In the drinking-water of 100 cities in the Federal Republic of Germany the average TCE concentration was 0.6 μg/litre, with a range of 0.1–5.9 μg/litre *(7)*. Concentrations in groundwater used for drinking-water in the Milan municipal area, Italy, have been found to be as high as 80 μg/litre *(4)*. The US National Organics Monitoring Survey observed TCE in drinking-water in three groups of cities in the USA. The mean concentrations were 11 μg/litre, 21 μg/litre and 1.3 μg/litre in 4 of 112 cities in March–April 1976, in 28 of 113 cities in May–July 1976 and in 19 of the cities in November 1976 – January 1977, respectively *(8)*. In drinking-water from locations near producer and user facilities the levels were higher: 19–32 μg/litre *(3)*.

Food

Residues of TCE found in food result from its use as a solvent or from environmental contamination. On the basis of investigations of 500 samples of representative food items in the Federal Republic of Germany, an average daily intake via solid food of 6.0 μg was calculated, whereas the intake via beverages was 1.2 μg/day. In another study in the Federal Republic of Germany the "theoretically possible intake" through food was 1.4 μg/day *(2,3)*.

Relative significance of different routes of exposure

Model calculations for the Netherlands show that, on average, the country's population is exposed to a TCE concentration of 0.8 μg/m^3 in the atmosphere. This would lead to a daily intake of 16 μg TCE via the ambient air. Specific groups may be exposed to higher levels *(2)*.

Available data indicate that daily intake via drinking-water is normally lower than 2 μg/day. Intake via water from areas where there is considerable TCE emission exceeds this level. In the Federal Republic of Germany the average daily intake via food was determined: values of 1.4 and 7.2 μg/day were obtained *(2)*. Possible maximum daily intakes from air and drinking-water by persons in industrial areas of Italy have been calculated to be as high as 300 μg/day *(9)*.

It may be concluded from the available data that respiratory intake is the most important route of exposure for the general population.

Kinetics and Metabolism

Absorption

Upon inhalation the initial uptake of TCE is rapid and the blood concentration during inhalation closely parallels the alveolar gas concentration. When equilibrium between the blood and alveolar gas concentrations is reached, uptake remains constant for the rest of the exposure *(1)*. The blood/air partition coefficient in humans is about 10 at 37 °C, and TCE retention varies according to the physical activity *(1)*. In rats TCE is absorbed almost completely after oral dosing. Dermal absorption in mice, guinea pigs and humans is slow and consequently dermal absorption will rarely be in toxic amounts *(1)*.

Distribution

Trichloroethylene is distributed to all body tissues and crosses the blood–brain barrier; it also crosses the placenta in animals and humans. About 3 hours are required for complete tissue equilibrium under continuous exposure. Considerable amounts are stored in lipid tissues *(1)*.

Biotransformation

Trichloroethylene is metabolized mostly in the liver, partly in the lungs and possibly in the kidneys. The compound is metabolized to a greater extent in mice than in rats and humans. In man an appreciable portion of absorbed TCE is exhaled unchanged. Numerous investigations of plasma and urine of animals and humans all demonstrated trichloroethanol, trichloroethanol-glucuronide and trichloroacetic acid as the principal metabolites *(1)*.

Biotransformation of TCE involves the formation of an epoxide, which is transformed into chloralhydrate. The latter is then converted partly to trichloroethanol and partly to trichloroacetic acid. Balance studies in man recover some 75% of TCE, so possibly minor metabolites exist *(1)*.

Elimination

After intravenous injection, elimination from the blood in rats was fairly rapid (half-time of 0.3–1 hour). For the clearance from lipids a half-time of 3.5 hours was found. This clearance involves two routes: 20–40% is exhaled unchanged, whereas the remainder is biotransformed into urinary metabolites. The half-time for urinary excretion of trichloroethanol and its conjugate is 7–14 hours; for trichloroacetic acid it is 37–70 hours. In humans, trichloroacetic acid accumulates in blood during repeated exposures to TCE *(1)*.

Health Effects

Effects on experimental animals and *in vitro* test systems

Toxicological effects

Acute toxicity studies have shown that the oral LD_{50} values in rodents range from 2400 to 4920 mg/kg body weight, and LC_{50} values range from 45 to 260 mg/litre. Acute toxic effects are related to a depressant action on the CNS, which can lead to coma and death *(1)*.

In liquid form, TCE has an irritant effect on the skin and eyes. In vapour form it is an irritant to the respiratory tract *(1)*.

Toxic effects on kidneys and liver after prolonged or chronic exposure to TCE were described in numerous studies *(1)*. Some histopathological changes in the kidneys of rats can occur after oral administration of 125 mg/kg body weight per day for 13 weeks. Nephrotic changes were found in mice following oral administration of 1000 mg/kg body weight per day for 2 years, and necrotic changes in the liver were found following an exposure of 6000 mg/kg body weight per day for 13 weeks *(10)*.

Continuous exposure of mice by inhalation to 810 mg/m^3 for 2 days resulted in increased liver weight, with significant changes in biochemical parameters *(11)*. Continuous exposure of rats, guinea pigs, rabbits and monkeys to 189 mg/m^3 for 90 days produced growth inhibition, but did not affect survival or behaviour, and no adverse effects were observed upon histological examination of a number of tissues (including liver and kidney tissues) *(12)*.

In most animal studies, embryotoxicity was observed, but in no case did a teratogenic effect occur. It appears that all developmental toxicity was shown at doses that are probably toxic to the mothers.

Mutagenic and carcinogenic effects

Quite often, mutagenicity tests were performed using TCE stabilized with compounds (e.g. epichlorohydrin and 1,2-epoxybutane) known to have mutagenic properties. Unfortunately, reports frequently do not specify the degree of purity of the TCE used; only in a relatively small number of studies was it evident that purified TCE was tested. When the available data are examined it becomes clear that in some cases positive responses occurred, but in many tests negative findings were observed. Thus, the results are conflicting. *In vitro* DNA-binding studies showed strong binding capacity after metabolic activation only. *In vivo* studies, however, revealed only very weak binding to DNA. Trichloroethylene induced cell transformation in rat embryo cells. On the basis of the studies available, it was concluded that the evidence for mutagenic effects is inadequate *(13)*.

The evidence for carcinogenicity of TCE was recently reviewed in two health criteria documents *(1,3)*. In a study conducted in 1976, in which TCE was stabilized with 0.09% epichlorohydrin and 0.19% epoxybutane, a significant increase was found in the incidence of hepatocellular carcinomas in mice at oral dose levels of 869–2339 mg/kg body weight per day for 78 weeks. No increase in tumour incidence was observed in rats orally exposed to 547 and 1097 mg/kg body weight per day, but increased mortality due to toxic nephropathy occurred *(14)*. In a US National Toxicology Program study carried out in 1982 *(10)*, epichlorohydrin-free TCE was applied by gavage. In mice (dose level 1000 mg/kg body weight per day) hepatocellular carcinomas again occurred. In male rats (dose levels 500 and 1000 mg/kg body weight per day) there was a minimal increase in the incidence of kidney tumours, and toxic nephrosis and cytomegaly occurred in almost all treated animals.

In another oral study, TCE was administered to Sprague-Dawley rats for 52 weeks (dose levels 50 and 250 mg/kg body weight per day) and the animals were kept until spontaneous death. No increase in tumour incidence was found *(15)*. In a further oral study in mice, using different samples of TCE (highly purified TCE, industrial TCE, and TCE combined with stabilizers) in doses of 2400 (males) or 1800 (females) mg/kg body weight per day, 5 days per week over a period of 78 weeks, no increase in tumour incidence was found in animals exposed to TCE without stabilizers *(16)*. Van Duuren et al. *(17)* reported oral, dermal and subcutaneous studies in mice, all of which yielded negative (no-response) results. Another oral study in mice was also negative *(18)*. These mouse experiments, however, showed various deficiencies (small number of animals, one dose group only, limited histopathology, etc.). In a mouse-skin bioassay to determine the carcinogenic activity of TCE, no response was found *(17)*.

Mice, rats and Syrian hamsters were exposed in inhalatory studies to 540 or 2700 mg/m^3 6 hours per day, 5 days per week for 18 months, and survivors were killed 12–18 months after the last exposure *(19)*. In rats and hamsters there was no increase in tumour incidence. In mice the incidence of malignant lymphomas was elevated in females. Because the spontaneous frequency of tumours is very high in female mice of this particular strain (NMRI mice), the significance of the observed increase (which lacked an evident dose–response relationship) is not clear *(19)*.

In a chronic inhalation study, mice exposed to levels of 270, 810 and 2430 mg TCE per m^3 for 7 hours per day, 5 days per week for 104 weeks showed a higher incidence of lung adenocarcinomas, with significant elevation at TCE concentrations of 810 and 2430 mg/m^3 *(20)*. The purity of the TCE used was 99%, with 0.128% carbon tetrachloride, 0.019% benzene,

0.019% epichlorohydrin and 0.01% 1,1,2-trichloroethane. In a similar study in rats no increase in tumour incidence was found *(20)*.

In 1982 IARC concluded that there was limited evidence for carcinogenicity of TCE in animals *(13)*. Recent animal data are believed not to contradict this assessment.

Effects on humans

Toxicological effects

Numerous case studies and studies among persons working with TCE have been reported. These are summarized in the relevant WHO *Environmental Health Criteria* publication *(1)*. The main effects of TCE are on the CNS (e.g. subjective complaints and signs indicating impairment of the psychomotor function). Observed effects also include skin and eye irritation. In some cases severe injury to the liver (centrilobular necrosis) and kidney changes are noted. There are also reports of cranial nerve damage following TCE intoxication. However, workers were often exposed to other chemicals as well and there were large fluctuations in the TCE concentrations: both factors hamper interpretation of the results.

Several controlled short-term neurobehavioural studies were conducted, with test concentrations of 116–5400 mg/m^3 and exposure times of up to 3 hours. At concentrations above 1080 mg/m^3 decreased psychomotor performance, visual-motor disturbances and subjective complaints were noted *(21–25)*. At 810 mg/m^3 there were slight disturbances in cardiac rhythm, indicative of greater functional exertion in performing mental tasks *(21,22)*. After two exposures to 594 mg/m^3, performance in tests for perceptivity, memory, reaction time, manual ability and dexterity was decreased *(26)*. At lower levels the results did not show a consistent pattern. In some studies there were effects on psychomotor/visual-motor performance at levels of 540 mg/m^3 *(25,27)*, but these results were not confirmed in some other studies *(1)*. At exposure levels ranging between 270 and 540 mg/m^3 for 3.5 and 7.5 hours, the visual and auditory potentials were affected *(28)*.

Several studies demonstrated that the urinary excretion of trichloroacetic acid may well be correlated with the observed toxic effects *(1,29)*. One of the first studies *(30)* showed that excretion of up to 20 mg of trichloroacetic acid per litre of urine was not associated with any health impairment. However, with urinary values between 40 and 75 mg/litre, headache, fatigue, somnolence, irritability and alcohol intolerance were reported in 50% of workers. When urinary levels of trichloroacetic acid exceeded 300 mg/litre, these symptoms were found in 100% of subjects. These early findings were confirmed later in many other studies *(1,29)*. Andersson *(31)* described some neurasthenic symptoms in 40% of workers excreting trichloroacetic acid in urine in amounts of less than 20 mg/litre, in 60% of those excreting 21–75 mg/litre, and in 80% of those excreting 76–760 mg/litre. However, adverse health effects related mainly to the nervous system were reported in the majority of studies when the urinary excretion of trichloroacetic acid was at a level of over 50 mg/litre. An average urinary trichloroacetic acid excretion of 50 mg/litre seems to correspond best to the time-weighted average TCE concentration of 135 mg/m^3 *(29)*.

Mutagenic and carcinogenic effects

In a study whose results have not been confirmed, an increased incidence of sister chromatid exchange in the lymphocytes of 2 out of 6 TCE workers is reported following chronic exposure *(32,33)*. In another study, however, no chromosome anomalies were found in 28 workers exposed to 324–408 mg/m^3, but pathological rates of hypodiploid cells occurred in 8 of 28 subjects *(34)*.

A number of epidemiological studies were performed to examine the possible carcinogenic effects of TCE in exposed workers. Major findings were reviewed and summarized in two recent documents *(1,3)*. In some cases there was an excess of cancer incidence in several sites, including skin, lung, genitals, bladder, liver, oesophagus and lymphatic system, but the increases did not show a consistent pattern, with the possible exception of those relating to urinary tract tumours and lymphomas. Other studies yielded negative results. In all cases simultaneous exposure to other compounds (e.g. tetrachloroethylene, carbon chloride, chromium, nickel, copper, iron, lead, corrosive acids/caustic solutions) could not be excluded *(1,3)*.

Evaluation of Human Health Risks

Exposure

Average concentrations in the ambient air of urban areas vary between a few μg/m^3 and 50 μg/m^3 (maximum value). Indoor air levels are generally in the same range, but may occasionally be higher (up to 100 μg/m^3). Occupational levels of exposure may be two or three orders of magnitude higher.

Health risk evaluation

The results of the available mutagenicity studies do not show a consistent pattern: on several points they seem conflicting, and, to complicate interpretation even further, in quite a few cases mutagenic impurities were present, or else the presence of these cannot be excluded. Recent data from animal studies are considered not to change the basis for the earlier conclusion that the evidence for carcinogenicity of TCE in animals is limited *(13)*. In the animal carcinogenicity studies tumour incidences were elevated in some cases, but these increases cannot be regarded as conclusive evidence of a carcinogenic effect. In particular, the development of liver tumours in mice only is not a sufficient basis for considering TCE to be carcinogenic to animals. The available epidemiological studies also do not provide valid data regarding a carcinogenic effect of TCE in humans; evidence for its carcinogenicity in humans must therefore be considered inadequate *(13)*.

Trichloroethylene has been categorized by IARC in Group 3, in which chemicals cannot be classified as to their carcinogenicity to humans.

In humans and animals, exposure to higher concentrations causes neurobehavioural, liver and kidney effects.

Available data from occupational studies suggest that adverse health effects related to the nervous system appear when the urinary level of trichloroacetic acid exceeds 50 mg/litre. Since such a level corresponds best to the TCE concentration of 135 mg/m^3 *(29)*, this can be regarded as the lowest-observed-adverse-effect level. The available data on the toxicity of TCE in humans do not allow a no-effect level in man to be indicated.

Guidelines

Data on occupational exposure, together with corroborative evidence from animal studies, suggest that the lowest-observed-adverse-effect concentration of TCE can be assumed to be around 135 mg/m^3. A protection (safety) factor of 100 is considered appropriate in extrapolation of data from occupational to environmental exposure. In selecting the size of the protection factor, the differing exposure pattern (continuous versus intermittent exposure), the variability and validity of the findings around the concentration of 135 mg/m^3, and the fact that a no-effect level in man cannot be established were of decisive importance.

An air quality guideline value of 1 mg/m^3 (24-hour average) is recommended. Below this value, adverse health effects of environmental exposure to TCE (outdoor or indoor) are not likely to occur. The level is considerably above the current environmental exposure.

References

1. *Trichloroethylene.* Geneva, World Health Organization, 1985 (Environmental Health Criteria, No. 50).
2. **Besemer, A.C. et al.** *Criteriadocument over trichloroetheen* [Trichloroethylene criteria document]. The Hague, Ministerie van Volkshuisvesting, Ruimtelijke Ordening en Milieubeheer, 1984 (Publikatiereeks Lucht, No. 33).
3. *Health assessment document for trichloroethylene.* Washington, DC, US Environmental Protection Agency, 1985 (Final report No. EPA-600/8-82-006F).
4. **Ziglio, G. et al.** Human environmental exposure to trichloro- and tetrachloroethylene from water and air in Milan, Italy. *Archives of environmental contamination and toxicology,* **12**: 57–64 (1983).
5. **de Bortoli, M. et al.** *Measurements of indoor air quality and comparison with ambient air: a study of 15 homes in Northern Italy.* Luxembourg, Commission of the European Communities, 1985.
6. **Andelman, J.B.** Inhalation exposure in the home to volatile organic contaminants of drinking water. *Science of the total environment,* **47**: 443–460 (1985).
7. **Bauer, U.** Belastung des Menschen durch Schadstoffe in der Umwelt. Untersuchungen über leicht flüchtige organische Halogen-Verbindungen in Wasser, Luft, Lebensmitteln und im menschlichen Gewebe. III Mitteilung [Human exposure to environmental chemicals. Investigations on volatile organic halogenated compounds in water, air, food, and human tissues. III]. *Zentralblatt für Bakteriologie, Mikrobiologie und Hygiene, I Abt. Orig. B,* **174**: 200–237 (1981).

8. *Guidelines for drinking-water quality. Vol. 2. Health criteria and other supporting information.* Geneva, World Health Organization, 1984.
9. **Ziglio, G. et al.** Esposizione ambientale a solventi clorurati in popolazioni studentesche di un comune del Nord Italia [Environmental exposure to chlorinated solvents in a population of schoolchildren of a northern Italian town]. *L'igiene moderna,* **82**: 133–161 (1984).
10. *Technical report on the carcinogenesis studies of trichloroethylene (without epichlorohydrin) in F344/n rats and BCF mice.* Research Triangle Park, NC, National Toxicology Program, 1983 (NIH Publication No. 83-1979, NTP TR 243).
11. **Kjellstrand, P. et al.** Trichloroethylene: further studies of the effects on body and organ weights and plasma butyrycholinesterase activity in mice. *Acta pharmacologica et toxicologica,* **53**: 375–384 (1983).
12. **Prendergast, J.A. et al.** Effects on experimental animals of long-term inhalation of trichloroethylene, carbon tetrachloride, 1,1,1-trichloroethane, dichlorodifluoromethane, and 1,1-dichloroethylene. *Toxicology and applied pharmacology,* **10**: 270–289 (1967).
13. Trichloroethylene. *In: Chemicals, industrial processes, and industries associated with cancer in humans. IARC Monographs, Volumes 1 to 29.* Lyon, International Agency for Research on Cancer, 1982, pp. 247–249 (IARC Monographs on the Evaluation of the Carcinogenic Risk of Chemicals to Humans, Supplement 4).
14. **US National Cancer Institute.** *Carcinogenesis bioassay of trichloroethylene, CAS No. 79-01-6.* Bethesda, MD, US Department of Health, Education, and Welfare,1976 (DHEW Publication No. (NIH) 76-802).
15. **Maltoni, C. & Gotti, G.** *Preliminary report on long-term carcinogenicity bioassays on trichloroethylene performed at the Bologna Institute of Oncology.* International Conference on Organic Solvent Toxicity, 15–17 October 1984, Stockholm. Unpublished paper.
16. **Henschler, D. et al.** Carcinogenicity study of trichloroethylene, with and without epoxide stabilizers, in mice. *Journal of cancer research and clinical oncology,* **107**: 149–156 (1984).
17. **Van Duuren, B.L. et al.** Carcinogenicity of halogenated olefinic and aliphatic hydrocarbons in mice. *Journal of the National Cancer Institute,* **63**: 1433–1439 (1979).
18. **Rudali, G.** A propos de l'activité oncogène de quelques hydrocarbures halogènes utilisés en thérapeutique [The oncogenic action of some halogenous hydrocarbons used in therapeutics]. *UICC Monographs,* **7**: 138–143 (1967).
19. **Henschler, D. et al.** Carcinogenicity study of trichloroethylene by long term inhalation in three animal species. *Archives of toxicology,* **43**: 237–248 (1980).
20. **Fukuda, K. et al.** Inhalation carcinogenicity of trichloroethylene in mice and rats. *Industrial health,* **21**: 243–254 (1983).
21. **Ettema, J.H. et al.** Effects of alcohol, carbon monoxide and trichloroethylene exposure on mental capacity. *International archives of occupational and environmental health,* **35**: 117–132 (1975).

22. **Ettema, J.H. et al.** Study of mental stresses during short-term inhalation of trichloroethylene. *Staub, Reinhaltung der Luft,* **35**: 409–410 (1975).
23. **Stewart, R.D. et al.** Experimental human exposure to trichloroethylene. *Archives of environmental health,* **20**: 64–71 (1970).
24. **Stopps, D.J. & McLaughlin, M.** Psychophysiological testing of human subjects exposed to solvent vapors. *American Industrial Hygiene Association journal,* **28**: 43–50 (1967).
25. **Vernon, R.J. & Ferguson, R.K.** Effects of trichloroethylene on visual-motor performance. *Archives of environmental health,* **18**: 894–900 (1969).
26. **Salvini, M. et al.** Evaluation of the psychophysiological functions in humans exposed to trichloroethylene. *British journal of industrial medicine,* **28**: 293–295 (1971).
27. **Nakaaki, K. et al.** An experimental study on the effect of exposure to trichloroethylene vapour in man. *Journal of science of labour,* **49**: 449–463 (1973).
28. **Winneke, G.** Acute behavioural effects of exposure to some organic solvents — psychophysiological aspects. *Acta neurologica scandinavica,* **66**(Suppl. 92): 117–129 (1982).
29. WHO Technical Report Series, No. 664, 1981 (*Recommended health-based limits in occupational exposure to selected organic solvents:* report of a WHO Study Group).
30. **Ahlmark, A & Forssman, S.** Evaluating trichloroethylene exposures by urinalyses for trichloroacetic acid. *Archives of industrial hygiene and occupational medicine,* **3**: 386–398 (1951).
31. **Andersson, A.** Gesundheitliche Gefahren in der Industrie bei Exposition für Trichloräthylen [Health hazards in industry due to exposure to trichloroethylene]. *Acta medica scandinavica,* **157**(Suppl. 323): 1–220 (1957).
32. **Gu, Z.W. et al.** Effets du trichloroethylene et de ses metabolites sur le taux d'échanges de chromatides sœurs. Etude *in vivo* et *in vitro* sur les lymphocytes humains [Effects of trichloroethylene and its metabolites on the sister chromatid exchange level: *in vivo* and *in vitro* study of human lymphocytes]. *Annales de génétique,* **24**: 105–106 (1981).
33. **Gu, Z.W. et al.** Induction d'échanges entre les chromatides sœurs (SCE) par le trichloroethylene et ses metabolites [Induction of sister chromatid exchange (SCE) by trichloroethylene and its metabolites]. *Toxicological European research,* **3**: 63–67 (1981).
34. **Konietzko, H. et al.** Cytogenetische Untersuchungen an Trichloräthylen-Arbeitern [Cytogenic investigations of trichloroethylene workers]. *Archiv für Toxikologie,* **40**: 201–206 (1978).

16

Vinyl chloride

General Description

At standard temperature and pressure, vinyl chloride (VC) is a nonirritating, colourless gas. It is generally odourless below 10 000 mg/m^3 (3900 ppm), but a sweetish odour may be detected by some sensitive individuals between 200 and 500 mg/m^3. The gas is easily liquefied under pressure and is usually stored or shipped as a liquid.

Vinyl chloride is highly stable in the absence of sunlight or oxygen. Above 400 °C, it dissociates into acetylene and hydrochlorine. In the atmosphere VC reacts with hydroxyl radicals and ozone, ultimately forming formaldehyde, carbon monoxide, hydrochloric acid and formic acid. On the basis of measured reaction rates with hydroxyl radicals and their concentration in air, it is estimated that the half-time of VC in the atmosphere is about 20 hours *(1)*.

Sources

The principal emission sources, in order of importance, are VC production plants, polyvinyl chloride (PVC) polymerization facilities, and plants where PVC products are fabricated. Minor sources include storage and handling facilities for VC and PVC and plants producing ethylene diamine or ethylene dichloride. In the United States, VC emissions have been reported from municipal landfills, but the exact source of emission is unclear and systematic survey data are unavailable.

Approximately 5 million tonnes of VC were produced in the whole of Europe in 1981. The levels of emission from VC and PVC production facilities depend upon the processes and control technology employed. The use of the best available technology can reduce emissions to less than 1% of production volume, but emissions from facilities in some countries exceed this value *(1)*.

Occurrence in air

The general background of VC in western Europe resulting from known production sources is estimated from dispersion model calculations to range from 0.1 μg/m^3 to 0.5 μg/m^3 *(1)*. These levels are lower than the detection limit (0.8 $\mu g/m^3$) of the best analytical procedure (gas chromatography/ mass spectroscopy) *(2)*. The same models would predict that average annual concentrations around well controlled sources would range from 1 to 10 $\mu g/m^3$ at distances of 1–5 km from the source and exceed 10 $\mu g/m^3$ only within 1 km. The 99-percentile 24-hour concentrations around such sources would be about 10 times higher than the above averages *(1)*. Dispersion models reasonably predict source-related, average concentrations. However, measurements over limited time periods can differ considerably, owing to fluctuating meteorological conditions. Poorly controlled, source-related environmental concentrations are reflected by measurements made in the United States in the mid-1970s near VC and PVC production facilities. Here, plant boundary concentrations often exceeded 1 mg/m^3 *(3)*.

Conversion factors

1 ppm = 2.589 mg/m^3

1 mg/m^3 = 0.386 ppm

Routes of Exposure

Air

Currently, general population exposure comes overwhelmingly from industrial production sources, the primary route of entry being inhalation. Assuming a daily inhalation of 20 m^3 air, the vast majority of the population would inhale 2–10 μg of VC daily. Individuals living within 5 km of well controlled production sources could be exposed to 10–100 times as much. Previously, VC was produced as a propellant in aerosol cans and episodic indoor air pollution from this source was considerable *(4)*. However, this type of use has been discontinued.

Occupational exposure

Approximately 10 000 individuals are occupationally exposed to VC during monomer or polymer production. Generally, exposures are lower than 10 mg/m^3. The yearly average for the entire occupationally exposed population is considerably less than this value.

Smoking

Vinyl chloride has been found in the smoke of cigarettes (1.3–16 ng/cigarette) and of small cigars (14–27 ng) *(5)*. Charcoal filter tips reduce VC in cigarette smoke.

Drinking-water

There is very little information on current concentrations of VC in water systems. Because of its volatility and reactivity, VC would not be expected to remain in significant concentrations in drinking-water. It has been detected only occasionally in samples of drinking-water taken in 100 cities of the Federal Republic of Germany. The highest level, 1.7 μg/litre, was tentatively ascribed to dissolution from PVC tubing *(1)*.

Food

In the mid-1970s VC was also identified as a contaminant of foods and liquids packaged in PVC material *(6)*. However, with the implementation of more stringent manufacturing specifications for PVC, such contamination decreased substantially and it is estimated that the maximum intake per person in foods and liquids would now be less than 0.1 μg/day *(7)*.

Relative significance of different routes of exposure

The most important exposure route is air contamination from VC and PVC production facilities. General environmental levels of VC from all sources in Europe are likely to lead to average exposures of 2–10 μg/day. Exposures from food and water are less than 0.1 μg/day. Heavy smokers may inhale additionally up to 0.5 μg/day.

Kinetics and Metabolism

Vinyl chloride is rapidly absorbed through the lungs and is carried by the blood stream to all organs. The highest concentrations of metabolites are found in the liver, kidneys and spleen *(8,9)*. Studies in rats indicate that this process saturates, with proportionately less VC being metabolized at concentrations exceeding 1000 mg/m^3 than at lower concentrations *(10)*.

First-order kinetics describes the metabolism of VC up to about 200 mg/m^3. The metabolism is believed to proceed through the microsomal mixed-function oxidase system, forming chloroethylene oxide which rapidly rearranges to chloroacetaldehyde *(11)*. The activated metabolite binds to cellular macromolecules *(12)* or nucleotides *(13)*. Detoxification occurs primarily by oxidation in the liver to polar compounds, which can be conjugated to glutathione and/or cysteine and excreted in the urine *(14)*. No significant accumulation of VC occurs in the body. From studies of its metabolism in rats VC is estimated to have a biological half-time of 20 minutes *(9)*.

Health Effects

Reviews of the health effects of VC include that of IARC, the Dutch criteria document *(1)*, the clinically oriented review by Lelbach & Marsteller *(15)* and a review of VC mortality by Nicholson et al. *(16)*.

Effects on experimental animals and *in vitro* test systems

Toxicological effects

The acute toxicity of VC is low; at higher concentrations a narcotic effect occurs. Two-hour LC_{50} values for different animal species vary from 300 to 600 g/m^3 *(1)*. In chronic exposure VC can induce a variety of toxic effects. These are mainly related to the liver, the CNS and the cardiovascular system.

Teratology studies, after VC inhalation, have been carried out in mice, rats and rabbits. No significant effects on malformations or anomaly rates resulted from exposures to VC at 130–6470 mg/m^3 (50–2500 ppm) for up to 24 hours per day for up to 12 days during different periods of pregnancy *(17–19)*. Other experiments have suggested some signs of embryotoxicity of VC in rats *(20)* and in mice *(19)*. Vinyl chloride has been shown to be a transplacental carcinogen in the rat at exposures of 15 000 or 26 000 mg/m^3 for 4 hours per day on days 12–18 of pregnancy *(21)*.

Mutagenic and carcinogenic effects

Vinyl chloride is one of the best studied chemicals in relation to animal species. Studies by Maltoni et al. *(21)* on nearly 7000 animals over a 10-year period provide a data base virtually unmatched in experimental carcinogenesis. In these studies, VC was found to produce a statistically significant excess of Zymbal gland carcinomas, liver angiosarcomas, nephroblastomas, neuroblastomas, mammary gland adenocarcinomas and forestomach papillomas in concentrations ranging from 13 to 77 000 mg/m^3. Liver haemangiosarcomas were found in experiments at concentrations as low as 25 mg/m^3. The risk increased linearly up to approximately 1300 mg/m^3, at which point 10% of the animals developed haemangiosarcoma. Thereafter, the rate of increase in risk lessened, but 30% of the animals developed haemangiosarcoma at 77 000 mg/m^3. The saturation of the haemangiosarcoma risk seen in the experimental animal data may be attributed to the shortened lifespan due to toxic effects in the heavily exposed animals (they did not live long enough to fully show their cancer risk). Further, the nonlinear metabolism of VC would affect the carcinogenic dose–response relationship.

The effect of age was investigated by Maltoni and co-workers, who exposed 21-week-old breeders and newborn Sprague-Dawley rats to 25 600 and 15 000 mg VC per m^3 for 4 hours per day, 5 days per week, for 5 weeks. Malignancies were found only among the newborn animals. Fifteen liver haemangiosarcomas and 20 hepatomas were found among 44 rats exposed to 25 600 mg/m^3; 17 haemangiosarcomas and 20 hepatomas were found among 42 exposed at 15 400 mg/m^3. In contrast to these results, Groth et al. *(22)* found that the incidence of angiosarcoma increased with age at the start of exposure in an experiment in which Sprague-Dawley rats were exposed to 2400 mg/m^3 for 7 hours per day, 5 days per week, for 24 weeks. The difference between the two studies may indicate a greatly increased sensitivity of newborns compared with older animals.

Interaction effects in the incidence of neoplasms in the rat have been demonstrated by Radike et al. *(23)*, who administered VC, ethanol, and VC with ethanol to groups of 80 male Sprague-Dawley rats. Forty of 80 rats exposed to VC and ethanol developed angiosarcomas, compared with 18 in the groups exposed to VC and none in those exposed to ethanol.

Vinyl chloride has been shown to be carcinogenic in mice *(21)*, hamsters *(21)* and rabbits *(24)*. In some strains, the haemangiosarcoma incidence in mice is similar to that in rats, but hamsters are much less sensitive. In contrast to rats, some strains of mice are highly susceptible to an increase in lung malignancies caused by VC exposure. In other mice strains, VC induces haemangiosarcomas in many tissues *(21,25)*. In fact, liver haemangiosarcomas are in the minority.

Vinyl chloride and its metabolites 2-chloroethylene oxide, 2-chloroethylene aldehyde and 2-chloroethanol are mutagenic in various test systems *(26)*.

Effects on humans

Toxicological effects

Between 1949 and 1974, a variety of effects of exposure to VC were documented. The symptoms observed included Raynaud's phenomenon, a painful vasospastic disorder of the hands, acro-osteolysis, primarily of the terminal phalanges of the hands, and pseudoscleroderma. Hepatomegaly and noncirrhotic portal fibrosis with portal hypertension and splenomegaly were also noted among individuals exposed to VC in polymerization facilities. The above symptoms were largely confined to people very heavily exposed to VC during reactor cleaning and were not found among individuals exposed to lower VC concentrations in the PVC processing industry. None of the above manifestations is of concern for individuals exposed in environmental circumstances.

VC is a narcotic agent and loss of consciousness can occur from exposures approaching 25 000 mg/m^3. This was the case among 4.5% of workers examined at a PVC polymerization facility *(27)*. At lesser concentrations (about 2300 mg/m^3), euphoria, dizziness, somnolence and narcosis were commonly reported *(28)*. At approximately 100 mg/m^3, 10% of workers exposed in a workshift experienced dizziness and 17% somnolence; lesser percentages of other CNS disorders were reported.

Several investigators reported a higher incidence of chromosomal aberrations in peripheral lymphocytes cultured from workers exposed to high levels of VC *(26)*. Three studies of communities close to PVC plants suggested an association between such locations and an increased risk of malformations, particularly of the CNS *(29–31)*. However, none of the studies produced a clearcut association, and other uncontrolled variables, including other industrial pollutants, may account for the differences observed. A single study *(32)* of workers exposed to VC, showing an increase in fetal death rates in their wives, was more convincing, but it had a number of methodological problems. Further studies are required before any definite conclusions can be reached. Sanockij et al. reported that, among wives of workers exposed to VC, the number of miscarriages increased, while the number of spermatozoa in the ejaculate of exposed males decreased *(33)*.

Mutagenic and carcinogenic effects

Twelve cohort studies of workers exposed to VC have been published. The size of the cohorts varied greatly, from 304 to 9677, in a large industry-sponsored study in the United States *(34)*. A notable feature of all the studies is that the populations followed were relatively young or recently employed, even though many plants in the studies started production in the 1940s. Most workers were hired after 1950, when United States and western European production increased sixfold in 10 years. Thus, data on effects 25 or more years after the beginning of exposure are limited.

Increased incidence of cancer in all sites is reported in most of the studies, although it does not achieve a 0.05 level of significance except in the study by Waxweiler et al. *(35)*. In the study by Ott et al. *(36)*, a highly exposed subgroup with 15 years' latency had 8 cancer deaths compared with 3.2 expected ($P < 0.05$). The absence of significant findings in other studies may be attributed to their low power or to the inclusion of a large number of individuals with very short or recent exposures.

One remarkable finding in virtually all the studies is the absence of significantly elevated mortality from chronic liver disease. The generally benign results reported in other studies contrast sharply with the severe liver disease from VC exposure documented in clinical studies *(15)*. Hepatomegaly, hepatic fibrosis, portal hypertension and bleeding oesophageal varices have commonly been found in individuals heavily exposed to VC, even without exposure to alcohol.

In the case of liver cancer, the overall data are consistent and striking. Haemangiosarcomas of the liver were reported in 8 of the 12 studies. In each of these, a very large and highly significant standard mortality rate (SMR) for liver cancer was seen. Methodological limitations can account for negative data in the other four studies. The large SMRs observed, however, result largely from low values for the expected number of cases rather than from a high incidence of observed cases. In all 12 studies, only 35 separate liver haemangiosarcomas were identified. As the overall excess number of deaths from liver and biliary cancer in all studies was 54, some haemangiosarcomas may not have been identified. The low numbers must also be seen in the light of the limited follow-up times in most studies.

The evidence for lung cancer is less clear. Some studies indicate an increase in lung cancer, but at a level that does not achieve statistical significance, except in the 15-year-latency population of Waxweiler et al. *(35)*. This, in part, may be the result of the low power of many of the studies. Only two have an 80% power to detect an overall risk of 1.5. Of significance, however, are the very low SMRs in the groups studied by Theriault & Allard *(37)*, Reinl et al. *(38)*, and Nicholson et al. *(16)*, cohorts in which many haemangiosarcomas were found. The four largest studies, although in some cases limited by the inclusion of short-term and recently employed workers, are also noteworthy for SMRs close to 100. Where available, data on subcohorts with longer latency (> 15 years) suggest some increased risk.

In a number of studies, cancers of the brain and the CNS were found to be significantly elevated, although the results differed considerably from study to study. Again, negative data may be simply the result of limited long-term follow-up or the low power of the study. In contrast to lung cancer, however, the largest study group had a significantly elevated risk of malignant tumours of the brain and the CNS. The human data are also modified by the recent finding of brain and CNS tumours resulting from various types of chemical plant exposure *(39)*. Excess brain malignancies, but not the etiological agents, have been identified in workers at several chemical/petrochemical plants in Texas and Louisiana, USA. Exposure to VC was documented for some cases, but it did not account for the overall findings. Since individuals in many of the VC studies considered here were exposed to other chemicals and petrochemicals, the possible role of these agents cannot be excluded.

Similar results are obtained for malignancies of the lymphatic and haematopoietic system. Here again, the analysis is limited by the few deaths and disparate results reported in different studies. Overall, there would appear to be an elevated risk, but the influence of confounding exposures precludes any definitive statement to this effect.

Animal data show VC to be a multi-site carcinogen *(21)*. While the epidemiological data are somewhat equivocal, VC should be considered potentially carcinogenic in humans in the lung, brain, and lymphatic and haematopoietic system, as well as the liver. Nevertheless, in all of the epidemiological studies reported to date, the number of excess cases at these nonhepatic sites is no more than the number of liver haemangiosarcomas found.

Nicholson et al. *(16)* have recently completed a follow-up (to the end of 1981) of a group of 491 workers at two polymerization plants. Eighty deaths occurred during follow-up beginning 10 or more years after employment. Of the 80 deaths, 9 were from haemangiosarcoma of the liver, 6 of which occurred after 1974. From an analysis of this very limited group, it would appear that haemangiosarcoma risk increases as approximately the square of time from the start of exposure in individuals exposed for 5 or more years.

The exposures that led to the currently observed mortality from VC were very high. Prior to 1955, it was estimated *(40)* that industrywide average exposures were approximately 2500 mg/m^3 and decreased to about 800 mg/m^3 by 1970. After 1974, substantial reductions in workplace exposures occurred following the identification of human cancer risk.

In a United States mortality study of PVC fabricators, including around 4300 deaths, an overrisk in gastrointestinal cancer was found in both sexes *(41)*. A statistically insignificant trend towards increased risk of tumours of the digestive organs was also found in a Swedish study *(42)*. These studies indicate that even low levels of VC exposure might represent a cancer risk to humans.

Nicholson et al. *(43)* have made projections of the future mortality that might occur from all VC exposures prior to 1975 in the United States and western European industry. These projections were made by assuming that the time course of VC risk follows a power law relationship with age *(44,45)*: $R = bt^k$, where R is the incidence rate of cancer at a specific site, t is age, and b and k are constants specific to site. Incidence data according to year of first exposure, number of years of exposure, calendar year of death, and number of years from onset of exposure, were matched to results calculated using a range of values for b and k. The values of b and k that best fit the data were then utilized to calculate the future mortality, assuming that the time course of risk continues throughout the life of the exposed population. The result suggests that 200–600 individuals in the United States and 550–2800 in western Europe may die of haemangiosarcoma from all occupational exposures to VC prior to 1975. Because of the assumption of a continued increasing risk, the estimates are likely to be high.

Vinyl chloride also causes chromosome breaks *(46)*, fragmentations and rearrangements *(47)* in the peripheral lymphocytes of PVC workers.

Evaluation of Human Health Risks

Exposure

Calculations based on dispersion models indicate that 24-hour average concentrations of 0.1–0.5μg/m^3 exist as background levels in much of western Europe, but such concentrations are below the current detection limit (approximately 1.0μg/m^3). In the vicinity of VC and PVC production facilities 24-hour ir concentrations can exceed 100μg/m^3, but are generally less than 10μg/m^3 at distances greater than 1 km from plants. The half-time of VC in the air is calculated to be 20 hours; this figure is based on measured rates of reaction with hydroxyl radicals and their concentrations in the air *(1)*.

Health risk evaluation

There is sufficient evidence of carcinogenicity of VC in humans and experimental animals *(26)*. Extrapolation (or rather interpolation) to lower exposure levels can be made, based on knowledge or assumptions about the dose and time-dependence of risk. As seen in the low exposure data of Maltoni et al. *(21)*, a linear dose–response relationship accords well with the animal data for haemangiosarcoma. The finding of at least three cases of haemangiosarcoma in PVC processors as compared with about 100 in VC or PVC production workers is compatible with a linear relationship. The average exposures in the production industry were about 100 times lower than those in the polymerization industry, but the workforce was 10 times larger.

Data from a cohort study *(16)* and an analysis of the incidence of haemangiosarcoma in the USA and western Europe *(43)* suggest that the risk of haemangiosarcoma increases as the second or third power of time from onset of exposure. Using a model in which the risk increases as t^3 during exposure and as t^2 subsequently, estimates of the relative risk in various exposure circumstances can be calculated and used to convert limited-duration exposure risks into lifetime exposure risks.

Estimates of cancer risk can be made from the data relating to the cohort studied by Nicholson et al. *(16)*. A group of 491 workers at two long-established PVC production plants was studied. One plant began operations in 1936 and the other in 1946. Each cohort member had a minimum of 5 years' employment; the average work duration was 18 years. It is estimated that the average VC exposure was 2050 mg/m^3. The overall SMR for cancer was 142 (28 observed; 19.7 expected); that for liver and biliary cancer was 2380 (10 observed; 0.42 expected). Using the liver cancer data, the estimated lifetime risk of death from VC exposure is 3.6×10^{-4} per mg/m^3, or $[(23.8 - 1) \times 0.003/(2050 \text{ mg/m}^3) \times 2.8 \times 70/18]$, where 0.003 is the lifetime risk of death from liver biliary cancer in white American males, 2.8 is the working week–total week conversion and 70/18 the work period–lifetime conversion. Since there are an equal number of cancers at other sites (averaging over 12 cohorts), the excess cancer risk is 7.2×10^{-4} per mg/m^3. If the total cancer SMR is used directly, the risk is 4.5×10^{-4} per mg/m^3, or $[(1.42 - 1) \times 0.2/(2050 \text{ mg/m}^3) \times 7.8 \times 70/18]$, which is in good agreement with the above. The average of the two estimates indicates that a 10^{-6} cancer risk occurs at exposures of 1.7μg/m^3.

The risk of cancer from VC can be calculated from data on the United States population exposed in the Equitable Environmental Health study *(34)*. This study identified 10 173 workers who were employed for one or more years in 37 (of 43) VC and PVC production plants. The average duration of employment before 1973 was 8.7 years. Using the data of Barnes *(40)*, a weighted exposure of 650 ppm (1665 mg/m^3) was estimated. Considering the total population at risk to be 12 000, the unit exposure lifetime risk from an average exposure of 9 years is 0.75×10^{-5} per mg/m^3, or $[(150/12\,000) \times (1/1665)]$.

Using a linear dose–response relationship converting to a lifetime exposure (assuming that one half of the workers began exposure at the age of 20 and one half at the age of 30), the continuous lifetime haemangiosarcoma risk is 4.7×10^{-4} per mg/m^3, or $[0.75 \times 10^{-5} \times 2.8 \times 22.4]$, where 2.8 is the ratio of the air volume inhaled in a full week (20 m^3 $\times$ 7) to that in a working week (10 m^3 $\times$ 5) and 22.4 is the average conversion to a lifetime for a ten-year exposure beginning at an average age of 25 years, taking into account the time course of haemangiosarcoma. (Without explicit consideration of the time course, the multiplier would be 70/9 = 7.8.) A 10^{-6} risk occurs at a concentration of 2.1 μg/m^3.

Assuming that the number of cancers in other sites may equal that of haemangiosarcomas, the best estimate for excess cancer risk is that a 10^{-6} risk occurs as a result of continuous lifetime exposure to 1.0 μg/m^3.

The risks estimated from epidemiological studies are the most relevant for human exposures. The above estimate from human angiosarcoma incidences is a conservative one, from the point of view of health, because of the use of a model that assumes that the haemangiosarcoma risk continues to increase throughout the lifetime of an exposed individual.

These risk estimates are in agreement with those made by others. The US Environmental Protection Agency has estimated that 11 cancer deaths per year would result from 4.6×10^6 people being exposed to 0.017 ppm (43 μg/m^3) *(48)*: this translates to a 10^{-6} lifetime risk at 0.25 μg/m^3. A Dutch criteria document, on the basis of animal data, estimates that a 10^{-6} risk occurs at 0.035 μg/m^3 *(1)*.

One cautionary note should be sounded: the particular sensitivity of newborn rats to VC, referred to above, suggests that risks may be much greater in childhood than those estimated from adult exposures. However, by the age of 10 years the latter risks should prevail.

Guidelines

Vinyl chloride is a human carcinogen and the critical concern with regard to environmental exposures to VC is the risk of malignancy. No safe level can be indicated. Estimates based on human studies indicate a lifetime risk from exposure to 1 μg/m^3 to be 1×10^{-6}.

References

1. *Criteriadocument over vinylchloride* [Vinyl chloride criteria document]. The Hague, Ministerie van Volkshuisvesting, Ruimtelijke Ordening en Milieubeheer, 1984 (Publikatiereeks Lucht, No. 34).

2. **Krost, K.J. et al.** Collection and analysis of hazardous organic emissions. *Analytical chemistry,* **54**: 810–817 (1982).
3. **Dimmick, W.F.** EPA programs of vinyl chloride monitoring in ambient air. *Environmental health perspectives,* **41**: 203–206 (1981).
4. **Gay, B.W. et al.** Measurements of vinyl chloride from aerosol sprays. *Annals of the New York Academy of Sciences,* **246**: 286–295 (1975).
5. *Tobacco smoking.* Lyon, International Agency for Research on Cancer, 1986 (IARC Monographs on the Evaluation of the Carcinogenic Risk of Chemicals to Humans, Vol. 38).
6. **Rösli, M. et al.** Rückstände von Vinylchlorid-Monomer in Speiseölen [Residues of vinyl chloride in edible oils]. *Mitteilungen aus der Gebiete der Lebensmitteluntersuchung und Hygiene,* **66**: 507–511 (1975).
7. Vinyl chloride. *British food journal,* **80**: 149–150 (1978).
8. **Bolt, H.M. et al.** Disposition of [1,2-^{14}C] vinyl chloride in the rat. *Archives of toxicology,* **35**: 153–162 (1976).
9. **Watanabe, P.G. et al.** Fate of [^{14}C] vinyl chloride following inhalation exposure in rats. *Toxicology and applied pharmacology,* **37**: 49–59 (1976).
10. **Gehring, P.J. et al.** Resolution of dose–response toxicity data for chemicals requiring metabolic activation. Example — vinyl chloride. *Toxicology and applied pharmacology,* **49**: 581–591 (1978).
11. **Bonse, G. & Henschler, D.** Chemical reactivity, biotransformation, and toxicity of polychlorinated aliphatic compounds. *CRC critical reviews in toxicology,* **4**: 395–409 (1976).
12. **Watanabe, P.G. et al.** Hepatic macromolecular binding following exposure to vinyl chloride. *Toxicology and applied pharmacology,* **44**: 571–579 (1978).
13. **Laib, R.J. & Bolt, H.M.** Alkylation of RNA by vinyl chloride metabolites *in vitro* and *in vivo*: formation of 1-N^6-etheno-adenosine. *Toxicology,* **8**: 185–195 (1977).
14. **Green, T. & Hathway, D.E.** The chemistry and biogenesis of the S-containing metabolites of vinyl chloride in rats. *Chemico-biological interactions,* **17**: 137–150 (1977).
15. **Lelbach, W.K. & Marsteller, H.J.** Vinyl chloride associated disease. *In:* Frick, P. et al., ed. *Ergebnisse der Inneren Medizin und Kinderheilkunde* [Advances in internal medicine and pediatrics]. Berlin, Springer-Verlag, 1981, Vol. 47.
16. **Nicholson, W.J. et al.** Occupational hazards in the VC-PVC industry. *In:* Jarvisalo, P. et al., ed. *Industrial hazards of plastics and synthetic elastomers.* New York, Alan R. Liss, 1984, pp. 155–176 (Progress in clinical and biological research, Vol. 141).
17. **John, J.A. et al.** The effects of maternally inhaled vinyl chloride on embryonal and fetal development in mice, rats and rabbits. *Toxicology and applied pharmacology,* **39**: 497–513 (1977).
18. **Ungváry, G. et al.** Effects of vinyl chloride exposure alone and in combination with trypan blue — applied systematically during all thirds of pregnancy on the fetuses of CFY rats. *Toxicology,* **11**: 45–54 (1978).

19. **Ungváry, G.** Studies on the teratogenicity of PVC. *Acta morphologica academiae scientiarum hungaricae,* **28**: 159–164 (1980).
20. **Salnikova, L.S. & Kicovskaja, J.A.** [Influence of vinyl chloride on embryogenesis in rats]. *Gigiena truda i professional'nye zabolevanija,* **3**: 46–47 (1980) (in Russian).
21. **Maltoni, C. et al.** Carcinogenicity bioassays of vinyl chloride monomer: a model of risk assessment on an experimental basis. *Environmental health perspectives,* **41**: 3–29 (1981).
22. **Groth, D.H. et al.** Effects of aging on the induction of angiosarcoma. *Environmental health perspectives,* **41**: 53–57 (1981).
23. **Radike, M.J. et al.** Effect of ethanol on vinyl chloride carcinogenesis. *Environmental health perspectives,* **41**: 59–62 (1981).
24. **Caputo, A. et al.** Oncogenicity of vinyl chloride at low concentrations in rats and rabbits. *International research communication,* **2**: 1582 (1974).
25. **Holmberg, B. et al.** The pathology of vinyl chloride exposed mice. *Acta veterinaria scandinavica,* **17**: 328–342 (1976).
26. *Some monomers, plastics and synthetic elastomers, and acrolein.* Lyon, International Agency for Research on Cancer, 1979 (IARC Monographs on the Evaluation of the Carcinogenic Risk of Chemicals to Humans, Vol. 19).
27. **Lilis, R. et al.** Prevalence of disease among vinyl chloride and polyvinyl chloride workers. *Annals of the New York Academy of Sciences,* **246**: 22–41 (1975).
28. **Suciu, I. et al.** Clinical manifestations in vinyl chloride poisoning. *Annals of the New York Academy of Sciences,* **246**: 53–69 (1975).
29. **Infante, P.F. et al.** Genetic risk of vinyl chloride. *Lancet,* **1**: 734–735 (1976).
30. **Edmonds, L.D. et al.** Congenital malformations and vinyl chloride. *Lancet,* **2**: 1098 (1975).
31. **Edmonds, L.D. et al.** Congenital central nervous system malformations and vinyl chloride monomer exposure. *Teratology,* **17**: 137–143 (1978).
32. **Infante, P.F.** Oncogenic and mutagenic risks in communities with polyvinyl chloride production facilities. *Annals of the New York Academy of Sciences,* **271**: 49–57 (1976).
33. **Sanockij, J.V. et al.** [A study of male reproductive function as affected by some chemicals]. *Gigiena truda i professional'nye zabolevanija,* **3**: 28–32 (1980) (in Russian).
34. *Epidemiological study of vinyl chloride workers.* Rockville, MD, Equitable Environmental Health, Inc., 1978.
35. **Waxweiler, R.J. et al.** Neoplastic risk among workers exposed to vinyl chloride. *Annals of the New York Academy of Sciences,* **271**: 40–48 (1976).
36. **Ott, M.G. et al.** Vinyl chloride exposure in a controlled industrial environment. A long-term mortality experience in 594 employees. *Archives of environmental health,* **30**: 333–339 (1975).
37. **Theriault, G. & Allard, P.** Cancer mortality of a group of Canadian workers exposed to vinyl chloride monomer. *Journal of occupational medicine,* **23**: 671–676 (1981).

38. **Reinl, W. et al.** The mortality of German vinyl chloride (VC) and polyvinyl chloride (PVC) workers. *Arhiv za higijenu rada i toksikologiju,* **30**(Suppl.): 399–402 (1979).
39. **Alexander, V. et al.** Brain cancer in petrochemical workers: a case series report. *American journal of industrial medicine,* **1**: 115–123 (1980).
40. **Barnes, A.W.** Vinyl chloride and the production of PVC. *Proceedings of the Royal Society of Medicine,* **69**: 277–281 (1976).
41. **Chiazze, L. et al.** Mortality among employees of PVC fabricators. *Journal of occupational medicine,* **19**: 623–628 (1977).
42. **Molina, G. et al.** Mortality and cancer rates among workers in the Swedish PVC processing industry. *Environmental health perspectives,* **41**: 145–151 (1981).
43. **Nicholson, W.J. et al.** Trends in cancer mortality among workers in the synthetic polymers industry. *In:* Jarvisalo, P. et al., ed. *Industrial hazards of plastics and synthetic elastomers.* New York, Alan R. Liss, 1984, pp. 65–78 (Progress in clinical and biological research, Vol. 141).
44. **Armitage, P. & Doll, R.** Stochastic models for carcinogenesis. *In:* Neyman, J., ed. *Proceedings of the Fourth Berkeley Symposium on Mathematical Statistics and Probability.* Berkeley, University of California Press, 1961, pp. 19–38.
45. **Cook, P.J. et al.** A mathematical model for the age distribution of cancer in man. *International journal of cancer,* **4**: 93–112 (1969).
46. **Funes-Cravioto, F. et al.** Chromosome aberrations in workers exposed to vinyl chloride. *Lancet,* **1**: 459 (1975).
47. **Ducatman, A. et al.** Vinyl chloride exposure and human chromosome aberrations. *Mutation research,* **31**: 163–168 (1975).
48. **Kuzmach, A.M. & McGaughy, R.E.** *Quantitative risk assessment for community exposure to vinyl chloride.* Washington, DC, US Environmental Protection Agency, 1975.

Part III

Inorganic substances

17

Arsenic

General Description

Arsenic (As) and its compounds are ubiquitous in nature and exhibit both metallic and nonmetallic properties. The trivalent and pentavalent forms are the most common oxidation states of arsenic. At least six groups of compounds can be distinguished in the many different arsenic compounds present in the environment:

(*a*) inorganic water-soluble compounds: arsenic trioxide and arsenic pentoxide; soluble arsenite and arsenate salts;

(*b*) inorganic compounds (low or no water solubility): various arsenite and arsenate salts, arsenides, arsenic selenide and arsenic sulfide;

(*c*) organic arsenic compounds: methylated arsenic compounds occurring naturally in the environment as a result of biological activity, or as pesticides, e.g. cacodylic acid;

(*d*) organic arsenic compounds occurring naturally in marine organisms, e.g. arsenobetaine;

(*e*) organic compounds used as feed additives, e.g. arsanilic acid;

(*f*) gaseous inorganic and organic arsenic compounds , e.g. arsine.

Sources

Arsenic appears in nature primarily in the form of sulfides in association with sulfides of ores of silver, lead, copper, nickel, antimony, cobalt and iron. Trace amounts of arsenic are found in soils and other environmental media. Uncontaminated soil generally contains less than 40μg arsenic per gram of soil; an average level of 7μg/g is suggested. However, levels of 100–2500μg/g have been found in the vicinity of copper smelters. Pesticides, herbicides and defoliants may cause arsenic levels to reach 700μg/g in agricultural soils *(1)*. Soil microorganisms can transform arsenic into volatile organic arsines, which may then be volatilized to air.

Arsenic is released to the atmosphere from both natural and anthropogenic sources. The principal natural source is volcanic activity, with minor contributions by exudates from vegetation and wind-blown dusts. Man-made emissions to air arise from the smelting of metals, the combustion of fuels, especially of low-grade brown coal, and the use of pesticides *(2)*.

The arsenic concentration of rivers and lakes varies greatly: it is generally less than 10 μg/litre but sometimes as high as 1 mg/litre. The concentrations in groundwater depend on the arsenic content of the bed-rock; unusually high levels have been reported in carbonate spring waters in New Zealand, Romania, the USSR and the USA (0.4–1.3 mg/litre), in artesian wells in Taiwan (up to 1.8 mg/litre) and in groundwater in Cordoba, Argentina (up to 3.4 mg/litre). In oxygenated water arsenic occurs in pentavalent form, but under reducing conditions the trivalent form predominates *(3)*.

Global natural emissions have been estimated to be 7900 tonnes per year, whereas anthropogenic emissions are about three times higher, i.e. 23 600 tonnes per year *(4)*.

The relatively high concentrations of arsenic in coal may result in substantial emissions to air on combustion. White arsenic (arsenic [III] oxide) is produced as a byproduct from the roasting of sulfide ores, collected on electrofilters, and serves as a basis for the manufacture of virtually all arsenicals. The world production of arsenic kept rising until about the mid-1940s (in 1943 it was estimated at some 70 000 tonnes annually). As arsenic pesticides, specifically insecticides, were gradually replaced by other preparations, the production of arsenic declined. The world production of arsenic trioxide is about 40 000–70 000 tonnes per year *(1)*. The largest amount of arsenic is still used in the production of agricultural chemicals (herbicides and pesticides), although the amounts produced vary among countries according to the different restrictions in force on this use. Arsenic is an active component of antifungal wood preservatives (e.g. Wolman's salt, which contains 25% sodium arsenite). It is also used in the pharmaceutical and glass industries, in the manufacture of sheep-dips, leather preservatives and poisonous baits. Arsenicals are used in the manufacture of pigments. Metallic arsenic is used in the manufacture of alloys. Crystals of gallium arsenide are used as light quantum generators in laser devices.

Arsanilic acid and its derivatives 4-aminophenylarsonic and 3-nitro-4-hydroxyphenylarsonic acids are, in some countries, added to cattle and poultry feed at 25–45 mg/kg and used as growth-stimulating agents *(5)*.

According to the different uses of arsenic and arsenicals, there is a wide spectrum of situations in which humans may be exposed to arsenic.

Occurrence in air

Representative background levels of arsenic in air are 1–10 ng/m^3 in rural areas. Concentrations can reach several hundred ng/m^3 in some cities and exceed 1000 ng/m^3 near nonferrous metal smelters and some power plants, depending on the arsenic content in the burnt coal.

Results from an air monitoring network in the USA give annual means ranging from 2.6 to 11 ng/m^3 during 1977–1981 *(5)*. Uncertainties are primarily due to the large numbers of analyses below the detection limits, which for the neutron activation technique used were 4–7 ng/m^3.

Higher values are also reported from other networks in cities where coal-burning sources are prevalent. For example, in Prague, airborne arsenic concentrations were found to be 450 ng/m^3 on average in winter and 70 ng/m^3 in summer *(6)*.

Arsenic in air is present mainly in particulate form as inorganic arsenic. It is assumed that methylated arsenic is a minor component in the air of suburban, urban and industrial areas, and that the major inorganic portion is a variable mixture of the trivalent and pentavalent forms *(5)*, the latter form being predominant.

Routes of Exposure

Air

Particulate arsenic compounds may be inhaled, deposited in the respiratory tract and absorbed in the blood. Assuming a breathing rate of 20 m^3 per day, and retention and absorption of the order of 30% of the intake amount, the uptake rates may be estimated as follows:

rural areas (1–10 ng/m^3) = 0.006–0.06 μg/day
urban areas (10–200 ng/m^3) = 0.06–1 μg/day.

Tobacco smoke may contain arsenic, especially when the tobacco plants have been treated with lead arsenate insecticide. Although the use of arsenic pesticides is now prohibited in most countries, the natural content of arsenic in tobacco may still result in some exposure. At present, it is estimated that about 6 μg of arsenic may be inhaled per pack of cigarettes smoked, of which about 2 μg would be retained in the lungs *(5)*.

Occupational exposure to arsenic occurs primarily among workers in metal-smelting industries *(7)*, among workers at some coal-burning power plants *(8)*, and among workers using or producing pesticides containing arsenic *(9)*. Air concentrations of arsenic in smelters have been reported to range from a few micrograms to milligrams per m^3 *(3,5)*. Workers may also be exposed to airborne arsenic from cutting and sawing wood treated with preservatives containing arsenic *(3)*.

Drinking-water

Drinking-water may contribute significantly to oral intake in certain regions where there are high arsenic concentrations in well-water or in mine drainage areas. More common drinking-water sources generally contain less than 10 μg/litre of arsenic. Flocculation treatment using either aluminium or ferric salts removes a high proportion, at least, of pentavalent arsenic *(3)*.

Food

Arsenic levels in foods are generally well below 1 mg/kg wet weight (see Table 1). The use of organic arsenic compounds as feed additives for poultry and pigs may lead to increased levels of arsenic in meat. Wine made from

Table 1. Relative significance of different routes of exposure to arsenic

Route	Arsenic concentration
Air	
Rural	0.2–10 ng/m³
Urban	10–750 ng/m³
Food	
Plants	0.4 μg/g
Fish	1–10 μg/g
Other	0.25 μg/g

grapes sprayed with arsenic insecticides or fungicides may contain appreciable levels of arsenic (up to 0.5 mg/litre) in the trivalent inorganic form *(5)*.

Much higher levels of arsenic, but in various organic forms (e.g. arsenobetaine), are present in seafood, particularly marine fish (1–10 mg/kg), and values over 100 mg/kg have been observed in certain bottom-feeding fish and crustaceans.

A representative intake of arsenic is of the order of 40 μg per day in foods of terrestrial origin and 80 μg per day in seafoods *(10)*. Both inorganic and organic arsenic compounds are readily absorbed from the gastrointestinal tract.

Kinetics and Metabolism

Absorption

The major routes of arsenic absorption in the general population are inhalation and ingestion. Factors affecting the extent of absorption from the lungs include chemical forms, particle size and solubility. Limited data from human subjects suggest that about 40% of inhaled arsenic is deposited in the lungs, yielding a net absorption of approximately 30% of the inhaled amount *(3)*.

Soluble inorganic arsenic is almost totally absorbed from the gastrointestinal tract. Less soluble forms have a much lower absorption.

Distribution

Blood is the main vehicle for the transport of arsenic following absorption, and arsenic is cleared relatively rapidly from it. Arsenic movement from the blood appears to conform to a three-compartment model which must reflect in part the biomethylation of inorganic arsenic noted above.

In man, tissue-partitioning data are mainly available from autopsy data. The muscles, bones, kidneys, liver and lungs have the highest absolute amounts, but skin and excretory/storage organs such as nails and hair have the highest levels on a concentration basis *(3)*.

The transplacental transfer of arsenic appears to occur in man. This finding is based on autopsy data and on reports showing that blood levels in the cords of neonates approximate those of the mothers *(3)*.

Recent data on valency and exposure level effects on the tissue distribution of arsenic indicate that levels of arsenic in the kidneys, liver, bile, brain, skeleton, skin and blood are 2–25 times higher for the trivalent form than for the pentavalent state and are greatly increased at higher dosing. This may be due to differences in the methylation rate of either form as well as to the level of exposure *(5)*.

Autopsy data from retired metal-smelter workers, obtained several years after cessation of occupational exposure, showed that arsenic levels in the lung were 8 times higher than in a control group *(11)*. This suggests the existence of arsenic compounds of very low solubility in the smelter environment.

Biotransformation

Extensive recent literature documents the *in vivo* methylation of inorganic arsenic to monomethyl and dimethyl arsenic (the latter being the major methylated metabolite) in man and animals *(5)*.

In man, dimethyl arsenic represents approximately 75% of total arsenic excretion, monomethyl arsenic being excreted in lesser amounts *(5)*. This biotransformation is dose-dependent and high exposure will give relatively more of the monomethyl form *(12)*.

The demonstration of interconversion of the two valency forms of inorganic arsenic claimed in earlier literature must be considered in the light of biomethylation, and only the more recent studies using chemical speciation techniques in addressing this problem can be considered reliable *(12)*. A recent study using speciation methods and human subjects also establishes the *in vivo* reduction of pentavalent to trivalent arsenic in humans *(5)*.

Elimination

Renal clearance appears to be the major route of excretion of absorbed arsenic in man and animals. Biliary transport of the element leads to enteric reabsorption, with little carriage in faeces.

Health Effects

Effects on experimental animals and *in vitro* test systems

A number of animal carcinogenicity studies are available on exposure to inorganic arsenic compounds via the respiratory tract. Pulmonary carcinomas and/or benign tumours were obtained after intratracheal instillation with arsenic trioxide and other inorganic compounds *(13,14)*. Two studies suggest a positive interaction between arsenic trioxide and benzo[*a*]pyrene in relation to pulmonary tumours, but the evidence is not conclusive *(14,15)*.

Arsenic is clastogenic and induces sister chromatid exchanges in a variety of mammalian cells *in vitro (16)*; trivalent arsenic is approximately one order of magnitude more potent than pentavalent arsenic *(17)*. Sodium arsenite caused a slight increase in chromosomal aberrations in bone-marrow cells of mice treated *in vivo (18)*.

Several studies have suggested that inorganic arsenic affects DNA repair mechanisms and acts as co-mutagen in bacterial test systems by inhibiting the repair of damage to DNA caused by another agent *(19)*.

Arsenic at relatively high exposure levels is teratogenic in a number of animal species, including hamster, rat and mouse *(5)*. Such effects have generally been observed after parenteral administration of either arsenite or arsenate. Oral exposures have not produced any notable effects on reproduction and development.

Effects on humans

Toxicological effects

The clinical picture of chronic poisoning with arsenic varies widely. It is usually dominated by changes in the skin and mucous membranes and by neurological vascular and haematological lesions. Involvement of the gastrointestinal tract, increased salivation, irregular dyspepsia, abdominal cramps and loss of weight may also occur. The neurological symptomatology may also include signs of optic nerve degeneration resulting in slight to severe visual impairment, including amaurosis. Alteration of vestibular functions may also occur. Reports of diminished sexual activity in persons with chronic arsenic exposure are frequent *(3,5)*.

Allergic contact dermatitis may occur in chronic arsenical poisoning. Chronic dermatological disorders may manifest themselves as eczematous, follicular, erythematous or even ulcerative dermatitis *(3)*.

Increased mortality from cardiovascular diseases has been observed in epidemiological investigations of smelter workers exposed to high levels of airborne arsenic. Peripheral vascular lesions, such as symptoms of *endangitis obliterans* and *acrodermatitis atrophicans* or so-called "blackfoot disease" (peripheral gangrene), have been reported in different exposure situations *(3,5)*.

Inorganic arsenic has an inhibitory effect on haematopoiesis, giving rise to anaemia, most commonly of the hypoplastic type. In severe cases of arsenical poisoning agranulocytosis or thrombopenia may develop *(3)*.

An increased rate of spontaneous abortions and lower mean birth weights has been reported among Swedish smelter workers and among subjects living in the vicinity of the smelter *(5)*. The rate of congenital malformations in the offspring of women working at the smelter was also higher. It is not possible to link these effects with exposure to any specific compound in the smelter environment.

Mutagenic and carcinogenic effects

Inorganic arsenic compounds are established human carcinogens *(20)*. Lung cancer is considered as the critical effect following exposure via inhalation *(3)*; consequently, cancer at other sites, e.g. skin cancer, will not be discussed in detail here.

Several studies show that exposure to inorganic arsenic compounds can increase the risk of lung cancer in smelter workers and persons engaged in the production and use of arsenic-containing pesticides. The data often indicate positive dose–response relationships. Both trivalent and pentavalent arsenic compounds have occurred in these exposure situations and at present the possibility cannot be ruled out that any form of inorganic arsenic may be carcinogenic. Study results on the interaction between inorganic arsenic and smoking are conflicting *(5)*, but at least one study has provided evidence of a multiplicative interaction *(21)*.

Some investigations of populations living near copper smelters and other point sources of arsenic emission to the air have revealed moderate increases in lung cancer mortality *(22–25)*. Other studies have failed to detect an effect in such situations *(26,27)*. No detailed exposure data were given in the studies, which makes it difficult to evaluate the findings quantitatively.

An increased frequency of chromosomal aberrations has been found in peripheral blood lymphocytes of wine-growers exposed to arsenic, in psoriatic patients treated with arsenic, and in arsenic-exposed workers at a copper smelter *(3,5)*. Sodium arsenate inhibits DNA repair in human skin biopsy cells and in lymphocytes *(3)*.

Evaluation of Human Health Risks

Exposure

There are many arsenic compounds, both organic and inorganic, in the environment. Airborne concentrations of arsenic range from 1 ng/m^3 to 10 ng/m^3 in rural areas and from a few nanograms to a few hundred nanograms per m^3 in urban areas. Near emission sources, such as non-ferrous metal smelters and coal-burning power plants, concentrations of airborne arsenic can exceed 1 μg/m^3.

Health risk evaluation

Inorganic arsenic can have acute, subacute and chronic effects which may be either local or systemic. Lung cancer is considered to be the critical effect following inhalation. An increased incidence of lung cancer has been seen in several occupational groups exposed to inorganic arsenic compounds. Some studies also show that populations near emission sources of inorganic arsenic, such as smelters, have a moderately elevated risk of lung cancer *(4)*. Information on the carcinogenicity of arsenic compounds in experimental animals was considered inadequate to make an evaluation *(4)*.

A significant number of studies concerning occupational exposure to arsenic and the occurrence of cancer have been described. As shown in Table 2, unit risks derived by the Carcinogen Assessment Group of the US Environmental Protection Agency (EPA) *(5)* from five sets of data involving two independently exposed worker populations are quite consistent, ranging from 1.25 to 7.6×10^{-3}, a weighted average of these five estimates giving a composite estimate of 4.29. The potency of airborne arsenic compared to other carcinogens lies in the first quartile of the

Table 2. US EPA unit risk estimates based on absolute risk linear models

Source of exposure	Study	Unit risk	Geometric mean unit risk
Anaconda smelter	Brown & Chu	1.25×10^{-3}	
	Lee-Feldstein	2.80×10^{-3}	2.56×10^{-3}
	Higgins	4.90×10^{-3}	
Tacoma smelter	Enterline & Marsh	6.81×10^{-3}	7.19×10^{-3}
		7.60×10^{-3}	

Source: US Environmental Protection Agency *(5)*.

52 carcinogens that have been evaluated by the US EPA Carcinogen Assessment Group *(5)*.

For the quantitative risk assessment, two epidemiological studies were evaluated by WHO working groups *(7,28)*. Although the absolute risk linear model conforms better with the available data, estimates calculated using the average relative risk model do not differ significantly from the US EPA estimates based on the absolute risk linear models *(5)*. However, since arsenic has a dose-dependent metabolism, it is uncertain if linear extrapolation is appropriate for estimating risks at lower levels.

A WHO Working Group on Arsenic *(3)* conducted a quantitative risk assessment for arsenic, assuming a linear relationship between the cumulative arsenic dose and the relative risk of developing lung cancer. Risk estimates for lung cancer from inorganic arsenic exposure were based on the study by Pinto et al. *(7)* of workers at the Tacoma smelter, WA, USA, in which there was a relative risk (R) of about 3 at an estimated average air concentration of 50 $\mu g/m^3$ for an average exposure duration of more than 25 years. The lifetime risk of lung cancer was calculated to be 7.5×10^{-3} per microgram of airborne arsenic per cubic metre.

The second study *(28)* relating to quantitative risk assessment included a large number of the 8047 males employed as smelting workers at the Anaconda copper smelter, MT, USA. The expected number of cancer deaths was calculated on an age-adjusted basis. Exposure to airborne arsenic levels were estimated to average 11.27, 0.58 and 0.27 mg/m^3 in the heavy, medium and light exposure areas, respectively *(5)*. Respirators were used with a varying degree of constancy in the areas of highest exposure. Therefore, the average individual exposure in the high-exposure areas was reduced by a factor of 10 to 1127 $\mu g/m^3$, assuming that the use of respirators would roughly reduce the risk by this factor *(5)*. Based on this division, the workers were categorized into heavy, medium or light exposure groups. The average duration of exposure was estimated to be 15 years for all three groups. From the lower, medium and higher estimation of time-weighted average exposures, the average lifetime daily exposures (X) were calculated to be 13.6, 27.2 and 52.9 $\mu g/m^3$, respectively [$X = \mu g/m^3 \times 8/24 \times 240/365 \times 15/70$]. The estimated relative risks were 2.3 (136/58.9), 4.5 (93/20.9) and 5.1 (33/6.5) for the light, medium and high exposure groups respectively, and risk estimates were then calculated to be 3.9×10^{-3}, 5.1×10^{-3} and 3.1×10^{-3}, respectively [$UR = P_0(R - 1)/X$]. The geometric mean of unit risks calculated from this study would be 4×10^{-3}.

If it is assumed that the risk estimates based on the Tacoma study are slightly higher because the urine measurements made may have underestimated the actual inhalation exposure, the lung cancer risk due to lifetime exposure to 1 $\mu g/m^3$ of inorganic arsenic can be considered to be 4×10^{-3}.

Guidelines

Because arsenic is carcinogenic and there is no known safe threshold, no safe level for arsenic can be recommended. At an air concentration of 1 μg arsenic per m^3, a conservative estimate of lifetime risk is 3×10^{-3}.

References

1. *Effects of arsenic in the Canadian environment.* Ottawa, National Research Council Canada, 1978 (Publication No. 15391).
2. **Merian, E.** Introduction on environmental chemistry and global cycles of chromium, nickel, cobalt, beryllium, arsenic, cadmium and selenium and their derivatives. *Toxicological and environmental chemistry,* **8**: 9–38 (1984).
3. *Arsenic.* Geneva, World Health Organization, 1981 (Environmental Health Criteria, No. 18).
4. **Woolson, E.A.** Man's perturbation of the arsenic cycle. *In:* Lederer, W.H. & Fensterheim, R.J., ed. *Arsenic: industrial, biomedical and environmental perspectives. Proceedings of the Arsenic Symposium, Gaithersburg, MD.* New York, Van Nostrand Reinhold, 1983.
5. *Health assessment document for inorganic arsenic.* Research Triangle Park, NC, US Environmental Protection Agency, 1984, p. 351 (Final report No. EPA-600/8-83-021F).
6. **Vondráček, V.** Koncentrace 3,4-benzpyrenu a sloučenin arzénu v pražském ovzduší [Concentration of 3,4-benzpyrene and arsenic compounds in the Prague atmosphere]. *Československà hygiena,* **8**: 333–339 (1963).
7. **Pinto, S.S. et al.** Mortality experience in relation to a measured arsenic trioxide exposure. *Environmental health perspectives,* **19**: 127–130 (1977).
8. **Bencko, V. et al.** Rate of malignant tumor mortality among coal-burning power plant workers occupationally exposed to arsenic. *Journal of hygiene, epidemiology, microbiology and immunology (Prague),* **24**: 278–284 (1980).

9. **Horiguchi, S. et al.** A long-term observation of the environmental conditions of a factory manufacturing lead arsenate as an insecticide. (Studies on lead arsenate poisoning, part I). *Osaka City medical journal,* **22**: 43–46 (1976).
10. **Bennett, B.G.** Exposure commitment concepts and application; summary exposure assessments for lead, cadmium and arsenic. *In: Exposure commitment assessments of environmental pollutants.* London, University of London Monitoring and Assessment Research Centre, 1981, Vol. 1 (MARC Report No. 23).
11. **Brune, D. et al.** Distribution of 23 elements in the kidney, liver and lungs of workers from a smeltery and refinery in North Sweden exposed to a number of elements and of a control group. *Science of the total environment,* **16**: 13–35 (1980).
12. **Lovell, M.A. & Farmer, J.G.** Arsenic speciation in urine from humans intoxicated by inorganic arsenic compounds. *Human toxicology,* **4**: 203–214 (1985).
13. **Pershagen, G. & Björklund, N.-E.** On the pulmonary tumorigenicity of arsenic trisulfide and calcium arsenate in hamsters. *Cancer letters,* **27**: 99–104 (1985).
14. **Pershagen, G. et al.** Carcinomas of the respiratory tract in hamsters given arsenic trioxide and/or benzo[*a*]pyrene by the pulmonary route. *Environmental research,* **34**: 227–241 (1984).
15. **Ishinishi, N. et al.** Preliminary experimental study on carcinogenicity of arsenic trioxide in rat lung. *Environmental health perspectives,* **19**: 191–196 (1977).
16. **Vainio, H. & Sorsa, M.** Chromosome aberrations and their relevance to metal carcinogenesis. *Environmental health perspectives,* **40**: 173–180 (1981).
17. **Wan, B. et al.** Studies of cytogenetic effects of sodium arsenicals on mammalian cells *in vitro. Environmental mutagenesis,* **4**: 493–498 (1982).
18. *Some metals and metallic compounds.* Lyon, International Agency for Research on Cancer, 1980 (IARC Monographs on the Evaluation of the Carcinogenic Risk of Chemicals to Humans, Vol. 23).
19. **Rossman, T.G.** Enhancement of UV-mutagenesis by low concentrations of arsenite in *E. coli. Mutation research,* **91**: 207–211 (1981).
20. *Chemicals, industrial processes and industries associated with cancer in humans. IARC Monographs, Volumes 1 to 29.* Lyon, International Agency for Research on Cancer, 1982 (IARC Monographs on the Evaluation of the Carcinogenic Risk of Chemicals to Humans, Supplement 4).
21. **Pershagen, G. et al.** On the interaction between occupational arsenic exposure and its relationship to lung cancer. *Scandinavian journal of work, environment & health,* **7**: 302–309 (1981).
22. **Blot, W.J. & Fraumeni, J.F., Jr.** Arsenical air pollution and lung cancer. *Lancet,* **2**: 142–144 (1975).

23. **Newman, J.A. et al.** Occupational carcinogenesis. Histologic types of bronchogenic carcinoma among members of coppermining and smelting communities. *Annals of the New York Academy of Sciences,* **271**: 260–268 (1976).
24. **Pershagen, G.** Lung cancer mortality among men living near an arsenic emitting smelter. *American journal of epidemiology,* **122**: 684–694 (1985).
25. **Matanoski, G. et al.** Cancer mortality in an industrial area of Baltimore. *Environmental research,* **25**: 8–28 (1981).
26. **Greaves, W.W. et al.** Relationship between lung cancer and distance of residence from nonferrous smelter stack effluent. *American journal of industrial medicine,* **2**: 15–23 (1981).
27. **Rom, W.N. et al.** Lung cancer mortality among residents living near the El Paso smelter. *British journal of industrial medicine,* **39**: 269–272 (1982).
28. **Lee-Feldstein, A.** Arsenic and respiratory cancer in man: follow-up of an occupational study. *In*: Lederer, W.H. & Fensterheim, R.J., ed. *Arsenic: industrial, biomedical and environmental perspectives. Proceedings of the Arsenic Symposium, Gaithersburg, MD.* New York, Van Nostrand Reinhold, 1983, pp. 245–254.

18

Asbestos

General Description

The term "asbestos" designates a group of naturally occurring fibrous serpentine or amphibole minerals that have extraordinary tensile strength, conduct heat poorly and are relatively resistant to chemical attack. The principal varieties of asbestos are chrysotile, a serpentine mineral, and crocidolite, amosite, anthophyllite, tremolite and actinolite, all of which are amphiboles.[a]

Chrysotile fibres consist of aggregates of long, thin, flexible fibrils that resemble scrolls or cylinders of uniform chemical composition. Although chrysotile is a reasonably well defined mineral, the five amphibole asbestiform fibres have such variable chemical compositions and physical properties that positive identification is sometimes troublesome.

The macroscopic asbestos fibres are actually bundles of thinner fibres made up of fibrils which, in the case of chrysotile, have a diameter of 20–25 nm. Each macroscopic fibre is highly anisotropic and tends to decompose into its thinner constituents under industrial handling or from weathering, giving rise to a fibrous, partially respirable aerosol.

Sources

Natural sources are important, because asbestos minerals are widely spread throughout the earth's crust and are not restricted to the few mineable deposits. In particular, chrysotile is present in most serpentine rock formations. Emissions are due to natural weathering and can be enhanced by man's activities, such as quarrying or street building. Very little, however, is known about the amounts emitted from natural sources.

Man-made emissions originate from activities in the following categories:

(*a*) mining and milling

(*b*) manufacture of products

(*c*) construction activities

---

[a] Man-made mineral fibres and other natural mineral fibres such as fibrous zeolites, wollastonite, attapulgite and sepiolite will not be discussed here.

(*d*) transport and use of asbestos-containing products

(*e*) disposal.

There has been a steep rise in the production and use of asbestos in the last 100 years. Asbestos consumption has levelled off in recent years *(1)* to about 4 million tonnes (1983) and, because of the relative decline in crocidolite and amosite usage *(1)* the figure represents essentially chrysotile production (in 1982 only about 5% of the total asbestos produced was in the form of amphibole asbestos (crocidolite and amosite) *(2)*).

In many industrialized countries most of the asbestos used was in the building sector (70–90% in several western European countries) *(2)*, and the demand for asbestos in these countries had already reached a constant level before the health effects of asbestos were widely debated. Because of its specific technical properties, asbestos has found an extremely large variety of applications (in about 3000 different products). In the future, legislative restrictions and the success in finding substitutes for asbestos in fibre-cement, brake-linings, insulation and many other applications will most probably lead to declining asbestos consumption in the above-mentioned countries.

Asbestos emissions occur during processing. When air filtration is used, the dust emissions from processing can be kept below 100 g per tonne *(3)*.

Rain acidity (from carbon dioxide and air pollutants) is known to corrode asbestos-cement sheets, constituting a further source of emission *(4)*. Brake-linings in cars are another source in the urban environment. Only a few measurements have been made of the contribution made by these sources to fibre emissions in urban air *(5,6)*.

Indoor asbestos fibre concentrations can be considerably higher than outdoor concentrations *(7)*. Indoor asbestos dust originates from insulation material sprayed on steelwork or ceilings (such material may become highly friable after some years), asbestos plasters, low-weight insulation plates, etc. *(8,9)*. Sometimes such materials have been used in direct or close contact with air-conditioning equipment. Even though some of these materials, such as spray asbestos, are no longer used, they are still found in many public buildings. Until the mid-1970s electric storage-heaters and some other electrical household equipment contained asbestos. One of the main forms of use of asbestos is as asbestos-cement; in this case the release of fibres to the general environment is minimized, since the fibres are essentially "locked" in the cement matrix. Asbestos cement products, therefore, do not usually pose problems for indoor air quality.

Factors such as renovation and repair, maintenance, external vibrations and vandalism can considerably increase the emission of asbestos dust from existing indoor sources *(10,11)*. Increased emission is also possible as a result of changes in temperature and reduction of humidity.

Unlike levels of asbestos fibres in ambient air, fibre concentrations in the air of indoor environments can always be related to their source, thus offering the possibility of modifying or removing the source of emission. Because people in temperate climates live indoors most of the time, the most relevant source of inhalation exposure will often be asbestos fibre concentrations in buildings.

Occurrence in air

Asbestos fibres of respirable size form part of a range of fibrous aerosols in the lower atmosphere. Other fibres include man-made mineral fibres, fibrous silica and aluminium oxide, fibrous gypsum and, in some geographical areas, fibrous zeolite, attapulgite, sepiolite and wollastonite *(1)*.

Once emitted into the atmosphere, asbestos fibres may travel considerable distances owing to their aerodynamic properties. Because no chemical breakdown of the fibres occurs, washout by rain or snow is the only cleaning mechanism.

Asbestos fibres normally constitute only a relatively small fraction of the total fibrous aerosol in ambient air *(6,12)*. The biologically more important so-called "critical" fibres are those equal to or longer than 5 μm and having diameters up to 3 μm *(1,10,12)* with an aspect ratio equal to or greater than 3 : 1.

Although asbestos fibres can be readily detected and monitored in occupational situations by using phase-contrast optical microscopy, their assessment in the environment calls for an integrated method capable of microchemical analysis of single fibres, measurement of fibre length and diameter, and counts of fibre numbers in given air samples. Electron microscopy is the only method which can detect and identify asbestos fibres among the very wide range of other fibrous and nonfibrous particles, of greatly varying toxic potency, in the ambient air. Used in conjunction with each other, these methods allow the identification of samples. In order positively to identify asbestos fibres in environmental ambient air it is necessary to use selected-area electron diffraction and/or energy-dispersive X-ray diffraction analysers attached to an electron microscope. Such instrumentation is costly and highly-trained personnel are required in order to obtain reliable results. Although integrated electron microscopy methods have been developed during the last 15 years *(13)*, they are still not in widespread use. Therefore, all data measured by these methods are, strictly speaking, representative only for the location tested and the time interval chosen. However, if one compares various sets of such data and restricts quantification to orders of magnitude, the following concentration pattern emerges:[a]

Rural areas (remote from asbestos emission sources):

— below 100 F/m^3 *(6,10)*

[a] *Note.* If not otherwise stated, concentrations in the following text are given as numbers of critical asbestos fibres per m^3, i.e. all fibres of length $L > 5 \mu m$, diameter $D < 3 \mu m$ and aspect ratio $L:D > 3:1$ measured by electron microscopy methods.

All concentrations are expressed as fibres per m^3 air, although concentrations in terms of fibres per litre and fibres per ml are often reported. Some studies have been carried out where results have been expressed as ng or μg per m^3 air. Because only results such as fibre number concentrations are considered to be directly relevant to an index of exposure with regard to possible health implications *(1,12)* and factors in attempts to convert mass to number concentrations are so variable *(12)*, mass concentrations are not considered here.

Urban areas:

— general levels may vary from below 100 to 1000 F/m^3 *(6,12)*

Near various emission sources the following figures have been measured as yearly averages *(6,12,14)*:

— downwind from an asbestos-cement plant at 300 m: 2200 F/m^3; at 700 m: 800 F/m^3; at 1000 m: 600 F/m^3 *(6)*;

— at a street crossing with heavy traffic, 900 F/m^3 *(14)*;

— on an express-way, up to 3300 F/m^3 *(9)*

Indoor air:

— in buildings without specific asbestos sources, concentrations are generally below 1000 F/m^3 *(12)*;

— in buildings with friable asbestos, concentrations vary irregularly; usually less than 1000 F^*/m^3 are found, but in some cases exposure reaches 10 000 F^*/m^3 *(9)*, where F^* = fibres counted with an optical microscope.

Occupational levels are orders of magnitude higher than those found in the environment, with values from 10^5 F^*/m^3 to more than 10^8 F^*/m^3 *(12,15)*, but are now being reduced to below 2×10^6 F^*/m^3 in most countries and to $(0.2–0.5) \times 10^6$ F^*/m^3 in some.

Routes of Exposure

Air

Various subgroups of the population are exposed to different fibre concentrations for varying lengths of time. It is usually assumed that the risks of any two persons are roughly comparable — other factors being equal — if their accumulated fibre burdens are the same. The fibre burden is the accumulated number of critical fibres (F) = fibre concentration (F/m^3) × number of years of exposure × air volume inhaled each year at place of exposure (m^3/year). In this context it is important to note that not all fibres inhaled are deposited and available for retention in the body. However, it is assumed that the fraction of fibres exhaled is more or less the same during all inhalation exposures. Assuming a breathing rate of 10 m^3 for an 8-hour working day and 200 working days per year, a worker inhales 2000 m^3 per year during working hours, while total inhalation for the general population is 7300 m^3 per year *(10)*.

Table 1 gives orders of magnitude of lifetime fibre burdens in various population subgroups, thus indicating the relative importance of different types of inhalation exposure *(10,16)*. Values for indoor air exposure have not been included because only a few are available and details are insufficient. Only for the USA has a calculation of indoor concentration been made in terms of average fibre concentration. The median values range from 400 F/m^3 *(10)* to 500 F/m^3 *(12)*. If these estimates were correct, they would

Table 1. Lifetime fibre burdens typical for industrial countries

Population[a] (%)	Subgroup	Fibre concentration (F/m³)[b]	Exposure time (years)	Inhaled volume (m³/year)	Accumulated critical fibres (F > 5 μm)
70	Urban population (moderate exposure)	30	70	7300	$\sim 1.5 \times 10^7$
25	Rural population	10	70	7300	10^5–10^6
5	Urban population (high exposure)	200	70	7300	$\sim 10^8$
1–2	General construction workers	10^3–10^5[c]	50	2000	10^8–10^{10}[c]
0.1	Asbestos workers	10^5–10^6[c]	50	2000	10^{10}–10^{11}[c]
	Irregular exposure (example)	10^4	0.7	7300	5×10^7

[a] These percentages reflect an assumed population distribution.

[b] F > 5 μm.

[c] Fibre count with optical microscope.

result in a lifetime fibre burden of up to 2×10^8 F. In this case outdoor exposure would be of minor importance to most of the US population.

There can also be more irregular exposure of the general public, with peak concentrations at specific sites. Table 1 shows that such exposure for 1% of a lifetime (0.24 hours per day) could contribute significantly to the total accumulated burden of fibres. Peak exposures in people passing building sites have been reported by several authors. These exposures occurred when asbestos-cement sheets were being cut *(16)* or spray insulation was being carried out *(17)*; in these cases no dust suppression measures were applied. Peak exposures may also occur in para-occupational situations (e.g. household activities *(18)*).

Drinking-water and food

Drinking-water and food may contain asbestos fibres from natural sources (e.g. rock) and man-made sources (e.g. asbestos-cement pipes). The total fibre content (fibres of all lengths) in drinking-water can vary from 10^4 F/litre to more than 10^8 F/litre *(3,4,19,20)*.

Relative significance of different routes of exposure

Inhalation is by far the most important route of exposure of humans to critical fibres, the amount of uptake by ingestion being questionable and, at the least, significantly lower.

Kinetics

Health effects due to asbestos exposure in the occupational environment have been clearly associated with inhalation. The relevance of the oral intake of asbestos fibres for human health is unclear; in any case, ingestion is far less important than inhalation. For this reason, only the deposition, retention and clearance of fibres from the human lungs are described here.

Deposition

Fibres with a diameter greater than 3 μm are not respirable *(10)*. Symmetrical fibres have a lower probability of deposition in the upper airways than nonsymmetrical fibres such as the "curly" chrysotile. A considerable number of fibres are deposited in the upper ciliated airways, where fibres can be removed by cleaning mechanisms *(10)*. Chrysotile fibres can split into fibrils and undergo partial dissolution within the lungs. The amphiboles do not subdivide into fibrils of smaller diameter or break up by length. They are much less soluble in lung fluids, and they have long residence times in the lungs *(10)*.

Clearance

Several mechanisms are involved in the clearing of fibres from their site of deposition, i.e. mucociliary clearance, translocation of alveolar macrophages containing small fibres, and uptake by epithelial cells lining the airways *(17)*. Overall, the clearing mechanisms are very effective (95–98%), although in the alveolar regions some fibres can remain. It should be noted, however, that one of the most effective clearance mechanisms (mucociliary clearance) is impaired by smoking *(1)*. Deposited fibres less than 5–10 μm long may often be engulfed by a single macrophage and thereafter be translocated; fibres longer than this are more difficult to clear. Long fibres may, however, become chemically dissolved or even mechanically broken down in the lung tissue. Durability seems to be greatest for amphiboles, less for chrysotiles *(9,12,21,22)*; chrysotile can, in fact, split readily into fibrils and undergo partial dissolution in lung fluids *(1)*.

The number of fibres retained in the lungs has been investigated very intensively in post-mortem lung material. All sorts of fibres known to exist in the atmospheric aerosol, including asbestos, have been found. The number of asbestos fibres per cm^3 of lung sample is related to past exposure, the ratio of lung burdens for asbestos workers and controls being roughly the same as that of their respective exposures *(10,12)*.

Health Effects

There is an intensive and well documented series of case studies and epidemiological observations which link past occupational exposure to asbestos with asbestosis, lung cancer and mesothelioma *(1,3,9,10,12,23)*. Another series of studies has mainly examined subgroups of the population in particular exposure situations, mostly indoor exposure or para-occupational exposure (see pp. 185–187).

Asbestosis is a slowly developing fibrosis of the lung caused by the inhalation of high concentrations of asbestos dust and/or long exposure *(24)*. Its severity depends both on the length of time since onset of exposure and on the intensity of the latter *(23)*. Although nonmalignant in itself, advanced asbestosis is often associated with lung cancer, especially among smokers. Hypoxia with cor pulmonale does occur in severe cases, while mild forms may not necessarily be associated with marked disablement *(10)*.

Lung cancer (bronchial carcinoma) is the most frequent kind of cancer in the male population (accounting for about 10% of all male deaths in many industrialized countries) and is clearly related to external factors such as smoking, ionizing radiation (e.g. radon) and occupational exposures to certain substances, including asbestos inhalation, in the latter case even without co-existing asbestosis *(9)*. Many studies have shown that smokers have a higher risk of developing lung cancer than nonsmokers when exposed to asbestos *(23)*.

Mesothelioma is a malignant tumour of the pleura or peritoneum *(9,23)*. It is a rare type of cancer (less than 0.04% of all deaths in the general population of the USA) *(10)*. A higher incidence of mesothelioma has nearly always been related to the inhalation of mineral fibres, and in the majority of cases to occupational asbestos exposure *(9)*. Causing few symptoms initially, mesothelioma is incurable when diagnosed *(10)*.

Effects on experimental animals and *in vitro* test systems

Experimental toxicology has developed models to establish the relative carcinogenic potency of different types and sizes of fibres and to examine possible interactive effects with other airborne pollutants *(3)*. This approach has proved helpful in the interpretation of epidemiological studies. Besides inhalation studies, ingestion and implantation studies have also been carried out; these have additionally been used to explore the association between asbestos and other forms of cancer in man, e.g. cancers of the gastrointestinal tract, which are epidemiologically less well established *(9,23)*.

Species vary in their response to asbestos. Effects in rats seem to resemble those in man more closely than those in other animals *(9)*; if not otherwise stated, the results reported refer to this species.

Inhalation

Lung cancer, mesothelioma and asbestosis have been observed in several studies regarding various asbestos and other fibres *(10)*. In three inhalation experiments with 10 mg chrysotile per m^3 over a period of 1 year, at least 20% of exposed rats developed lung tumours *(25–27)*.

Intrapleural and intraperitoneal injection

The direct injection of a specific amount of dust with a given particle-size distribution into the pleura or the peritoneum makes it possible to study carcinogenic potency as a function of fibre type and shape without interference from variable factors such as deposition, clearance and disintegration, which are difficult to control in inhalation experiments.

The injection of asbestos produced mesotheliomas, the resulting incidence being highly fibre-specific and sensitive *(3,9)*. Intraperitoneal injection of 1 mg chrysotile (UICC Canada) or 0.3 mg Actinolite resulted in a tumour rate of about 90% and 80%, respectively *(28)*. Experiments show that fibre length and diameter are the most important factors in causing mesothelioma. As shown by Pott in his graph of the carcinogenicity factor *(29)*, fibres shorter than 5 μm and thicker than 2 μm elicit little response, while those longer than 10 μm and thinner than 0.5 μm yield the highest response. Nonasbestos fibres of the same critical dimensions also caused mesothelioma in animals, while experiments with fibres of noncritical dimensions are negative, irrespective of fibre category *(29)*. This finding is important considering the presence of many other fibre types in ambient air.

Ingestion

In several studies, rats and other laboratory animals were exposed to asbestos in diet or drinking-water. Only a single — usually very high — dose was applied in each investigation. The results are inconclusive, if not negative *(1,20)*.

Effects on humans

Asbestosis

While early stages of asbestosis are still observed at asbestos-cement plants under modern regulations *(24)*, asbestosis has never been observed in relation to nonoccupational asbestos exposure. It is therefore concluded that environmental concentrations of asbestos are not sufficient to induce asbestosis *(1,9)*.

Mesothelioma

Without doubt, exposure to all kinds of asbestos is closely linked to mesothelioma. Several epidemiological studies indicate that amphibole asbestos is more potent in inducing mesothelioma than chrysotile *(1)*, which seems to contradict the findings from intraperitoneal injection experiments *(9)*. The contradiction is, however, to a great extent resolved by the observation that technical processes involving amphiboles produce dust with a higher proportion of long, thin — i.e. more dangerous — fibres *(9)*. Furthermore, amphibole fibres are more resistant in the body and have a lower lung clearance than chrysotile *(9)*.

The time elapsing between first exposure to asbestos and the clinical manifestation of tumours ranges from 20 to 50 years for mesothelioma in the populations of workers studied *(23)*. Dose–response relationships are derived from retrospective epidemiological studies of exposure data relating to situations several decades ago. Clearly these exposure data can only be approximations *(9,15)*. Increased incidence rates seen in nonoccupationally exposed people living in the same household as asbestos workers or in the vicinity of strong asbestos emission sources have been attributed to this exposure *(30)*.

Lung cancer

Clarifying the relationship between asbestos exposure and lung cancer is a much more complicated task. Both the synergism of smoking and asbestos inhalation and the high level of lung cancer in the general population (background exposure) must be taken into account. The synergistic effect can be described by a multiplicative model *(31)* which is used below in connection with the extrapolation of lung cancer risk. The risk of lung cancer seems to rise consistently from mining and milling operations, through branches such as asbestos-cement production, to asbestos textile and insulation work, according to the increasing portion of more dangerously shaped fibres in these processes (see Table 2).

Other cancers

Recent evaluations *(9,10,20,23)* of epidemiological occupational studies provide little evidence for the induction of other cancers, including gastrointestinal cancers, although recently it has been suggested that laryngeal cancer may be related to heavy occupational exposure to asbestos *(23)*. Clearly the risk, if any, for the general population from "other" cancers is very small *(9,20,45)*.

Evaluation of Human Health Risks

Exposure

Actual indoor and outdoor concentrations in air range from below one hundred to several thousand fibres per m^3.

Table 2. Increase in the relative risk of lung cancer, as shown by different studies

K_L per 100 F*year/ml	Type of activity	Reference
0.04	mining and milling	*(33)*
0.045	mining and milling	*(34)*
0.06	friction material	*(35)*
0.1	factory processes	*(36,37)*
(M) 0.4–1.1	factory processes	*(38)*[b]
(F) 2.7[a]	factory processes	
0.2	asbestos-cement	*(39)*
0.07	textiles (before 1951)	*(40)*
0.8[a]	textiles (after 1950)	
6(M) 1.6[a]	textiles	*(41)*
1.6	textiles	*(42)*[c]
1.1	insulation products	*(43)*[b]
1.5	insulation	*(44)*[b]

[a] Fewer than 10 cases of lung cancer expected (i.e. small cohort).

[b] Inadequate knowledge of actual fibre concentrations.

[c] Same factory as in *(41)*, but larger cohort.

Source: based on a table by Liddell *(32)*.

Health risk evaluation

On the basis of the evidence from both experimental and epidemiological studies, it is clear that asbestos inhalation can cause asbestosis, lung cancer and mesothelioma. The evidence that ingested asbestos causes gastrointestinal or other cancers is insufficient. Furthermore, the carcinogenic properties of asbestos are most probably due to its fibre geometry and remarkable integrity. Other fibres with the same characteristics may also be carcinogenic.

Current environmental concentrations of asbestos are not considered a hazard with respect to asbestosis. However, a risk of mesothelioma and lung cancer from the current concentrations cannot be excluded.

A WHO Task Group recently expressed reservations about the reliability of risk assessment models applied to asbestos risk. Its members suggested that such models can only be used to obtain a broad approximation of the lung cancer risk of environmental exposures to asbestos and "that any number generated will carry a variation over many orders of magnitude". The same was found to be true for estimates of the risk of mesothelioma. The same document stated: "In the general population the risks of mesothelioma and lung cancer attributable to asbestos cannot be quantified reliably and probably are undetectably low." *(1)*.

The following estimates of risk are based on the relatively large amount of evidence from epidemiological studies concerning occupational exposure. Data from these studies have been conservatively extrapolated to the much lower concentrations found in the general environment. Although there is evidence that chrysotile is less potent than amphiboles, as a precaution chrysotile has been attributed the same risk in these estimates.

Mesothelioma

A formula by which the excess incidence of mesothelioma can be approximated has been derived by Peto *(46)*. Fibre concentration, duration of exposure and time since first exposure are parameters incorporated in this model, which assumes a linear dose–response relationship. Peto verified this model from data on an urban population exposed for its whole life and on workers exposed for many decades. In both cases, duration of exposure is assumed to be equal or close to time since first exposure. The data show that the incidence of mesothelioma is proportional to the fibre concentration to which the workers were exposed and to $time^{3.5}$ since first exposure for both workers and the general population. Starting from this relationship, one may calculate the risk of lifetime exposure to environmental concentrations from the incidence of mesothelioma in occupational populations exposed to much higher concentrations, but for a shorter time.

Apart from incomplete knowledge about the true workplace exposure, a further complication arises from the fact that workplace concentrations were measured by means of an optical microscope, counting only fibres longer than 5 μm and thicker than, say, 0.5 μm.[a]

Several studies have been performed to calculate the risk of mesothelioma resulting from nonoccupational exposure to asbestos. Lifetime exposure to 100 F^*/m^3 has been estimated by various authors to carry differing degrees of mesothelioma risk (see Table 3).

The risk estimates in Table 3 differ by a factor of 4. A "best" estimate may be 2×10^{-5} for 100 F^*/m^3.

An independent check of this risk estimate can be made by calculating the incidence of mesothelioma in the general population, based on a hypothetical average asbestos exposure 30–40 years ago *(49)*. If the latter had been 200–500 F^*/m^3 (corresponding to about 400–1000 F/m^3 as measured today), the resulting lifetime risk of mesothelioma would be $(4–10) \times 10^{-5}$. With the average United States death rate of 9000×10^{-6} per year, this would give 0.4–0.9 mesothelioma cases each year per million persons from past

[a] In this chapter all fibre concentrations based on optical microscopy are marked F^*/m^3 and risk estimates will be based on F^*/m^3. If concentrations measured by optical microscope are to be compared with environmental fibre concentrations measured by a scanning electron microscope, a conversion factor has to be used, i.e. $2\ F/m^3 = 1\ F^*/m^3$.

Table 3. Estimates of mesothelioma risk resulting from lifetime exposure to asbestos

Risk of mesothelioma from 100 F^*/m^3	Values in original publication (risk for fibre concentration indicated)	Reference
1.0×10^{-5}	1.0×10^{-4} for 1000 F^*/m^3	*(45)*
$\sim 2.0 \times 10^{-5}$	1.0×10^{-4} for (130–800) F^*/m^3	*(47)*
$\sim 3.9 \times 10^{-5}$	1.56×10^{-4} for 400 F^*/m^3	*(10,48)*
$\sim 2.4 \times 10^{-5}$	2.75×10^{-3} (females), 1.92×10^{-3} (males) } for 0.01 F/ml	*(17)*

environmental asbestos exposure. The reported mesothelioma incidence in the USA ranges from 1.4×10^{-6} per year to 2.5×10^{-6} per year according to various authors *(10,49)*. Thus, the calculated risk figures would account for only part of the observed incidence. However, other factors which may account for this discrepancy must be considered.

- Uncertainties in the risk extrapolations result from the lack of reliable exposure data in the cohort studies, errors in the medical reports, and necessary simplifications in the extrapolation model itself *(17)*. Furthermore, the amount of past ambient exposure can only be an educated guess.

- The incidence of nonoccupational mesotheliomas is calculated from the difference between the total of observed cases and the number of those probably related to occupational exposure. Neither of these two figures is exactly known. Moreover, the influence of other environmental factors in the generation of mesothelioma is unknown.

In the light of these uncertainties, the result obtained by using the risk estimate can be considered to be in relatively good agreement with the annual mesothelioma death rate based on national statistical data.

Lung cancer

Unlike mesothelioma, lung cancer is one of the most common forms of cancer. As several exogenous noxious agents can be etiologically responsible for bronchial carcinoma, the extrapolation of risk and comparison between different studies is considerably complicated. In many epidemiological studies, in particular, the crucial effect of smoking has not been properly taken into account. Differentiation of the observed risks according to smoking habits has been carried out, however, in the cohort of North American insulation workers studied by Hammond et al. *(31)*.

This study suggests that the relative risk at a given time is approximately proportional to the cumulative amount of fine asbestos dust received up to this point, for both smokers and nonsmokers. The risks for non-asbestos-exposed nonsmokers and smokers must therefore be multiplied by a factor which increases in proportion to the cumulative exposure.

The dose–response relationship in the case of asbestos-induced lung cancer can be described by the following equation *(17)*.

$$I_L \text{ (age, smoking, fibre dose)} = I_L^o\text{(age, smoking)}[1 + K_L \times C_f \times d]$$

This equation could also be written as:

$$K_L = [(I_L/I_L^o) - 1]/C_f \times d = \text{(relative risk} - 1\text{)/(cumulative exposure)}$$

where:

K_L = a proportionality constant, which is a measure of the carcinogenic potency of asbestos

C_f = fibre concentration

d = duration of exposure, in years

I_L = lung cancer incidence, observed or projected, in a population exposed to asbestos concentration C_f during time d

I_L^o = lung cancer incidence expected in a group without asbestos exposure but with the same age and smoking habits (this factor includes age dependence).

There are several studies which allow the calculation of K_L. Liddell *(32,50)* has done this in an interesting and consistent manner. The results are given in Table 2 (p. 189).

Taking the data in Table 2 as a basis, a reasonable estimate for K_L is 1.0 per 100 F*years/ml. For a given asbestos exposure the risk for smokers is about 10 times that for nonsmokers *(31)*. In extrapolating from workers to the general public a factor of 4 for correction of exposure time has to be applied to K_L.

The incidence of lung cancer in the general population exposed to 100 F*/m³ is calculated as follows:

$$I_L = I_L^o(1 + 4 \times 0.01 \times 10^{-4} \text{ F}^*\text{/ml} \times 50 \text{ years})$$

or

$$I_L = I_L^o(1 + 2 \times 10^{-4} \text{ F}^*\text{/ml})$$

The extra risk is $I_L - I_L^o$. Values for I_L^o are about 0.1 for male workers and 0.01 for male nonsmokers *(10)*.

Lifetime exposure to 100 F*/m^3 (lifetime assumed to be 50 years; since, in a lifetime of 70 years, the first 20 years without smoking probably do not make a large contribution) is therefore estimated as follows.

Status	Risk of lung cancer per 100 000	Range (using the highest and lowest values of K_L from Table 2)
Smokers	2.0	0.08–3.2
Nonsmokers	0.2	0.008–0.32

This risk estimate can be compared, when adjusted to 100 F*/m^3, with estimates for male smokers made by other authors or groups:

Breslow (National Research Council) *(48)*: 7.3 × 10^{-5};

Schneiderman et al. *(47)*: (14–1.4) × 10^{-5};

US Environmental Protection Agency *(17)*: 2.3 × 10^{-5}.

A fibre concentration of 100 F*/m^3 (about 200 F/m^3 as seen by scanning electron microscope) thus gives a total risk of (2 + 2) × 10^{-5} for smokers or 2.2 × 10^{-5} for nonsmokers.

Guidelines

Asbestos is a proven human carcinogen (IARC Group 1). No safe level can be proposed for asbestos because a threshold is not known to exist. Exposure therefore should be kept as low as possible.

Several authors and working groups have produced estimates indicating that, with a lifetime exposure to 1000 F/m^3 (0.0005 F*/ml or 500 F*/m^3, optically measured) in a population of whom 30% are smokers, the excess risk due to lung cancer would be in the order of 10^{-6}–10^{-5}. For the same lifetime exposure, the mesothelioma risk for the general population would be in the range of 10^{-5}–10^{-4}. These ranges are proposed with a view to providing adequate health protection, but their validity is difficult to judge. An attempt to calculate a "best" estimate for the lung cancer and mesothelioma risk is described above, beginning on p. 191.

References

1. *Asbestos and other natural mineral fibres.* Geneva, World Health Organization, 1986 (Environmental Health Criteria, No. 53).
2. *Asbestos.* Paris, OECD, 1984, p. 21 (Env/Air 81.18; 2nd Rev.).
3. **Umweltbundesamt** [Federal Office for the Environment]. *Umweltbelastung durch Asbest und andere faserige Feinstäube* [Environmental pollution by asbestos and other fibrous fine dusts]. Berlin (West), Erich Schmidt Verlag, 1980 (Berichte 7/80).

4. **Meyer, E.** Vorkommen von Asbestfasern im Trinkwasser [Occurrence of asbestos fibres in drinking-water]. *In:* Fischer, M. & Meyer, E., ed. *Zur Beurteilung der Krebsgefahr durch Asbest* [Assessment of the cancer risk of asbestos]. Munich, Medizin Verlag, 1984, pp. 62–78.
5. **Fischer, M. & Meyer, E.** The assessment of the health risk from asbestos fibres by the Federal Health Office of the Federal Republic of Germany. *VDI-Berichte,* **475**: 325–330 (1983).
6. **Marfels, H. et al.** Imissionsmessungen von faserigen Stäuben in der Bundesrepublik Deutschland — I. Messungen in der Nähe einer Industriequelle [Measurements of fibrous dusts in ambient air of the Federal Republic of Germany. I. Measurements in the vicinity of an industrial source]. *Staub, Reinhaltung der Luft,* **44**: 259–263 (1984).
7. **Lohrer, W.** Asbestbelastete Innenräume — Analyse und Bewertung des Gefahrenpotentials [Asbestos exposed interiors — analysis and evaluation of risk potentials]. *Staub, Reinhaltung der Luft,* **43**: 434–438 (1983).
8. **Nicholson, W.J., et al.** *Asbestos contamination of the air in public buildings.* Research Triangle Park, NC, US Environmental Protection Agency, 1975 (Final report, contract 68-02-1346).
9. *Report of the Royal Commission on Matters of Health and Safety Arising from the Use of Asbestos in Ontario.* Toronto, Ontario Ministry of the Attorney General, 1984, Vol. 1–3.
10. **National Research Council.** *Asbestiform fibers: nonoccupational health risks.* Washington, DC, National Academy Press, 1984.
11. **Sawyer, R.N.** Indoor air pollution: application of hazard criteria. *Annals of the New York Academy of Sciences,* **330**: 579–586 (1979).
12. **Commins, B.T.** *The significance of asbestos and other mineral fibres in environmental ambient air.* Maidenhead, Commins Ass., 1985.
13. **Chatfield, E.J.** Measurement and interpretation of asbestos fibre concentrations in ambient air. *In: Proceedings of the 5th Colloquium on Dust Measuring Technique and Strategy, October 1984.* Johannesburg, South African Asbestos Producers Advisory Committee, 1985, pp. 269–296.
14. **Marfels, H. et al.** Imissionsmessungen von faserigen Stäuben in der Bundesrepublik Deutschland — II. Messungen an der Kreuzung einer Großstadt [Measurements of fibrous dusts in ambient air of the Federal Republic of Germany. II. Measurements on a busy crossing of a large town]. *Staub, Reinhaltung der Luft,* **44**: 410–414 (1984).
15. **Nicholson, W.J.** Asbestos and inorganic fibres. *Arbete och hälsa,* **17** (1981).
16. **Woitowitz, K.J. & Rödelsperger, K.** Asbestemissionen im Grenzbereich von Arbeitsplatz und Umwelt [Asbestos emissions in the borderline area of workplace and environment]. *In:* Fischer, M. & Meyer, E., ed. *Zur Beurteilung der Krebsgefahr durch Asbest* [Assessment of the cancer risk of asbestos]. Munich, Medizin Verlag, 1984.
17. *Airborne asbestos health assessment update.* Research Triangle Park, NC, US Environmental Protection Agency, 1985 (EPA-600/8-84-003F).

18. **Lilis, R. & Selikoff, I.J.** Asbestos disease in occupationally exposed workers and household contacts. *In:* Fischer, M. & Meyer, E., ed. *Zur Beurteilung der Krebsgefahr durch Asbest* [Assessment of the cancer risk of asbestos]. Munich, Medizin Verlag, 1984.
19. **Commins, B.T.** *Asbestos fibres in drinking water.* Maidenhead, Commins Ass., 1983.
20. **Toft, P. et al.** Asbestos in drinking water. *Critical reviews in environmental control,* **14**(2): 151–197 (1984).
21. **Friedberg, K.D. & Ulmer, S.** Deposition und Elimination von Staub aus faserförmigem Material in der Lunge von Versuchstieren [Deposition and elimination of fibrous material dust in the lungs of experimental animals]. *VDI Berichte,* **475**: 269–274 (1983).
22. **Lippmann, M.** Peer review: inhalation and elimination of MMMF — aids to the understanding of the effects of MMMF. *In: Biological effects of man-made mineral fibres.* Copenhagen, WHO Regional Office for Europe, 1984, Vol. 2, pp. 355–366.
23. **Doll, R. & Peto, J.** *Asbestos: effects on health of exposure to asbestos.* London, H.M. Stationery Office, 1985.
24. **Weicksel, P. & Ulrich, E.** Asbeststaubgefährdung in der modernen asbestverarbeitenden Industrie [The hazard of asbestos dust in the modern asbestos-processing industry]. *In:* Fischer, M. & Meyer, E., ed. *Zur Beurteilung der Krebsgefahr durch asbest* [Assessment of the cancer risk of asbestos]. Munich, Medizin Verlag, 1984.
25. **Davis, J.M.G. et al.** Mass and number of fibres in the pathogenesis of asbestos-related lung disease in rats. *British journal of cancer,* **37**: 673–688 (1978).
26. **Wagner, J.C. et al.** Animal experiments with MMM(V)F fibres — effects of inhalation and intrapleural inoculation in rats. *In: Biological effects of man-made mineral fibres.* Copenhagen, WHO Regional Office for Europe, 1984, Vol. 2, pp. 209–233.
27. **McConnell, E.E. et al.** A comparable study of the fibrogenic and carcinogenic effects of UICC Canadian chrysotile B asbestos and microfibres (JM 100). *In: Biological effects of man-made mineral fibres.* Copenhagen, WHO Regional Office for Europe, 1984, Vol. 2, pp. 234–250.
28. **Pott, F. et al.** Carcinogenicity studies on fibres, metal compounds, and some other dusts in rats. *Experimental pathology,* **32**: 129–152 (1987).
29. **Pott, F.** Some aspects on the dosimetry of the carcinogen potency of asbestos and other fibrous dusts. *Staub, Reinhaltung der Luft,* **38**: 486–490 (1978).
30. **Hain, E.** Untersuchungen über gesundheitliche Asbestschäden in Hamburg (1969–1979) [Studies of health damage caused by asbestos in Hamburg (1969–1979)]. *In:* Fischer, M. & Meyer, E., ed. *Zur Beurteilung der Krebsgefahr durch Asbest* [Assessment of the cancer risk of asbestos]. Munich, Medizin Verlag, 1984.
31. **Hammond, E.C. et al.** Asbestos exposure, cigarette smoking and death rates. *Annals of the New York Academy of Sciences,* **330**: 473–490 (1979).

32. **Liddell, F.D.K.** Some new and revised risk extrapolations from epidemiological studies on asbestos workers. *In:* Fischer, M. & Meyer, E., ed. *Zur Beurteilung der Krebsgefahr durch Asbest* [Assessment of the cancer risk of asbestos]. Munich, Medizin Verlag, 1984.
33. **McDonald, J.C. et al.** Dust exposure and mortality in chrysotile mining, 1910–1975. *British journal of industrial medicine,* **37**: 11–24 (1980).
34. **Nicholson, W.J. et al.** Long-term mortality experience of chrysotile miners and millers in Thetford Mines, Quebec. *Annals of the New York Academy of Sciences,* **330**: 11–21 (1979).
35. **Berry, G. & Newhouse, M.L.** Mortality of workers manufacturing friction materials using asbestos. *British journal of industrial medicine,* **40**: 1–7 (1983).
36. **Henderson, V.L. & Enterline, P.E.** Asbestos exposure: factors associated with excess cancer and respiratory disease mortality. *Annals of the New York Academy of Sciences,* **330**: 117–126 (1979).
37. **Enterline, P. et al.** Mortality in relation to occupational exposure in the asbestos industry. *Journal of occupational medicine,* **14**: 897–903 (1972).
38. **Newhouse, M.L. & Berry, G.** Patterns of mortality in asbestos factory workers in London. *Annals of the New York Academy of Sciences,* **330**: 53–60 (1979).
39. **Weill, H. et al.** Influence of dose and fiber type on respiratory malignancy risk in asbestos cement manufacturing. *American review of respiratory diseases,* **120**: 345–354 (1979).
40. **Peto, J.** Lung cancer mortality in relation to measured dust levels in an asbestos textile factory. *In:* Wagner, J.C., ed. *Biological effects of mineral fibres.* Lyon, International Agency for Research on Cancer, 1980 (IARC Scientific Publications, No. 30).
41. **Dement, J.M. et al.** Estimates of dose–response for respiratory cancer among chrysotile asbestos textile workers. *Annals of occupational hygiene,* **26**: 869–887 (1982).
42. **Fry, J.S. et al.** Respiratory cancer in chrysotile production and textile manufacture. *Scandinavian journal of work, environment & health,* **9**: 68–70 (1983).
43. **Seidman, H. et al.** Short-term asbestos work exposure and long-term observation. *Annals of the New York Academy of Sciences,* **330**: 61–67 (1979).
44. **Selikoff, J.J. et al.** Mortality experience of insulation workers in the United States and Canada, 1943–1976. *Annals of the New York Academy of Sciences,* **330**: 91–116 (1979).
45. **Aurand, K. & Kierski, W.-S., ed.** *Gesundheitliche Risiken von Asbest.* Eine Stellungnahme des Bundesgesundheitsamtes Berlin [Health risks of asbestos. A position paper of the Federal Health Office, Berlin]. Berlin (West), Dietrich Reimer Verlag, 1981 (BgA-Berichte, No. 4/81).
46. **Peto, J.** Dose and time relationships for lung cancer and mesothelioma in relation to smoking and asbestos exposure. *In:* Fischer, M. & Meyer, E., ed. *Zur Beurteilung der Krebsgefahr durch Asbest* [Assessment of the cancer risk of absestos]. Munich, Medizin Verlag, 1984.

47. **Schneiderman, M.S. et al.** *Assessment of risks posed by exposure to low levels of asbestos in the general environment.* Berlin (West), Dietrich Reimer Verlag, 1981 (BgA-Bericht, No. 4/81).
48. **Breslow, L. et al.** Letter. *Science,* **234**: 923 (1986).
49. **Enterline, P.E.** Cancer produced by nonoccupational asbestos exposure in the United States. *Journal of the Air Pollution Control Association,* **33**: 318–322 (1983).
50. **Liddell, F.D.K. & Hanley, J.A.** Relations between asbestos exposure and lung cancer SMRs in occupational cohort studies. *British journal of industrial medicine,* **42**: 389–396 (1985).

19

Cadmium

General Description

Cadmium (Cd) is a soft, silver-white metal. It has a relatively high vapour pressure. In air the vapour is rapidly oxidized to cadmium oxide. Many inorganic cadmium compounds are soluble in water (e.g. cadmium sulfate, cadmium nitrate, cadmium chloride), whereas cadmium sulfide and cadmium oxide are almost insoluble in water.

Sources

Cadmium occurs together with zinc in nature, the ratio being generally 1 : 100–1 : 1000. Cadmium is obtained as a byproduct in the refining of zinc; in certain zinc ores the cadmium concentration may be about 5% *(1)*.

Since 95% of primary cadmium is obtained from zinc production, it is the level of production of zinc and not the demand for cadmium that governs the supply of cadmium *(2)*. Before the First World War cadmium was usually not recovered from zinc plants or other nonferrous plants; contamination of the environment by cadmium has thus been taking place for a long time. In this century there has been increasing interest in cadmium, and the average annual production throughout the world was about 12 000 tonnes in 1960–1969 *(1)* and 15 000–20 000 tonnes in 1980–1985 *(3)*.

Cadmium compounds are used in the electroplating of metals, as pigments or stabilizers in plastics, in alkaline batteries and in alloys with other metals such as copper.

In Europe estimates have been made of the atmospheric cadmium emissions from different sources in the Member States of the European Community *(4)*. The steel industry, waste incineration, volcanic action (Mount Etna) and zinc production seem to account for the largest emissions. As incineration is increasingly chosen as a method of refuse disposal in European countries, this source of atmospheric cadmium pollution is of growing concern. The attempt to ban cadmium in Sweden was mainly aimed at reducing cadmium emissions from such disposal.

Within the European Community there are 18 primary zinc production plants. Zinc plants, as well as other metal-producing plants, are distributed throughout Europe; in particular, eastern Belgium and the Ruhr area are examples of regions with a conglomerate of industries emitting cadmium.

Occurrence in air

In a number of reports *(1,4–6)*, data on cadmium in air have been compiled. Generally the yearly means in rural areas range from <1 to 5 ng/m^3, whereas in urban areas 5–15 ng/m^3 and in industrialized areas 15–50 ng/m^3 have been reported. Much higher concentrations have been measured near industries processing metals. Weekly means of 300 ng/m^3 have been reported *(6)*, as well as short-term values of 5–11 μg/m^3 *(4,5)*. A compilation of measurement data for Member States of the European Community gives ranges of 0.1–1 ng/m^3 for remote areas, 1–50 ng/m^3 for urban areas and 1–100 ng/m^3 for industrial areas *(7)*. The concentrations of cadmium in air will be reflected in dustfall measurements. In Denmark the average concentration in ambient air was found to be 3 ng/m^3 and the average precipitation of cadmium was 0.2 mg/m^2 per year *(6)*. In northern Europe the levels of deposited cadmium ranged from 0.0001 to 0.0006 mg/m^2 per month, whereas near cadmium-processing plants values of 0.5–5 mg/m^2 per month were found *(5)*. Daily deposition values in Member States of the European Community were in the range of 0.02–0.8 μg/m^2 in remote areas, 0.4–30 μg/m^2 in urban areas and 0.7–200 μg/m^2 in industrial areas *(7)*, corresponding to yearly deposition values of 0.007–0.29, 0.15–11 and 0.25–73 mg/m^2. In a pathway analysis it was estimated that 1 ng/m^3 would cause the deposition of 0.16 mg/m^2 per year *(8)*.

The distribution of cadmium pollution in northern Europe has been studied in detail by conducting analyses of cadmium in moss *(5)*. In southern Sweden and Norway concentrations were around 1 mg cadmium per kg dry weight, whereas above the Arctic Circle concentrations were around 0.1 mg cadmium per kg dry weight. Around cadmium-emitting point sources much higher concentrations have been found *(5)*.

Routes of Exposure

Air

Assuming an air concentration of 50 ng cadmium per m^3, a daily inhalation of 20 m^3 of air, and indoor concentrations similar to outdoor concentrations (even in industrialized areas), the average daily intake of cadmium via inhalation would not be more than 1 μg. Less than 50% of the inhaled amount is expected to be absorbed from the lungs. In the following an absorption factor of 25% will be used *(5)*.

Tobacco contains cadmium and smoking may contribute significantly to the uptake of cadmium. Cigarettes may contain from 0.5 to 3 μg cadmium per gram of tobacco, depending on the country of origin *(9,10)*. Smoking a pack of 20 cigarettes a day may result in the inhalation of 1–6 μg cadmium per day, since about 10% of the cadmium is in the mainstream smoke; 25–50% may be absorbed via the lungs, i.e. up to 3 μg per day. Cigarettes smoked in Europe generally contain 1–2 μg cadmium per gram.

Drinking-water

Drinking-water normally contains very low concentrations of cadmium, usually ranging between 0.1 and 2.0 μg/litre *(3)*; levels up to 5 μg/litre have occasionally been reported and on rare occasions levels up to 10 μg/litre have been detected. In some areas, well-water may contain elevated concentrations of cadmium *(11)*.

The estimated daily exposure to cadmium via water, based on a water consumption of 2 litres per day, ranges from substantially less than 1 μg to over 10 μg per day.

Food

The daily intake via food has been well documented *(4,5,12)*. In European countries and in North America the average intake is 10–30 μg/day, but there may be large individual variations depending on age and dietary habits. In Japan the average intake is generally 40–50 μg, but may be much higher in severely polluted areas. About 5% of ingested cadmium will normally be absorbed by adults, but higher absorption (up to 20%), has been found in women with severe iron deficiency *(13)*.

Relative significance of different routes of exposure

Assuming a gastrointestinal absorption rate of 5% and 10 μg cadmium ingested, this would result in an absorbed amount of 0.5 μg, i.e. twice the amount of cadmium absorbed from air in an industrialized area with relatively high cadmium concentrations in air. Since most people will be exposed to much lower air concentrations of cadmium, it is obvious that dietary exposure is more important.

However, cadmium in food comes partly from cadmium deposited on soils as fallout of atmospheric cadmium and thereby entering the food chain. This indirect exposure pathway can be quantified with much less certainty *(8)*. The parameters which describe the multimedia transfer are not well known; indeed, they may vary widely, depending on the various environmental conditions (e.g. soil properties and type of plant).

Estimates based on measurements in several European countries indicate that, at present levels in air in rural areas, fallout contributes relatively little to the background concentration of cadmium in soil. In industrial areas the deposition may be up to 50 times higher than in rural areas *(3)*.

A WHO Working Group *(14)*, by using models, attempted to estimate the increase in dietary cadmium caused by the application of sewage sludge to agricultural lands. A similar approach may be useful for estimating the impact of cadmium deposited from the air.

The present acidification of soils may increase the availability of metals like cadmium. A combination of increased cadmium levels in air and increased soil acidity may in the future contribute significantly to cadmium concentrations in food.

Concerning total exposure, the important contribution which smoking can make should be emphasized. Heavy smoking may contribute as much as or even more cadmium than dietary sources and much more than ambient air.

Kinetics and Metabolism

A major part of cadmium absorbed from the lungs or the intestine will initially be deposited in the liver. Cadmium will be stored there, bound to metallothionein, a low molecular weight protein which, in addition to a high cadmium-binding capacity, can also bind zinc and copper.

From the liver, cadmium is slowly transferred to the kidneys, where eventually the highest cadmium concentrations are found *(5)*. Also in the kidneys cadmium is bound to metallothionein. Excretion from this site is slow. The biological half-time of cadmium was estimated to be around 10 years in the liver and slightly longer in the kidneys *(15)*. Urine cadmium is regarded as a reliable indicator of the body-burden of cadmium *(16)*. An adult nonsmoker will excrete about 0.4 μg cadmium per gram creatinine.

The newborn child is virtually free from cadmium, but during its lifetime there will be a considerable accumulation of cadmium, especially in the kidneys. Among people aged 50 years in Europe and North America, the average cadmium concentrations are generally 10–20 mg/kg wet weight in the renal cortex in nonsmokers and 50–100% higher in smokers *(17,18)*. The highest values in Europe have been reported from the Liège area in Belgium. The average concentration in persons aged 40–59 years, smokers and nonsmokers, was 39 mg/kg wet weight *(18)*. The distributions are skewed and among smokers in some European countries concentrations in the renal cortex have at present upper limits of 100–150 mg/kg wet weight.

It is also of interest that horses accumulate large amounts of cadmium. Adult horses (15–25 years of age) may, even in nonpolluted areas, have renal levels of cadmium which are about five times higher than in adult humans who are nonsmokers *(12)*. The critical concentration in the horse renal cortex is probably about 200 mg/kg, a concentration that has been exceeded in some European horses *(12)*.

Health Effects

Toxicological effects

Acute respiratory effects (chemical pneumonitis) may be expected at cadmium fume concentrations in air above 1 mg/m^3 *(1,5)*. Chronic respiratory effects may be expected after occupational exposure to 20 μg cadmium per m^3 for about 20 years *(1)*.

In long-term low-level exposure the kidney is regarded as the critical organ and it has been estimated that renal dysfunction may appear when the cadmium concentration in the renal cortex is around 200 mg/kg wet weight *(5,19–21)*.

The proportion of populations, exposed to various levels of cadmium via air and food, that would be expected to have cadmium concentrations above their individual critical concentration, has been calculated on the basis of a more elaborate metabolic model and data on the statistical distribution of cadmium concentrations in the human kidney cortex *(21)*. A 10% prevalence of effects (after 45 years of exposure) was estimated at a daily intake of about 200μg cadmium via food only. The same response rate would occur after 10 years' occupational exposure to 50μg cadmium per m^3.

The renal accumulation of cadmium affects the reabsorption capacity of the renal tubules and the first sign of renal toxicity is an increased excretion of low molecular weight proteins, "tubular proteinuria". Further accumulation may cause disturbances in the reabsorption of amino acids, glucose and minerals *(1,5)*. The renal dysfunction is irreversible and the only way to avoid chronic renal dysfunction is to prevent cadmium concentrations in the renal cortex from reaching the critical concentrations of around 200 mg/kg wet weight.

Since urinary excretion of cadmium is related to body-burden or renal burden, monitoring urinary cadmium is an effective way of detecting risks of renal dysfunction. The critical concentration in the urine is 10μg cadmium per gram of creatinine *(5,19)*, but to be safe a limit of 5μg cadmium per gram of creatinine should be used.

Cadmium can also be determined in liver and kidneys by *in vivo* neutron activation *(19,22)*. Table 1 shows that there is a clear dose–response relationship between liver concentrations of cadmium and the prevalence of tubular proteinuria *(22)*.

Renal concentrations of cadmium did not show the same clear relationship *(22)* because renal cadmium concentrations will decrease when there is advanced tubular damage *(5)*.

As seen from Table 1, tubular proteinuria appears at liver concentrations above 30 mg/kg wet weight. The corresponding concentration found in the renal cortex was about 200 mg/kg wet weight. In occupational exposure the ratio of liver cadmium to renal cadmium is greater than that

Table 1. Relationship between liver cadmium level and prevalence of abnormal $_2$-microglobulinuria ($_2$-mU) in a group of 148 workers from two zinc-cadmium smelters with hepatic cadmium ≧ 10 ppm and renal cortical cadmium ≧ 50 ppm

<table>
<tr><th>Cadmium in liver (ppm)</th><th>No. of workers</th><th></th><th>Prevalence of abnormal $_2$-mU (%)</th><th></th></tr>
<tr><td>10–19</td><td>54</td><td></td><td>0</td><td></td></tr>
<tr><td>20–29</td><td>27</td><td></td><td>4</td><td></td></tr>
<tr><td>30–39</td><td>28</td><td></td><td>11</td><td></td></tr>
<tr><td>40–49</td><td>18</td><td></td><td>17</td><td></td></tr>
<tr><td>50–59</td><td>8</td><td></td><td>25</td><td></td></tr>
<tr><td>60–69</td><td>5</td><td rowspan="2">13</td><td rowspan="2"></td><td rowspan="2">77</td></tr>
<tr><td>70–160</td><td>8</td></tr>
</table>

Source: Roels, H.A. et al. *(22)*.

seen in lifetime low-level exposures *(5)*, and in the general population a no-effect level in the liver is probably around 10 mg/kg wet weight. Present levels of liver cadmium are, on an average, about 1 mg/kg wet weight.

Other effects, e.g. teratogenic, will be prevented if the renal burden of cadmium is kept low. There is no evidence that cadmium has a direct teratogenic effect. The placenta is an effective barrier and the newborn is virtually free from cadmium. However, an increase in the body-burden of cadmium will have an influence on zinc metabolism, which may cause zinc deficiency in the fetus *(5)*.

Carcinogenic effects

IARC has classified cadmium as a Group 2B carcinogen *(23,24)*, having concluded that there was sufficient evidence of cadmium being carcinogenic in animals. This was based on the production of local sarcomas in rats after intramuscular injections of cadmium powder or cadmium oxide, chloride, sulfate and sulfide, and the production of testicular tumours in rats and mice after subcutaneous injections of cadmium salts. It has not been possible to produce tumours by long-term oral exposures in rats, even after 2 years' exposure to large amounts of cadmium (50 mg/kg diet). Inhalation studies reported in 1983 have shown a dose-dependent, highly significant increase in lung cancer in rats which were exposed to cadmium chloride aerosols (12.5–50 μg cadmium per m^3) for 23 hours per day for 18 months and observed for another 13 months *(25)*.

In 1982, IARC confirmed its conclusion that there was limited evidence of cadmium being a human carcinogen *(24)*. An excess of deaths from prostatic cancer among cadmium workers has been found in several studies. The excess is generally small and seems to be confined to groups with very high exposures in the past *(26,27)*. Prostatic cancer has a very complex etiology and it is unlikely that cadmium exposures in the general population can play a role in the development of this type of cancer *(28)*. The US Environmental Protection Agency (EPA) has recently concluded that there is insufficient evidence of cadmium causing prostatic cancer in man *(29)*.

An excess of lung cancer among cadmium workers has also been noted in several studies *(27)*. In most studies the excess is not significant, but a study on cadmium production workers with high exposures in the past to cadmium oxide fume and dust reported a significant excess of lung cancer *(30)*. This and other studies have been evaluated by US EPA, which deemed the evidence to be limited with regard to lung cancer risks for humans *(29)*.

Evaluation of Human Health Risks

Exposure

Cadmium concentrations in rural areas of Europe are typically a few ng/m^3 (below 5 ng/m^3); urban values range between 5 and 50 ng/m^3, but are mostly not higher than 20 ng/m^3.

Health risk evaluation

Whereas the data on relationships between internal dose (liver cadmium, renal cadmium or urine cadmium) and response are reliable, it is more difficult to relate external exposure to effects. This has mainly been done by using models *(5,15)*.

If the only route of exposure were inhalation (excluding tobacco smoke as a source), the critical average air concentration to reach the critical renal cortex concentration can be calculated as follows. Assuming a whole-body biological half-time of 19 years and 25% absorption of inhaled cadmium from the lungs, then 50 years of continuous exposure to 2.9 $\mu g/m^3$ would lead to a renal cortex concentration of 200 mg cadmium per kg *(5)*. This calculation does not take into account cadmium absorbed from food and water. Assuming a dietary intake of 30 μg cadmium per day, and 5% absorption, about 1.5 μg would be absorbed daily, compared with 14.5 μg from air. Additional exposure may come from smoking and a reasonable estimate will thus be that about 2 μg cadmium per m^3 air might cause a renal accumulation of cadmium sufficient to cause dysfunction. Yearly means of that magnitude have not been reported in ambient air and are not likely to occur.

The use of other half-times or absorption factors in these calculations would give lower or higher estimates. However, the present average levels of cadmium in ambient air are below any of these estimates, even near point sources, and thus do not constitute any risk of direct health effects in humans. Applying a protection (safety) factor of 10 would result in an air level of 0.2 $\mu g/m^3$, which is a safe level with regard to renal effects through inhalation.

However, as was pointed out earlier, cadmium in ambient air is transferred to soil by wet or dry deposition, is taken up by plants and enters the food chain. The rate of transfer from soil to plants will depend on the type of soil, plant species, pH, use of fertilizers, meteorology, etc. This chain of events is at present impossible to quantify, but present levels of cadmium are, on an average, about 20 mg/kg wet weight in the renal cortex, which means that for a large segment of the population the safety factor is less than 10 (for horses there may be no safety factor). Any increase in dietary exposure to cadmium will thus narrow the gap to the critical level. Consequently, airborne cadmium and the precipitation of cadmium onto soils should not be allowed to increase. The safe level with regard to inhalation (0.2 $\mu g/m^3$) may cause a fallout of 10–30 mg/m^2 per year, which would have a significant impact on cadmium uptake in plants. By monitoring the precipitation of cadmium and cadmium concentrations in foodstuffs it should be possible to prevent unwanted increases in exposure to cadmium. The use of fertilizers with a low cadmium content will be of great importance.

Evidence of the carcinogenicity of cadmium compounds in animals has been found to be sufficient by IARC. With regard to the carcinogenic risk of cadmium for humans, the evidence has been deemed to be limited by IARC and US EPA.

The Carcinogen Assessment Group of US EPA *(29)* made an assessment of cadmium, with regard to its carcinogenicity, on the basis of human lung cancer data *(30)* and estimated the incremental cancer risk from continuous lifetime exposure to 1 μg cadmium per m^3 to be 1.8×10^{-3}. A lifetime risk based on animal data was estimated to be 9.2×10^{-2} *(29)*.

Existing levels of cadmium in air give a large margin of protection with regard to toxic effects of inhalation exposure on the lungs or the kidneys. However, since cadmium in air will be precipitated onto soil and contribute to cadmium uptake by plants and the food chain, there is concern about future risks through dietary exposure to cadmium, which will increase renal levels of cadmium and eventually cause renal dysfunction. There is thus a need to prevent further increases in ambient air levels of cadmium. At present, guidelines should primarily be based on risks for renal accumulation by dietary exposure, since the estimates of cancer risks for humans are very uncertain.

Guidelines

While there is sufficient evidence of its carcinogenicity in animals and limited evidence of its carcinogenicity in humans, cadmium has been classified by IARC as a Group 2B carcinogen. Because data based on human studies are still problematic, no risk estimate is indicated here.

The following guideline values based on noncarcinogenic effects are recommended. In rural areas present levels of $<$ 1–5 ng/m^3 should not be allowed to increase. In urban areas without agricultural activities, and in industrialized areas, levels of 10–20 ng/m^3 may be tolerated. An increase in existing levels should not be permitted.

References

1. **Commission of the European Communities.** *Criteria (dose/effect relationships) for cadmium.* Oxford, Pergamon Press, 1978.
2. **Stubbs, R.L.** Cadmium — markets and trends. *In: Cadmium 81. Edited Proceedings of the Third International Cadmium Conference, Miami.* Cadmium Association, London, Cadmium Council, New York, International Lead Zinc Research Organization, New York, 1982.
3. *Control of toxic substances in the atmosphere – cadmium.* Solna, National Swedish Environmental Protection Board, 1985 (Report No. 3062).
4. **Hutton, M.** *Cadmium in the European Community: a prospective assessment of sources, human exposure and environmental impact.* London, University of London Monitoring and Assessment Research Centre, 1982 (MARC Report No. 26).
5. **Friberg, L. et al.** *Cadmium in the environment,* 2nd ed. Cleveland, OH, CRC Press, 1974.
6. **Denmark, Ministry of the Environment.** *Cadmiumforurening* [Cadmium pollution]. Copenhagen, Storgaard Jensen, 1980.
7. **Lahmann, E. et al.** *Heavy metals: identification of air quality and environmental problems in the European Community.* Luxembourg, Commission of the European Communities, 1986, Vol. 1 & 2 (Report No. EUR 10678 EN/I and EUR 10678 EN/II).

8. **Bennett, B.G.** *Exposure commitment assessments of environmental pollutants.* Vol. 1. London, University of London Monitoring and Assessment Research Centre, 1981 (MARC Report No. 23).
9. **Elinder, C.-G. et al.** Cadmium exposure from smoking cigarettes: variations with time and country where purchased. *Environmental research,* **32**: 220–227 (1983).
10. **Scherer, G. & Barkemeyer, H.** Cadmium concentrations in tobacco and tobacco smoke. *Ecotoxicology and environmental safety,* **7**: 71–78 (1983).
11. *Guidelines for drinking-water quality. Vol. 2. Health criteria and other supporting information.* Geneva, World Health Organization, 1984, pp. 84–90.
12. **Piscator, M.** Dietary exposure to cadmium and health effects: impact of environmental changes. *Environmental health perspectives,* **63**: 127–132 (1985).
13. **Flanagan, P.R. et al.** Increased dietary cadmium absorption in mice and human subjects with iron deficiency. *Gastroenterology,* **74**: 841–846 (1978).
14. **Dean, R.B. & Suess, M.J., ed.** The risk to health of chemicals in sewage sludge applied to land. *Waste management and research,* **3**: 251–278 (1985).
15. **Kjellström, T. & Nordberg, G.F.** A kinetic model of cadmium metabolism in the human being. *Environmental research,* **16**: 248–269 (1978).
16. WHO Technical Report Series, No. 647, 1980 (*Recommended health-based limits in occupational exposure to heavy metals:* report of a WHO Study Group).
17. **Kjellström, T.** Exposure and accumulation of cadmium in populations from Japan, the United States, and Sweden. *Environmental health perspectives,* **28**: 169–197 (1979).
18. **Friberg, L. & Vahter, M.** Assessment of exposure to lead and cadmium through biological monitoring: results of a UNEP/WHO global study. *Environmental research,* **30**: 95–128 (1983).
19. **Roels, H. et al.** The critical level of cadmium in human renal cortex: a re-evaluation. *Toxicology letters,* **15**: 357–360 (1983).
20. **Friberg, L. et al.** Cadmium. *In:* Friberg, L. et al., ed. *Handbook on the toxicology of metals,* 2nd ed. Amsterdam, Elsevier, 1986, pp. 130–184.
21. **Kjellström, T.** Critical organs, critical concentrations, and whole body dose–response relationships. *In:* Friberg, L. et al., ed. *Cadmium and health: a toxicological and epidemiological appraisal.* Boca Raton, FL, CRC Press, 1986, Vol. II, pp. 231–246.
22. **Roels, H.A. et al.** *In vivo* measurement of liver and kidney cadmium in workers exposed to this metal: its significance with respect to cadmium in blood and urine. *Environmental research,* **26**: 217–240 (1981).
23. *Cadmium, nickel, some epoxides, miscellaneous industrial chemicals and general considerations on volatile anaesthetics.* Lyon, International Agency for Research on Cancer, 1976 (IARC Monographs on the Evaluation of the Carcinogenic Risk of Chemicals to Man, Vol. 11).

24. *Chemicals, industrial processes and industries associated with cancer in humans. IARC Monographs, Volumes 1 to 29.* Lyon, International Agency for Research on Cancer, 1982 (IARC Monographs on the Evaluation of the Carcinogenic Risk of Chemicals to Humans, Supplement 4).
25. **Takenaka, S. et al.** Carcinogenicity of cadmium chloride aerosols in rats. *Journal of the National Cancer Institute,* **70**: 367–373 (1983).
26. **Sorahan, T. & Waterhouse, J.A.H.** Cancer of prostate among nickel-cadmium battery workers. *Lancet,* **1**: 459 (1985).
27. **Elinder, C.-G. et al.** Cancer mortality of cadmium workers. *British journal of industrial medicine,* **42**: 651–655 (1985).
28. **Piscator, M.** Role of cadmium in carcinogenesis with special reference to cancer of the prostate. *Environmental health perspectives,* **40**: 107–120 (1981).
29. *Updated mutagenicity and carcinogenicity assessment of cadmium. Addendum to the health assessment document for cadmium (May 1981) EPA-600/8-81-023.* Washington, DC, US Environmental Protection Agency, 1985 (Final report No. EPA-600/8-83-025F).
30. **Thun, M.J. et al.** Mortality among a cohort of U.S. cadmium production workers — an update. *Journal of the National Cancer Institute,* **74**: 325–333 (1985).

20

Carbon monoxide

General Description

Carbon monoxide (CO) is a colourless, odourless, tasteless gas that is slightly lighter than air. It reacts with haemoglobin to form carboxyhaemoglobin (COHb). The affinity of haemoglobin is more than 200 times higher for carbon monoxide than for oxygen.

Sources

Carbon monoxide is one of the most common and widely distributed air pollutants. It is a product of the incomplete combustion of carbon-containing materials, but is also produced by some industrial and biological processes.

Total emissions of carbon monoxide to the atmosphere equal or even exceed those of all other air pollutants combined. It therefore continues to be a potential industrial as well as a general environmental hazard.

The largest sources of carbon monoxide emissions are incomplete combustion processes, such as in automobiles, industrial processes, heating facilities and incinerators. Estimates of man-made carbon monoxide emissions vary from 350 to 600 million tonnes per annum. Some widespread natural nonbiological and biological sources have also been identified. These natural sources may be important as far as background concentrations are concerned, but their influence on carbon monoxide concentrations in urban areas is negligible.

Occurrence in air

Natural background levels of carbon monoxide range between 0.01 and 0.23 mg/m^3 (0.01–0.20 ppm). Concentrations in urban areas depend on weather and traffic density. They vary greatly according to time and distance from the sources. The 8-hour mean concentrations are generally less than 20 mg/m^3 (17 ppm). However, maximum 8-hour mean concentrations of up to 60 mg/m^3 (53 ppm) have occasionally been recorded.

Urban carbon monoxide concentrations vary with the density of petrol-powered vehicles and most cities have carbon monoxide peak levels that coincide with the morning and evening rush-hours. Variations in these levels are also influenced by topography and weather conditions. The fluctuation in ambient concentrations is only slowly reflected in the carboxyhaemoglobin levels in humans, as it takes 4–12 hours for approximate equilibrium between air levels and blood levels to occur. Thus, environmental concentrations tend to be expressed in terms of 8-hour average concentrations. The presentation of such data as moving 8-hour averages over the day rather than as three consecutive nonoverlapping periods has the advantage of reflecting the human body's response under common environmental conditions.

Carbon monoxide is widely generated indoors by unvented combustion appliances, particularly if they are operated in poorly ventilated rooms. In the room air of kitchens short-time concentrations of 11.5–34.5 mg carbon monoxide per m^3 (10–30 ppm) have been measured in the Federal Republic of Germany *(1)* and in Dutch cities values of 57.5 mg carbon monoxide per m^3 (50 ppm) and higher were found in 17% of homes *(2)*.

The relation between carbon monoxide concentrations inside and outside of cars has been measured in Delft, Netherlands, and in Frankfurt am Main, Federal Republic of Germany *(3)*. The concentrations found in cars were higher than those measured at fixed monitoring stations in both cities. Maximum 30-minute values of 41 mg/m^3 (36 ppm) have been found.

Some individuals may also be exposed to carbon monoxide in the course of their work. Persons likely to be exposed include traffic policemen or wardens, garage workers, employees at metallurgical, petroleum, gas or chemical plants, and firemen. Such occupational exposure can be considerable. For example, levels of carbon monoxide in some workplace situations may exceed 115 mg/m^3 (100 ppm) and can, for example, reach levels as high as 570 mg/m^3 (500 ppm) in garages, so that carboxyhaemoglobin levels among nonsmoking workers in such places may be up to five times higher than normal. Highway inspectors have been reported to show carboxyhaemoglobin concentrations of 4–7.6% (smokers) and 1.4–3.8% (nonsmokers) during their daily work. By contrast, carboxyhaemoglobin levels in the general population rarely exceed 1%, although a study of 18 urban areas in North America showed that 45% of nonsmokers exposed to ambient carbon monoxide had carboxyhaemoglobin levels exceeding 1.5% *(4)*. In a more recent study *(5)*, carboxyhaemoglobin values among nonsmokers of 1.3–1.5% (males) and 1.2–1.4% (females) have been reported; 3–5% exceeded critical carboxyhaemoglobin values of 2.5%.

Tobacco smoking can also be an important source of indoor pollution. A review of indoor byproduct levels of tobacco smoke *(6)* reported that the difference in carbon monoxide values between smokers' and nonsmokers' sections of cafeterias ranged from 0.8 mg/m^3 (0.7 ppm) to 3.5–5.7 mg/m^3 (3–5 ppm), while the difference between indoor and outdoor values ranged from 0.0 to 9.5 ppm. Tobacco smoking usually makes the largest contribution to the carbon monoxide body-burden in those who do smoke.

Conversion factors

1 ppm = 1.145 mg/m^3

1 mg/m^3 = 0.873 ppm

Routes of Exposure

Since carbon monoxide is a gas and will not penetrate the skin, the only important route of human exposure is via inhalation.

Kinetics and Metabolism

Inhaled carbon monoxide reacts with the iron in protohaem of haemoglobin and forms strong bonds. The distribution of carbon monoxide in the organism is influenced by the perfusion of the organs, particularly at the beginning of carbon monoxide uptake or when carbon monoxide levels in environmental air increase. The resulting carboxyhaemoglobin is about 250 times more stable than oxyhaemoglobin in humans; however, the factor is lower in most animal species. The carboxyhaemoglobin concentration in blood increases very rapidly in the coronary and cerebral arteries, and more slowly in the periphery (extremities) *(7)*. After reaching a steady state, the distribution of carbon monoxide is determined by the partial pressure of oxygen and carbon monoxide in organs and tissues, and also by the different affinity in relation to the amount of individual haemoproteins.

Several important relationships between carboxyhaemoglobin levels and other physiological parameters, considered in the 1979 WHO *(4)* and US Environmental Protection Agency (EPA) *(8)* criteria documents, have continued to be the subject of discussion. The relationship between external carbon monoxide exposure levels and consequent increases in blood carboxyhaemoglobin levels is important. Many factors can affect the rate at which carboxyhaemoglobin increases above pre-existing endogenous levels in response to inhalation of exogenous carbon monoxide. These include, for example, the pattern of external carbon monoxide exposure, as in the case of acute short-term exposure to high carbon monoxide concentrations versus longer-term exposure to relatively low levels of carbon monoxide. Carbon monoxide exposure–carboxyhaemoglobin concentration relationships have been modelled by Coburn and co-workers *(9)*. Carboxyhaemoglobin levels predicted by the Coburn equations are widely accepted as the currently best available modelled estimates of carboxyhaemoglobin levels likely to result from varying carbon monoxide concentrations, exposure durations and exercise levels. It should be noted that some questions have been raised regarding the specific mathematical approach employed by Coburn in solving his equations to predict carboxyhaemoglobin concentration as a function of time, considering appropriate physiological parameters. A modified Coburn model *(10)* may be used for such predictions with respect to risk groups on the basis of specified physiological parameters.

The exact mechanisms responsible for carbon monoxide-induced hypoxia are not known in detail. The well founded and most widely accepted theory regarding the mechanism of carbon monoxide toxicity is that the preferential binding of carbon monoxide to haemoglobin produces hypoxia by reducing oxygen transport by red blood cells to the tissues and impedes the dissociation of oxygen from haemoglobin in the capillaries. However, other mechanisms have been postulated for the reduction of oxygen transport or utilization in the tissues. It is possible that carbon monoxide binds to intracellular haemoproteins such as myoglobin and cytochrome oxidase, which depends on the relationship of oxygen tension (PO_2) and carbon monoxide tension (PCO) to carbon monoxide-binding constants *(11)*. The affinity of myoglobin for carbon monoxide is roughly 30–50 times that of oxygen.

The binding of carbon monoxide to myoglobin in heart and skeletal muscle may be high enough to reduce intracellular oxygen transport in those tissues *(11,12)*. Using a computer simulation of a three-compartment model (arterial blood, venous capillary blood, and tissue myoglobin), Agostoni et al. *(12)* predicted, on the basis of theoretical considerations, that conditions would be favourable for the formation of carboxymyoglobin at carboxyhaemoglobin levels of 5–10%, particularly in areas where the oxygen tension (PO_2) was physiologically low (e.g. in subendocardium) and when conditions of hypoxia, ischaemia or increased metabolic demands were present. This could provide additional theoretical support for experimental evidence of myocardial ischaemia, such as electrocardiographic irregularities and decrements in work capacity. However, it is not known whether binding of carbon monoxide to myoglobin could affect health (e.g. decreases in maximal oxygen consumption during exercise) at carboxyhaemoglobin levels as low as 4–5%. Additional research is needed before this hypothesis can be more definitively evaluated.

The elimination of carbon monoxide from the organism basically follows the same physical and physicochemical laws as the carbon monoxide uptake. It can be described as an exponential function of second order by assuming a constant respiratory minute volume and cardiac minute output. The elimination half-time lies in the range 2–8 hours.

Health Effects

Four types of health effects are reported *(8)* to be associated with carbon monoxide exposures (especially those producing carboxyhaemoglobin levels of below 10%):

(*a*) cardiovascular effects *(13–16)*

(*b*) neurobehavioural effects *(17–25)*

(*c*) fibrinolysis effects *(26)*

(*d*) perinatal effects *(27)*.

Hypoxia caused by carbon monoxide leads to deficient function in sensitive organs and tissues like the brain, heart, the inner wall of blood vessels and platelets.

With regard to cardiovascular effects, a decreased oxygen uptake capacity and the resultant decreased work capacity under maximal exercise conditions have been clearly shown to occur in healthy young adults, starting at 5.0% COHb. In some studies, small but significant decreases in work time to exhaustion at carboxyhaemoglobin levels as low as 3.3–4.3% were observed, although maximum aerobic capacity was not diminished *(28,29)*. These cardiovascular effects may have health implications for the general population in terms of a potential curtailment of certain physically demanding occupational or recreational activities under circumstances of sufficiently high carbon monoxide exposure. However, of greater concern at more typical ambient carbon monoxide exposure levels are certain cardiovascular effects (e.g. aggravation of angina symptoms during exercise) likely to occur in a smaller but sizeable segment of the general population. This group, consisting of chronic angina patients, is at present regarded as the most sensitive risk group in relation to carbon monoxide exposure effects, on the basis of evidence of an aggravation of angina occurring in patients at carboxyhaemoglobin levels of 2.9–4.5% COHb *(16)*. Such aggravation of angina is thought to represent an adverse health effect. Dose–response relationships for cardiovascular effects in patients with severe coronary artery diseases remain to be more conclusively defined, and the possibility cannot be ruled out that such effects may sometimes occur at levels below 2.9% COHb.

Statistically significant impairment of vigilance tasks has been described at carboxyhaemoglobin levels above 5% COHb *(22–24)*. No reliable evidence demonstrating decrements in neurobehavioural function in healthy young adults has been reported at carboxyhaemoglobin levels below 5%.

Much of the research at 5% COHb showed no effect, even when neurobehavioural functions similar to those affected in other studies were investigated *(17–21)*. However, none of those studies which demonstrated significant effects on neurobehavioural functions examined carboxyhaemoglobin levels below 5%, i.e. possible effects at lower carboxyhaemoglobin levels would not have been found in these studies. Therefore, on the one hand the empirical evidence indicates that carboxyhaemoglobin levels as low as 5% produce decrements in neurobehavioural functions and, on the other hand, the possibility of decrements at carboxyhaemoglobin levels lower than 5% cannot be fully excluded *(8)*.

Higher levels of carboxyhaemoglobin may lead to secondary effects, for instance, decrease in blood pH and changes in fibrinolysis *(26)*. Perinatal effects such as reduced birth weight and retarded postnatal development have also been described.

There is general agreement that from levels of 10–15% COHb upwards many of these malfunctions may occur, as well as subjective symptoms such as headache or dizziness.

Table 1 summarizes studies relating human health effects to different low-level exposures to carbon monoxide.

Patients with angina pectoris, or those with obstructed coronary arteries but not yet manifesting overt symptoms of coronary artery disease, appear to be the best identified as a sensitive group within the general population, i.e. at increased risk of experiencing health effects (exacerbation of cardiovascular symptoms) of concern at ambient or near-ambient carbon monoxide exposure levels. Several other probable risk groups have been identified: (*a*) pregnant women and young infants; (*b*) the elderly (especially those with reduced cardiopulmonary function); (*c*) individuals with chronic bronchitis and emphysema; (*d*) younger individuals with severe cardiac or

Table 1. Human health effects associated with low-level carbon monoxide exposure: lowest-observed-effect levels[a]

Carboxyhaemoglobin concentration (%)	Effects
2.3– 4.3	Statistically significant decrease (3–7%) in the relation between work time and exhaustion in exercising young healthy men
2.9– 4.5	Statistically significant decrease in exercise capacity (i.e. shortened duration of exercise before onset of pain) in patients with angina pectoris and increase in duration of angina attacks
5 – 5.5	Statistically significant decrease in maximal oxygen consumption and exercise time in young healthy men during strenuous exercise
<5	No statistically significant vigilance decrements after exposure to carbon monoxide
5 – 7.6	Statistically significant impairment of vigilance tasks in healthy experimental subjects
5 –17	Statistically significant diminution of visual perception, manual dexterity, ability to learn, or performance in complex sensorimotor tasks (e.g. driving)
7 –20	Statistically significant decrease in maximal oxygen consumption during strenuous exercise in young healthy men

[a] Compiled on the basis of references *7, 8, 13–25, 28–30.*

acutely severe respiratory diseases; (*e*) individuals with haematological diseases (e.g. anaemia) which affect the oxygen-carrying capacity of the blood; (*f*) individuals with genetically unusual forms of haemoglobin associated with reduced oxygen-carrying capacity; (*g*) individuals using drugs that have depressant properties in relation to the CNS; (*h*) persons with high-altitude exposure. However, there is currently little empirical evidence to enable the carboxyhaemoglobin levels at which such individuals are likely to experience definite health effects to be specified.

Genetic differences in the effects of acute carbon monoxide intoxication have been assessed in various species of laboratory animal *(4)*. It may therefore be worthwhile investigating whether genetic differences can explain some of the discrepancies in carbon monoxide intoxication in humans.

Evaluation of Human Health Risks

Exposure

Natural background levels of carbon monoxide vary between 0.01 and 0.23 mg/m^3 (0.01–0.20 ppm). Eight-hour mean concentrations are generally less than 20 mg/m^3 (17 ppm). However, maximum 8-hour mean concentrations up to 60 mg/m^3 (53 ppm) have occasionally been recorded. In kitchens short-term concentrations of 11.5–34.5 mg/m^3 (10–30 ppm) have been found, occasionally reaching 57.5 mg/m^3 (50 ppm) or more.

Health risk evaluation

Average carboxyhaemoglobin levels in the general nonsmoking population are around 1.2–1.5% (in cigarette smokers around 3–4%). Heavy smokers may reach carboxyhaemoglobin levels of about 10%. It should be noted that healthy persons have endogenous carboxyhaemoglobin levels of 0.5–1.0%.

At low carboxyhaemoglobin levels (below 10% COHb), it is mainly cardiovascular and neurobehavioural effects that have to be evaluated. The aggravation of symptoms in angina pectoris patients which occurs at carboxyhaemoglobin levels of 2.9–4.5% is of great concern as an adverse health effect. Transient cardiovascular and neurobehavioural effects have been reported in the range of 2–3% COHb, but these findings could not be reproduced by other authors and furthermore severe methodological criticism has been voiced. Among healthy young adults, decreased oxygen uptake capacity and resultant decreased work capacity under maximal exercise conditions have been clearly shown starting at 5.0% COHb. However, some studies observed small, but significant, decreases in work time to exhaustion at carboxyhaemoglobin levels of 3.3–4.3%, although maximum aerobic capacity was not diminished. From the empirical evidence on neurobehavioural effects, it appears that carboxyhaemoglobin levels around 5% produce decrements in neurobehavioural function.

With regard to the no-observed-effect level, no adverse effects have been reported below 2% COHb, but findings in the range of 2.0–2.9% could not be reproduced by other authors and are subject to severe methodological criticism *(8,29)*.

In quantitative hazard evaluation using the Coburn formula, the carboxyhaemoglobin levels shown in Table 2 may be predicted for sedentary subjects, those engaged in light physical work and those performing heavy physical work.

When using these values one must bear in mind that the carboxyhaemoglobin levels are predicted for people who were previously unexposed. Other exposure patterns may be computed by using Coburn's formula or a revised model.

For longer exposure times, the low carboxyhaemoglobin values for 8-hour exposure to 11.5 mg/m^3 (10 ppm) may be said to be near the state of equilibrium. Thus, they could also be taken as 24-hour means and would still provide protection against adverse effects. Most countries have specified an 8-hour value of 10 mg/m^3 (9 ppm).

Table 2. Predicted carboxyhaemoglobin levels for subjects engaged in different types of work

Carbon monoxide concentration		Exposure time	Predicted COHb level for those engaged in:		
ppm	mg/m³		sedentary work	light work	heavy work
100	115	15 minutes	1.2	2.0	2.8
50	57	30 minutes	1.1	1.9	2.6
25	29	1 hour	1.1	1.7	2.2
10	11.5	8 hours	1.5	1.7	1.7

Some problems arise when there are shorter exposure times and higher carbon monoxide concentrations in the air. At a concentration of 29 mg/m^3 (25 ppm) the carboxyhaemoglobin level would be kept below or at 3%, even if the exposure lasted 2 hours, but at a level of 40 mg/m^3 (35 ppm) the predicted 2-hour carboxyhaemoglobin values would be between 2.0% and 4.3%, depending on the degree of physical activity. Therefore, if a 1-hour value of 40 mg/m^3 (35 ppm) were proposed, precautions would have to be taken to ensure that the exposure time of 1 hour was not exceeded. One way of doing this would be to take the 8-hour value as a moving average with 1-hour intervals. The moving average of 8 hours should also be used to deal with carbon monoxide fluctuations over time (for instance the diurnal fluctuations resulting from varying traffic volume).

In short-term exposures (5–15 minutes) the carboxyhaemoglobin content is much higher in coronary than in peripheral blood vessels. This may lead to an underestimation of the risk as predicted from the carboxyhaemoglobin level in the peripheral blood, shown in tables and nomograms.

If the 1-hour mean value of 29 mg/m^3 (25 ppm) were taken as a short-term guideline value, then very high carbon monoxide levels in the air could occur for shorter exposure times, for example a 5-minute exposure could reach 345 mg/m^3 (300 ppm).

Guidelines

As smoking is a major contributor to carboxyhaemoglobin levels in smokers, recommendations for exposure limits are designed to protect nonsmokers.

A carboxyhaemoglobin level of 2.5–3.0% is recommended for the protection of the general population, including sensitive groups.

The following guidelines to prevent carboxyhaemoglobin levels exceeding 2.5–3% in nonsmoking populations, based on the Coburn formulae, are proposed.

1. A maximum permitted exposure of 100 mg/m^3 for periods not exceeding 15 minutes.

2. Time-weighted average exposures at the following levels for the periods of exposure indicated (where the mg/m^3 are rounded off to the nearest 10):

— 60 mg/m^3 (50 ppm) for 30 minutes;

— 30 mg/m^3 (25 ppm) for 1 hour;

— 10 mg/m^3 (10 ppm) for 8 hours.

References

1. **Seifert, B.** Luftverunreinigungen in Wohnungen und anderen Innenräumen [Air pollution in dwellings and other indoor areas]. *Staub, Reinhaltung der Luft,* **44**: 377–382 (1984).
2. **Boleij, J. et al.** Innenluftverunreinigungen durch Kohlenmonoxid und Stickstoffoxide [Indoor air pollution by carbon monoxide and nitrogen oxides]. *In:* Aurand, K. et al., ed. *Luftqualität in Innenräumen* [Indoor air quality]. Stuttgart-New York, Gustav Fischer Verlag, 1982, pp. 199–208.
3. **den Tonkelaar, W.A.M. & Rudolf, W.** Luftqualität im Innern von Kraftfahrzengen [Air quality inside motor cars]. *In:* Aurand, K. et al., ed. *Luftqualität in Innenräumen* [Indoor air quality]. Stuttgart-New York, Gustav Fischer Verlag, 1982, pp. 219–233.
4. *Carbon monoxide.* Geneva, World Health Organization, 1979 (Environmental Health Criteria, No. 13).
5. **Roscovanu, A. et al.** Untersuchungen zum Carboxyhämoglobingehalt ausgewählter Bevölkerungsgruppen in unterschiedlich strukturierten und belasteten Gebieten Nordrhein-Westfalens [Carboxyhemoglobin levels of selected population segments living in urban and rural environments of North Rhine-Westphalia]. *Zentralblatt für Bakteriologie, Mikrobiologie und Hygiene, I Abt. Orig. B,* **180**: 359–380 (1985).
6. **Sterling, T.D. et al.** Indoor byproduct levels of tobacco smoke: a critical review of the literature. *Journal of the Air Pollution Control Association,* **32**: 250–259 (1982).
7. **Zorn, H.R. et al.** Evaluation of exposure limits for CO with consideration of other air pollutants. *In:* Hartmann, H.F., ed. *Proceedings of the Seventh World Clean Air Congress, Sydney, 25–29 August 1986.* Paris, The International Union of Air Pollution Prevention Associations, 1986.
8. *Air quality criteria for carbon monoxide* (Report No. EPA-600/8-79-022) and *Revised evaluation of health effects associated with carbon monoxide exposure: addendum to the 1979 air quality criteria document for carbon monoxide* (Report No. EPA-600/8-83-033F). Washington, DC, US Environmental Protection Agency, 1979 and 1984.

9. **Coburn, R.F. et al.** Considerations of the physiological variables that determine the blood carboxyhemoglobin concentration in man. *Journal of clinical investigation,* **44**: 1899–1910 (1965).
10. **Hauck, H. & Neuberger, M.** Carbon monoxide uptake and the resulting carboxyhaemoglobin in man. *European journal of applied physiology,* **53**: 186–190 (1984).
11. **Coburn, R.F.** Mechanisms of carbon monoxide toxicity. *Preventive medicine,* **8**: 310–322 (1979).
12. **Agostoni, A. et al.** Influence of capillary and tissue PO_2 on carbon monoxide binding to myoglobin: a theoretical evaluation. *Microvascular research,* **20**: 81–87 (1980).
13. **Horvath, S.M. et al.** Maximal aerobic capacity at different levels of carboxyhemoglobin. *Journal of applied physiology,* **38**: 300–303 (1975).
14. **Ekblom, B. & Huot, R.** Response to submaximal and maximal exercise at different levels of carboxyhaemoglobin. *Acta physiologica scandinavica,* **86**: 474–482 (1972).
15. **Ayres, S.M. et al.** The prevalence of carboxyhemoglobinemia in New Yorkers and its effects on the coronary and systemic circulation. *Preventive medicine,* **8**: 323–332 (1979).
16. **Anderson, E.W. et al.** Effect of low-level carbon monoxide exposure on onset and duration of angina pectoris: a study in ten patients with ischemic heart disease. *Annals of internal medicine,* **79**: 46–50 (1973).
17. **Haider, M. et al.** Effects of moderate CO dose on the central nervous system — electrophysiological and behaviour data, clinical relevance. *In:* Finkel, A.J. & Duel, W.C., ed. *Clinical implications of air pollution research. Air Pollution Medical Research Conference, American Medical Association, 5–6 December 1974.* Acton, MA, Publishing Sciences Group, 1976, pp. 217–232.
18. **Winneke, G.** Behavioral effects of methylene chloride and carbon monoxide as assessed by sensory and psychomotor performance. *In:* Xintaras, C. et al., ed. *Behavioral toxicology. Early detection of occupational hazards.* Washington, DC, US Department of Health, Education, and Welfare, 1974, pp. 130–144 (DHEW Publication (NIOSH) No. 74-126).
19. **Christensen, C.L. et al.** Effects of three kinds of hypoxias on vigilance performance. *Aviation, space and environmental medicine,* **48**: 491–496 (1977).
20. **Beningnus, V.A. et al.** Lack of effects of carbon monoxide on human vigilance. *Perceptual and motor skills,* **45**: 1007–1014 (1977).
21. **Putz, V.R. et al.** *Effects of CO on vigilance performance. Effects of low level carbon monoxide on divided attention, pitch discrimination, and the auditory evoked potential.* Cincinatti, OH, US Department of Health, Education, and Welfare, 1976 (DHEW Publication (NIOSH) No. 77-124).
22. **Horvath, S.M. et al.** Carbon monoxide and human vigilance: a deleterious effect of present urban concentrations. *Archives of environmental health,* **23**: 343–347 (1971).

23. **Groll-Knapp, E. et al.** Auswirking geringer Kohlenmonoxid-Konzentrationen auf Vigilanz und computeranalysierte Hirnpotentiale [Effects of low carbon monoxide concentrations on vigilance and computer-analysed brain potentials]. *Staub, Reinhaltung der Luft,* **32**: 185–188 (1972).
24. **Fodor, G.G. & Winneke, G.** Wirkung niedriger CO-Konzentrationen auf Monotonie-Resistenz und psychomotorisches Leistungsvermögen [Effect of low CO concentrations on resistance to monotony and on psychomotor capacity]. *Staub, Reinhaltung der Luft,* **32**: 169–175 (1972).
25. **Putz, V.R.** The effects of carbon monoxide on dual-task performance. *Human factors,* **21**: 13–24 (1979).
26. **Kalmaz, E.V. et al.** Effect of long-term low and moderate levels of carbon monoxide exposure on platelet counts of rabbits. *Journal of environmental pathology and toxicology,* **4**: 351–358 (1980).
27. **Longo, L.D.** The biological effects of carbon monoxide on the pregnant woman, fetus, and newborn infant. *American journal of obstetrics and gynecology,* **129**: 69–103 (1977).
28. **Weiser, P.C. et al.** Effects of low-level carbon monoxide exposure on the adaptation of healthy young men to aerobic work at an altitude of 1610 meters. *In:* Folinsbee, L.J. et al., ed. *Environmental stress. Individual human adaptations.* New York, Academic Press, 1978, pp. 101–110.
29. *Review of the NAAQS for carbon monoxide: reassessment of scientific and technical information.* Research Triangle Park, NC, US Environmental Protection Agency, 1984 (EPA-450/5-84-004).
30. **Klein, J.P. et al.** Hemoglobin affinity for oxygen during short-term exhaustive exercise. *Journal of applied physiology,* **48**: 236–242 (1980).

21

Chromium

General Description

Sources

Chromium (Cr) is a grey, hard metal most commonly found in the trivalent state in nature, but hexavalent compounds are found in small quantities. Chromite ($FeOCr_2O_3$) is the only important ore containing chromium. So far, the ore has not been found in the pure form; its highest grade contains about 55% chromic oxide.

In Europe, ferrochromium is produced mainly in Finland, France, Italy, Norway, Sweden and Yugoslavia. Potassium chromate is produced mainly in the Federal Republic of Germany, Italy, Switzerland and the United Kingdom. Sodium chromate and dichromate are now among the most important chromium products, and are used chiefly for manufacturing chromic acid, chromium pigments, in leather tanning, and for corrosion control *(1)*.

Chromium levels in soil vary according to area and the degree of contamination from anthropogenic chromium sources. Tests on soils have shown chromium concentrations ranging from 1 to 1000 mg/kg, with an average concentration ranging from 14 to about 70 mg/kg *(2)*.

As chromium is almost ubiquitous in nature, chromium in the air may originate from wind erosion of shales, clay and many other kinds of soil. In countries where chromite is mined, production processes may constitute a major source of airborne chromium. In Europe, endpoint production of chromium compounds probably constitutes the most important source of chromium in air.

Occurrence in air

Information on chromium concentrations in the atmosphere is limited. Measurements carried out above the North Atlantic, north of latitude 30 °N, several thousands of kilometres from major land masses, showed concentrations from 0.07 to 1.1 ng chromium per m^3 *(3)*. Concentrations above the South Pole were slightly lower.

In the Shetland Islands and Norway 0.7 ng chromium per m^3 has been found, 0.6 ng/m^3 in north-west Canada, 1–140 ng/m^3 in continental Europe,

20–70 ng/m^3 in Japan, and 45–67 ng/m^3 in Hawaii *(4)*. Monitoring of the ambient air during the period 1977–1980 in many urban and rural areas of the United States showed chromium concentrations to be in the range of 5.2 ng/m^3 (24-hour background level) to 156.8 ng/m^3 (urban annual average). The maximum concentration determined in the USA in any one measurement was about 684 ng/m^3 (24-hour average) *(2)*. Ranges of chromium levels in Member States of the European Community were given in a recent survey as follows: remote areas: 0–3 ng/m^3; urban areas: 4–70 ng/m^3; and industrial areas: 5–200 ng/m^3 *(5)*. Mass median diameters of chromium particles in the ambient air have been reported to be in the range 1.5–1.9 μm *(2)*.

Chromium levels in the industrial environment have only been reported in a few studies. Many of the studies referred to by IARC *(1)* were performed before 1960. Considering the great difficulties associated with the analysis of the various chromium compounds *(6)*, the results obtained during the last 5–10 years may be considerably more reliable than those in the early studies.

From the point of view of toxicity and carcinogenicity, the hexavalent chromium compounds are of much greater significance for workers and the general population than are trivalent and other valency states of chromium compounds *(7)*. Therefore, chromium(VI) and chromium(III) have to be considered separately.

Routes of Exposure

Air

The inhalation of chromium-containing aerosols is an important route of exposure to chromium compounds because the bronchial tree is the major target organ for carcinogenic effects of chromium(VI). Chromium intake from inhalation, based on a 24-hour respiratory volume of 20 m^3 in urban areas with an average chromium concentration of 0.05 μg/m^3, will be about 50 ng.

Chromium has been determined as a component of cigarette tobacco produced in the USA, its concentration varying from 0.24 to 6.3 mg/kg *(1)*.

Drinking-water

The concentration of chromium in water varies according to the type of the surrounding industrial sources and the nature of the underlying soils. An analysis of 3834 tap water samples in representative cities of the USA showed a chromium concentration ranging from 0.4 to 8 μg/litre *(2)*.

Food

The daily chromium intake from individual foods is difficult to assess because studies have used methods which are not easily comparable. The chromium intake from typical North American diets was found to be 60–90 μg/day *(2)* and may be generally in the range of 50–200 μg/day.

The chromium content of British commercial alcoholic beverages was reported to be slightly higher than that of USA-produced wines, namely, 0.45 mg/litre for wine, 0.30 mg/litre for beer and 0.135 mg/litre for spirits *(2)*.

Kinetics and Metabolism

Only limited experimental data are available on the biokinetics of chromium *(8)*. It can be assumed that the rate of uptake in the airways is to a great extent governed by the water solubility of the compounds *(1)* and by the size distribution of the inhaled particles *(8)*.

It has been suggested that chromium concentrations in human lungs increase with age *(9)*. Absorption by inhalation exposure appears to occur rapidly, although it is difficult to quantify the extent of absorption. A preliminary estimate of pulmonary absorption, following deposition of chromium(III) chloride in the lungs by instillation, indicates that approximately 5% is absorbed *(2)*. Four recent studies on the fate of trivalent and hexavalent chromium after intratracheal administration have given some information on pulmonary retention and absorption of chromium compounds *(10–13)*. Chromates with low solubility are mainly cleared to the gastrointestinal tract, whereas more soluble chromates are absorbed into the blood *(11)*. Trivalent chromium is retained to a greater extent in the lungs than hexavalent chromium *(12)*.

Following oral exposure, gastrointestinal absorption of chromium is also low, an estimated 5% or less being absorbed *(2)*. From studies on the uptake of chromium(VI) compounds in the gastrointestinal tract, it can be assumed that the rate of uptake is to a great extent governed by the water solubility of the compounds *(1)*. *In vitro* studies indicate that gastrointestinal juices are capable of reducing chromium(VI) to chromium(III); however, data from *in vivo* studies are insufficient to demonstrate whether this reduction process has the capacity to eliminate any differences in absorption between ingested chromium(VI) and chromium(III) compounds *(2)*. Pulmonary cells have been shown *in vitro* to have some capacity to reduce hexavalent chromium; however, this capacity is low compared to that of liver cells *(14)*.

After exposure of some experimental animals (dams), absorbed chromium(III) and chromium(VI) can be transported to a limited extent to the fetus *in utero,* although available data do not allow quantitative estimates of fetal exposure. In man the chromium concentration in the tissues of newborn babies has been found to be higher than that found later in life. There is a significant decline of concentration in children until about 10 years of age. After this age there is a slight increase in the lung tissue concentration, but a slight decrease in that in the other tissues. Chromium transported by the blood is distributed to other organs, the greatest retention being found in the spleen, liver, and bone marrow.

In both animals and humans, absorbed chromium is eliminated from the body in a rapid phase, representing clearance from the blood, and in a slower phase, representing clearance from tissues. The primary route of elimination is urinary excretion, which accounts for a little over 50% of the eliminated chromium, while faecal excretion accounts for only 5% of elimination from the blood. The remaining chromium is deposited into deep body compartments, such as bone and soft tissue. Elimination from these tissues proceeds very slowly. The estimated half-time for whole-body chromium elimination is 22 days for chromium(VI) and 92 days for chromium(III) *(2)*.

Health Effects

It has been generally agreed that certain trivalent chromium compounds are essential to man *(15)*. Chromium deficiency has been described in both man and animals, but a quantitative definition of the daily requirement of chromium in human nutrition has not been arrived at *(15)*.

Effects on experimental animals and *in vitro* test systems

Toxicological effects
Systemic effects of chromium compounds in animals have been reviewed by Tandon *(16)*, covering effects in the kidneys, liver and airways.

Mutagenic and carcinogenic effects
Animal experiments have not been very successful in identifying the compounds which increase the risk of bronchiogenic cancer so well documented in chromate workers. Hueper *(17)* reported 3 injection-site tumours in 31 rats. Baetjer et al. *(18)* were unable to induce bronchiogenic cancer in mice and rats by inhalation. They reported 4 cases of lymphosarcoma, but such effects have not been observed by other authors. Sintered calcium chromate has been found to induce injection-site sarcomas in mice *(19)*. After intratracheal injection of calcium chromate and strontium chromate in 218 rats, Hueper & Payne *(20)* found 3 fibrosarcomas after 2 years of observation, the individual chromate dose being 10–12.5 mg. Nettesheim et al. *(21)* induced 14 adenocarcinomas and adenomas in the bronchial tree among 131 mice during inhalation experiments using calcium chromate dust (13 mg/m^3, 25 hours per week for lifetime). Levy & Venitt made intrabronchial implantations of chromates in rats and found that only chromates with low solubility caused a significant increase in bronchial carcinomas *(22)*.

Evidence has been presented demonstrating the mutagenic capacity of a number of hexavalent chromium compounds *in vitro* and *in vivo*. An extensive review of the relevant studies is made by Levis & Bianchi *(23)*.

Effects on humans

Toxicological effects
Chrome ulcers, corrosive reaction on the nasal septum, acute irritative dermatitis and allergic eczematous dermatitis are recorded. Slight effects on the respiratory tract due to hexavalent chromium (as chromic acid) have been reported to occur at concentrations above 2 μg/m^3 *(24)*.

Mancuso & Hueper *(25)* described a spotty, moderately severe but not nodular pneumoconiosis in chromate workers. Sluis-Cremer & du Toit *(26)* reported fine nodular pneumoconiosis in a few chromite workers in the Republic of South Africa, but these results were not confirmed by other authors.

On the basis of the available literature, it is difficult to suggest any dose–response relationship between local exposure to hexavalent chromium compounds and the development of septal ulceration of the nose.

Sensitization dermatitis and asthma appear mainly after exposure to hexavalent chromium compounds. A dose–response relationship is difficult to establish and the effects are probably more dependent on individual immunological characteristics than on the actual dose *(27,28)*.

In humans, systemic effects have been reported to occur in the airways, the cardiovascular system, the kidneys and the liver *(29,30)*.

Mutagenic and carcinogenic effects

From the end of the 1940s onwards, a number of epidemiological studies of chromate-exposed workers have been published. The results vary from a slight excess of lung cancer, compared with the expected figures, to a 50-fold excess, dependent on definition of the study population, level and duration of exposure, and choice of methods *(31)*.

Alwens & Jonas *(32)* reported 20 cases of lung cancer in workers producing dichromates in Frankfurt am Main, Federal Republic of Germany, the time from first exposure to the development of cancer being 30 years. Machle & Gregorius *(33)* observed 156 deaths in their study group of approximately 1445 workers recruited from 7 major chromate-producing plants in the USA. Forty-six deaths were due to cancer; of these, 32 were due to lung cancer, a 16-fold excess compared with expected figures.

In a study on dichromate-producing workers in England, Bidstrup & Case *(34)* found 12 lung cancer cases versus 3.3 expected. When this population, plus workers employed at a later date, was followed up by Alderson et al. *(35)*, the relative risk had dropped to 1.8. Two cases of nasal cancer were also found in this study. In a study of two groups of lung cancer hospital patients in the USA, Baetjer *(36)* found a 28.6-fold excess risk in chromate production workers. Additional documentation on lung cancer risk in chromate production workers has been presented by Mancuso & Hueper *(25)*, Taylor *(37)*, Ohsaki et al. *(38)*, and Hayes et al. *(39)*.

Pokrovskaja & Šabynina *(40)* suggested that workers exposed during ferrochromium production had an excess risk of developing cancer in the lungs and the gastrointestinal tract. With regard to gastrointestinal tract cancer, this finding was not confirmed in two Scandinavian studies *(41,42)*. In one of these studies *(42)* the relative risk of lung cancer was between 2.3 and 8.5, depending on the choice of reference population. Hexavalent chromium was found in the working atmosphere in both study plants and could be the major cause of the excess lung cancer risk.

Lung cancer in chromium pigment workers was reported by Gross & Kölsch *(43)*, who described 8 lung cancer cases in 3 small chromium pigment-producing enterprises. In an epidemiological study of chromium pigment workers *(44)*, an excess lung cancer risk was found, where the relative risk was 38. However, this figure was only based on three cases of cancer. In a follow-up study, 3 additional cases had occurred in the same subpopulation (24 workers), which was considered to be at highest risk *(45)*. The lung cancer risk in chromium pigment producers has been confirmed in other studies, but the relative risk has been much lower *(46,47)*. There are indications that zinc chromate has a greater carcinogenic potency than lead chromate *(46)*.

Mancuso *(48)* presented a study on chromate-producing workers, using health data collected in 1949 for estimation of exposure. Weighted average exposures to soluble, insoluble and total chromium, combined with length of exposure, were applied to dose/time subdivisions. He concluded that a carcinogenic potential is present with all forms of chromium.

A slight excess of lung cancer risk in chromium pigment spray painters has been demonstrated in one study *(49)*, while no excess was found in another *(50)*.

In the chromium-plating industry exposure to chromium trioxide may represent a lung cancer hazard. Royle *(51)* found 17 lung cancer cases among 1238 platers, versus 10 in a matched control population. He also observed 9 cases of cancer of the gastrointestinal tract as compared with 4 in the reference population. Similar results were found by Franchini et al. *(52)*, who detected 3 lung cancer cases among hard-chromium platers versus 0.7 expected.

Welding fumes from stainless-steel welding contain hexavalent chromium *(53)*. It may therefore be assumed that stainless-steel welders are at increased risk of developing lung cancer. A number of epidemiological cancer studies have been carried out in welding populations *(31)*, but only a few studies have isolated stainless-steel welders as a separate study group. In a retrospective study of a group of 234 stainless-steel welders, the relative risk of lung cancer was 4.4 (3 cancers observed, 0.68 expected) *(54)*. In the future, stainless steel welding may remain one of the most important work operations involving exposure to hexavalent chromium.

Increased prevalence of chromosomal aberrations in lymphocytes has been found in workers exposed to chromates. The trivalent form seems to be inactive in *in vitro* systems unless the conditions of exposure allowed purified DNA to be directly exposed. Under such conditions trivalent chromium may cause modifications in the physicochemical properties as well as decreased fidelity of replication. The bulk of evidence on the genetic effects of chromium compounds is amply sufficient to form the assessment that hexavalent chromium is a powerful mutagen in humans.

Evaluation of Human Health Risks

Exposure

Chromium is present throughout nature. Available data, generally expressed as total chromium, show a concentration range of 5–100 ng/m^3. There are few data accepted as valid on the valency state or the availability of chromium in the ambient air.

Health risk evaluation

Trivalent chromium is recognized as a trace element that is essential to both man and animals. Hexavalent chromium compounds are toxic and carcinogenic, although the various compounds have a wide range of potencies. As the bronchial tree is the major target organ for carcinogenic effects of hexavalent chromium compounds, and cancer primarily occurs after inhalation exposure, uptake in the respiratory organs is of greatest importance with respect to the cancer risk in humans. IARC has stated that for chromium and certain chromium compounds there is sufficient evidence of carcinogenicity in humans (Group 1) *(1)*.

A large number of epidemiological studies have been carried out on the association between human exposure to chromates and the occurrence of cancer, particularly lung cancer, but only a few of these include measurements of exposure *(33,39,42,44,45,48,55)*. Measurements were made mainly when the epidemiological studies themselves were performed, whereas the carcinogenic response is caused by exposure dating back 15–25 years. There is therefore a great need for studies which include historical exposure data.

Three sets of data for chromate production workers can be used for the quantitative risk assessment of chromium(VI) lifetime exposure *(39,42, 44,45,55)*. The average relative risk model is used in the following to estimate the incremental unit risk.

Using the study performed by Hayes et al. on chromium production workers *(39)*, several cohorts were investigated for cumulative exposure to chromium(VI) in terms of $\mu g/m^3$-years by Braver et al. *(55)* (cumulative exposure = usual exposure level in $\mu g/m^3 \times$ average duration of exposure). Average lifetime exposures for two cohorts can be calculated from the cumulative exposures of 670 and 3647$\mu g/m^3$-years, as 2$\mu g/m^3$ and 11.4$\mu g/m^3$, respectively [$X = \mu g/m^3 \times 8/24 \times 240/365 \times$ (No. of years)/70].

The relative risk for these two cohorts, calculated from observed cases and expected lung cancers, was 1.75 and 3.04. Based on the vital statistics data, the background lifetime probability of death due to lung cancer (P_0) is assumed to be 0.04. The risks (unit risk = UR) associated with a lifetime exposure to 1$\mu g/m^3$, based on results from these two cohorts, can be calculated to be 1.5×10^{-2} and 7.2×10^{-3}, respectively [$UR = P_0(R - 1)/X$]. The arithmetic mean of these two risk estimates is 1.1×10^{-2}.

A risk assessment can also be made on the basis of the study carried out by Langård et al. on ferrochromium plant workers in Norway *(42)*. The chromium concentration to which the workers were exposed is not known, but measurements taken in 1975 showed a geometric mean value of about 530$\mu g/m^3$. Assuming that the hexavalent chromium content in the sample was 19% and previous concentrations were at least as high as in 1975, the ambient concentration would have been about 100μg chromium(VI) per m^3. On the assumption that occupational exposure lasted for about 22 years, the average lifetime exposure can be determined as 6.9$\mu g/m^3$ [$X = 100 \mu g/m^3 \times 8/24 \times 240/365 \times 22/70$].

When workers in the same plant who were not exposed to chromium were used as a control population, the relative risk (R) of lung cancer in chromium-exposed workers was calculated to be 8.5. Therefore, the lifetime unit risk is 4.3×10^{-2}.

Since earlier exposures must have been much higher than the values measured in 1975, the calculated unit risk of 4.3×10^{-2} can only be considered as an upper-bound estimate.

The highest relative incidence rate ever demonstrated in chromate workers in Norway is about 38, at an exposure level of about 0.5 mg chromium(VI) per m^3 *(44,45)*. This relative rate is based on the incidence of bronchial cancer of 0.079 in the total Norwegian male population, irrespective of smoking status. If the average exposure duration is about 7 years, the average lifetime daily exposure is calculated to be 11 $\mu g/m^3$ [$X = 500 \mu g/m^3 \times 8/24 \times 240/365 \times 7/70$]. The incremental unit risk was calculated to be 1.3×10^{-1}.

The very high lifetime risk of 1.3×10^{-1} found in this study may be due to the relatively small working population.

Differences in the epidemiological studies cited may suggest that the different hexavalent chromium compounds have varying degrees of carcinogenic potency.

The estimated lifetime risks based on various epidemiological data sets, in the range of 1.3×10^{-1} to 1.1×10^{-2} are relatively consistent. As a best estimate, the geometric mean of the risk estimates of 4×10^{-2} may be taken as the incremental unit risk resulting from a lifetime exposure to 1 μg chromium(VI) per m^3.

Using some other studies and different risk assessment models, the US Environmental Protection Agency (EPA) estimated the lifetime cancer risk due to exposure to hexavalent chromium to be 1.2×10^{-2}. This estimate placed chromium(VI) in the first quartile of the 53 compounds evaluated by the US EPA Carcinogen Assessment Group for relative carcinogenic potency *(2)*.

Guidelines

Because hexavalent chromium is carcinogenic and there is no known safe threshold, no safe level for hexavalent chromium can be recommended. At an air concentration of 1 μg hexavalent chromium per m^3, the lifetime risk is estimated to be 4×10^{-2}.

References

1. *Some metals and metallic compounds.* Lyon, International Agency for Research on Cancer, 1980 (IARC Monographs on the Evaluation of the Carcinogenic Risk of Chemicals to Humans, Vol. 23).
2. *Health assessment document for chromium.* Research Triangle Park, NC, US Environmental Protection Agency, 1984 (Final report No. EPA-600/8-83-014F).
3. **Duce, R.A et al.** Atmospheric trace metals at remote northern and southern hemisphere sites: pollution or natural? *Science,* **187**: 59–61 (1975).
4. **Bowen, H.J.M.** *Environmental chemistry of the elements.* London, Academic Press, 1979.
5. **Lahmann, E. et al.** *Heavy metals: identification of air quality and environmental problems in the European Community.* Luxembourg, Commission of the European Communities, 1986, Vol. 1 & 2 (Report No. EUR 10678 EN/I and EUR 10678 EN/II).

6. **Torgrimsen, T.** Analysis of chromium. *In:* Langård, S., ed. *Biological and environmental aspects of chromium.* Amsterdam, Elsevier Biomedical Press, 1982, pp. 65–99.
7. **Hayes, R.B.** Cancer and occupational exposure to chromium chemicals. *In:* Lilienfeld, A.M., ed. *Reviews in cancer epidemiology,* Vol. 1. New York, Elsevier-North Holland, 1980, pp. 293–333.
8. **Langård, S.** Absorption, transport and excretion of chromium in man and animals. *In:* Langård, S., ed. *Biological and environmental aspects of chromium.* Amsterdam, Elsevier Biomedical Press, 1982, pp. 149–169.
9. **Schroeder, H.A. et al.** Abnormal trace metals in man — chromium. *Journal of chronic diseases,* **15**: 941–964 (1962).
10. **Weber, H.** Long-term study of the distribution of soluble chromate-51 in the rat after a single intratracheal administration. *Journal of toxicology and environmental health,* **11**: 749–764 (1983).
11. **Bragt, P.C. & van Dura, E.A.** Toxicokinetics of hexavalent chromium in the rat after intratracheal administration of chromates of different solubilities. *Annals of occupational hygiene,* **27**: 315–322 (1983).
12. **Edel, J. & Sabbioni, E.** Pathways of Cr(III) and Cr(VI) in the rat after intratracheal administration. *Human toxicology,* **4**: 409–416 (1985).
13. **Wiegand, H.J. et al.** Disposition of intratracheally administered chromium(III) and chromium(VI) in rabbits. *Toxicology letters,* **22**: 273–276 (1984).
14. **Petrilli, F.L. & de Flora, A.** Metabolic deactivation of hexavalent chromium mutagenicity. *Mutation research,* **54**: 137–147 (1978).
15. **Mertz, W.** Biological role of chromium. *Federation proceedings,* **26**: 186–193 (1967).
16. **Tandon, S.K.** Organ toxicity of chromium in animals. *In:* Langård, S., ed. *Biological and environmental aspects of chromium.* Amsterdam, Elsevier Biomedical Press, 1982, pp. 209–220.
17. **Hueper, W.C.** Experimental studies in metal cancerogenesis. X. Cancerogenic effects of chromite ore roast deposited in muscle tissue and pleural cavity of rats. *AMA archives of industrial health,* **18**: 284–291 (1958).
18. **Baetjer, A.M. et al.** Effect of chromium on incidence of lung tumors in mice and rats. *AMA archives of industrial health,* **20**: 124–135 (1959).
19. **Payne, W.W.** Production of cancers in mice and rats by chromium compounds. *AMA archives of industrial health,* **21**: 530–535 (1960).
20. **Hueper, W.C. & Payne, W.W.** Experimental cancers in rats produced by chromium compounds and their significance to industry and public health. *American Industrial Hygiene Association journal,* **20**: 274–280 (1959).
21. **Nettesheim, P. et al.** Effect of calcium chromate dust, influenza virus, and 100R whole-body X radiation on lung tumor incidence in mice. *Journal of the National Cancer Institute,* **47**: 1129–1138 (1971).
22. **Levy, L.S. & Venitt, S.** Carcinogenicity and mutagenicity of chromium compounds: the association between bronchial metaplasia and neoplasia. *Carcinogenesis,* **7**: 831–835 (1986).

23. **Levis, A.G. & Bianchi, V.** Mutagenic and cytogenetic effects of chromium compounds. *In:* Langård, S., ed. *Biological and environmental aspects of chromium.* Amsterdam, Elsevier Biomedical Press, 1982, pp. 171–208.
24. **Lindberg, K. & Hedenstierna, G.** Chrome plating: symptoms, findings in the upper airways, and effects on lung function. *Archives of environmental health,* **38**: 367–374 (1983).
25. **Mancuso, T.F. & Hueper, W.S.** Occupational cancer and other health hazards in a chromate plant: a medical appraisal. I. Lung cancers in chromate workers. *Industrial medicine and surgery,* **20**: 358–363 (1951).
26. **Sluis-Cremer, G.K. & du Toit, R.S.J.** Pneumoconiosis in chromite miners in South Africa. *British journal of industrial medicine,* **25**: 63–67 (1968).
27. **Polak, L. et al.** Studies on contact hypersensitivity to chromium compounds. *In:* Kallos, P. et al., ed. *Progress in allergy,* Vol. 17. Basle, Karger, 1973, pp. 145–226.
28. **Pedersen, N.B.** The effects of chromium on the skin. *In:* Langård, S., ed. *Biological and environmental aspects of chromium.* Amsterdam, Elsevier Biomedical Press, 1982, pp. 249–275.
29. **Langård, S.** Chromium. *In:* Waldron, H.A., ed. *Metals in the environment.* London, Academic Press, 1980, p. 111–132.
30. **Norseth, T.** Health effects of nickel and chromium. *In:* Di Ferrante, E., ed. *Trace metals: exposure and health effects.* Oxford, Pergamon Press, 1978, pp. 135–146.
31. **Langård, S.** The carcinogenicity of chromium compounds in man and animals. *In:* Burrows, D., ed. *Chromium: metabolism and toxicity.* Boca Raton, FL, CRC Press, 1983, pp. 13–30.
32. **Alwens, W. & Jonas, W.** Der Chromat-Lungenkrebs [Lung cancer caused by chromate]. *Acta unionis internationalis contra cancrum,* **3**: 103–118 (1938).
33. **Machle, W. & Gregorius, F.** Cancer of the respiratory system in the United States chromate-producing industry. *Public health reports,* **63**: 1114–1127 (1948).
34. **Bidstrup, P.L. & Case, R.A.M.** Carcinoma of the lung in workmen in the bichromates-producing industry in Great Britain. *British journal of industrial medicine,* **13**: 260–264 (1956).
35. **Alderson, M.R. et al.** Health of workmen in the chromate-producing industry in Britain. *British journal of industrial medicine,* **38**: 117–124 (1981).
36. **Baetjer, A.M.** Pulmonary carcinoma in chromate workers. I. A review of the literature and report of cases. *AMA archives of industrial hygiene and occupational medicine,* **2**: 487–504 (1950).
37. **Taylor, F.H.** The relationship of mortality and duration of employment as reflected by a cohort of chromate workers. *American journal of public health,* **56**: 218–229 (1966).

38. **Ohsaki, Y. et al.** Lung cancer in Japanese chromate workers. *Thorax,* **33**: 372–374 (1978).
39. **Hayes, R.B. et al.** Mortality in chromium chemical production workers: a prospective study. *International journal of epidemiology,* **8**: 365–374 (1979).
40. **Pokrovskaja, L.V. & Šabynina, N.K.** [Carcinogenous hazard in the production of chromium ferroalloys]. *Gigiena truda i professional'nye zabolevanija,* **10**: 23–26 (1975) (in Russian).
41. **Axelsson, G. et al.** Mortality and incidence of tumours among ferrochromium workers. *British journal of industrial medicine,* **37**: 121–127 (1980).
42. **Langård, S. et al.** Incidence of cancer among ferrochromium and ferrosilicon workers. *British journal of industrial medicine,* **37**: 114–120 (1980).
43. **Gross, E. & Kölsch, F.** Über den Lungenkrebs in der Chromfarbenindustrie [Lung cancer in the chromate dye industry]. *Archiv für Gewerbepathologie und Gewerbehygiene,* **12**: 164–170 (1943).
44. **Langård, S. & Norseth, T.** A cohort study of bronchial carcinomas in workers producing chromate pigments. *British journal of industrial medicine,* **32**: 62–65 (1975).
45. **Langård, S. & Vigander, T.** Occurrence of lung cancer in workers producing chromium pigments. *British journal of industrial medicine,* **40**: 71–74 (1983).
46. **Davies, J.M.** Lung cancer mortality among workers making lead chromate and zinc chromate pigments at three English factories. *British journal of industrial medicine,* **41**: 158–169 (1984).
47. **Frentzel-Beyme, R.** Lung cancer mortality of workers employed in chromate pigment factories: a multicentric European epidemiological study. *Journal of cancer research and clinical oncology,* **105**: 183–188 (1983).
48. **Mancuso, T.F.** Consideration of chromium as an industrial carcinogen. *In:* Hutchinson, T.C., ed. *Proceedings of the International Conference on Heavy Metals in the Environment, Toronto, 1975.* Toronto, Institute for Environmental Studies, 1975, pp. 343–356.
49. **Dalager, N.A. et al.** Cancer mortality among workers exposed to zinc chromate paints. *Journal of occupational medicine,* **22**: 25–29 (1980).
50. **Chiazze, L. et al.** Mortality among automobile assembly workers. I. Spray painters. *Journal of occupational medicine,* **22**: 520–526 (1980).
51. **Royle, H.** Toxicity of chromic acid in the chromium plating industry (1). *Environmental research,* **10**: 39–53 (1975).
52. **Franchini, I. et al.** Mortality experience among chromeplating workers: initial findings. *Scandinavian journal of work, environment & health,* **9**: 247–252 (1983).
53. **Stern, R.M.** Chromium compounds: production and occupational exposure. *In:* Langård, S., ed. *Biological and environmental aspects of chromium.* Amsterdam, Elsevier Biomedical Press, 1982, pp. 5–47.

54. **Sjögren, B.** A retrospective cohort study of mortality among stainless steel welders. *Sçandinavian journal of work, environment & health,* **6**: 197–200 (1980).
55. **Braver, E.R. et al.** An analysis of lung cancer risk from exposure to hexavalent chromium. *Teratogenesis, carcinogenesis and mutagenesis,* **5**: 365–378 (1985).

22

Hydrogen sulfide

General Description

Sources

Hydrogen sulfide (H_2S) is a colourless gas, soluble in various liquids including water and alcohol. It can be formed under conditions of deficient oxygen, in the presence of organic material and sulfate. Most of the atmospheric hydrogen sulfide has natural origins. Hydrogen sulfide occurs around sulfur springs and lakes, and is an air contaminant in geothermally active areas. Saline marshes can also produce sulfide *(1)*. The estimated global release of hydrogen sulfide from saline marshes into the atmosphere is 8.3×10^5 tonnes per year.

Human activities can release naturally occurring hydrogen sulfide into ambient air. For instance, some natural gas deposits contain up to 42% hydrogen sulfide *(2)*. In industry, hydrogen sulfide can be formed whenever elemental sulfur or sulfur-containing compounds come into contact with organic materials at high temperatures. Hydrogen sulfide is formed, for instance, during coke production, in viscose rayon production, in waste-water treatment plants, in wood pulp production using the sulfate method, in sulfur extraction processes, in oil refining and in the tanning industry. In Canada, in 1978, the kraft pulping industry was estimated to be responsible for 97% of the country's total anthropogenic hydrogen sulfide emissions *(3)*. However, only 10% of the total global emissions of this compound are of anthropogenic origin.

Occurrence in air

In one report *(2)*, the average ambient air hydrogen sulfide level was estimated to be 0.3 $\mu g/m^3$ (0.0002 ppm). In north-west London, over a period of 2.5 years, air levels of hydrogen sulfide were generally below 0.15 $\mu g/m^3$ (0.0001 ppm) under clear conditions *(2)*. In and around the city of Rotorua, New Zealand, where there is geothermal activity, there is usually a sufficient hydrogen sulfide concentration to cause odours. During continuous monitoring in Rotorua a hydrogen sulfide concentration of 0.08 mg/m^3 (0.05 ppm) was exceeded more than 55% of the time in the mid-winter months *(2)*. Rather high concentrations of hydrogen sulfide have been measured near point sources. Near a pulp and paper-mill in California, peak concentrations of up to 0.20 mg/m^3 (0.13 ppm) were measured *(2)*. In a Finnish town with two sulfate pulp mills (annual emissions of 1993 tonnes and 794 tonnes of hydrogen sulfide, respectively) concentrations of hydrogen sulfide near the mills were estimated, using the emission data and a dispersion model for the spread of gaseous sulfur compounds *(4)*. The average annual concentrations were calculated to be up to 55 μg/m^3, monthly average concentrations up to 100 μg/m^3, 24-hour concentrations up to 540 μg/m^3, and 1-hour concentrations up to 1600 μg/m^3.

In another Finnish town, hydrogen sulfide concentrations near a viscose rayon mill were partly measured and partly estimated by using a dispersion model *(5)*. When the smokestack of the mill was only 55 m high, the average annual concentrations exceeded 10 μg/m^3, 24-hour concentrations were approximately 200 μg/m^3 and short-term concentrations were up to 450 μg/m^3. When a higher smokestack was installed the annual concentrations were reduced to 4 μg/m^3, 24-hour concentrations to 35 μg/m^3 and 1-hour concentrations to a maximum of 80 μg/m^3.

During accidental exposures, concentrations from 150 mg/m^3 (100 ppm) to 18 000 mg/m^3 (12 000 ppm) have been reported *(2)*. In a Finnish study *(6)*, a health survey for hydrogen sulfide and other sulfides was carried out at six kraft mills. The hydrogen sulfide concentrations varied from less than 0.075 mg/m^3 (0.05 ppm) to 30 mg/m^3 (20 ppm), the highest concentrations being found near vacuum pumps. A Japanese study in 18 viscose rayon plants showed occupational exposure levels of hydrogen sulfide ranging from 0.45 to 11.7 mg/m^3 (0.3–7.8 ppm), with a mean of 4.5 mg/m^3 (3 ppm) *(7)*.

Hydrogen sulfide is the main toxic substance involved in livestock rearing systems with liquid manure storage *(8)*. It is also a hazard at waste treatment facilities.

Conversion factors

1 ppm = 1.5 mg/m^3

1 mg/m^3 = 0.670 ppm

Routes of Exposure

The respiratory system is the main route of human exposure to hydrogen sulfide both in workplaces and in the ambient air. A recent report *(9)* criticizes the earlier belief that hydrogen sulfide can enter the body via tympanic membrane defects in workplace concentrations. There is very limited data on the sulfide levels of drinking-water *(10)*.

Kinetics and Metabolism

Absorption

There are no exact data about the absorption of hydrogen sulfide through the lungs, but the absorbed proportion is probably large. Hydrogen sulfide is dissociated at physiological pH values to a hydrogen sulfide anion, in which form it is most probably absorbed *(11)*.

The rate of absorption in the gastrointestinal tract is not known. However, hydrogen sulfide poisoning can be produced by infusing soluble sulfide salts in the gastrointestinal tract *(11)*. Penetration through the skin is insignificant *(2)*.

Distribution

Sodium sulfide, when administered parenterally, concentrates in the liver, but small amounts can also be found in the kidneys and lungs *(11)*. In five lethal cases of hydrogen sulfide poisoning, sulfide concentrations in blood were between 1.7 and 3.75 mg/litre *(11)*; this was due to an acute lethal hydrogen sulfide concentration of at least 750–1400 mg/m^3. Because of the metabolic process of the sulfide anion after the intoxication, it is difficult to determine the air concentration of hydrogen sulfide on the basis of blood sulfide levels. Moreover, the sulfide concentrations in other organs have not been determined.

Biotransformation

The sulfide anion is transformed by the sulfide oxidase system in the liver and kidneys mainly into thiosulfate and sulfate *(11)*. In the mucosa of the gastrointestinal tract a thiol-S-methyltransferase system can also detoxify hydrogen sulfide *(11)*.

Elimination

When sodium sulfide was injected intraperitoneally into mice no metabolites were found in the exhaled air *(11)*.

When sodium (35)S-sulfide was administered orally to rats, 50% of the dose was found as sulfate in the urine after 24 hours *(11)*. When administered parenterally, 91% of the dose was found in urine (74%) and faeces (17%).

Health Effects

Effects on experimental animals

The main toxicological effect at the cellular level in the brain is the inhibition of the enzyme cytochrome oxidase at the end of the mitochondrial respiratory chain *(11)*. Because of abnormal mitochondrial function, many secondary changes appear in cells which have a great energy demand.

In acute hydrogen sulfide intoxications brain oedema, degeneration and necrosis of the cerebral cortex and the basal ganglia have been reported in rhesus apes *(11)*. A rhesus monkey developed brain damage after a nonlethal hydrogen sulfide intoxication *(11)*. An exposure of hydrogen sulfide at a concentration of 150 mg/m^3 for 2 hours caused inhibition of the brain protein synthesis in mice in 48 hours; this was normalized in 72 hours *(11)*. It has been reported that the cerebral biochemical effects caused by repeated subclinical hydrogen sulfide intoxications are cumulative in mice *(11)*.

Information on the effects of low-level concentrations of hydrogen sulfide on experimental animals is limited. The following effects have been reported on the basis of animal experiments: an increase of the reticulocytes, an increase of the blood cell volume in rats, a diminished leucocyte count, an increased lymphocyte count, and an increased thymol turbidity in rabbits *(11)*. Injection of hydrogen sulfide water solution (265 mg/litre) into the ear vein of the rabbit caused prolongation of the diastole and lowered the heart frequency. There were changes in the T-wave *(11)*. A decrease of the heart frequency has been reported in other mammals that were given intravenous injections of 0.5–10 mg/litre of hydrogen sulfide *(11)*.

Findings in the canary, cat, dog, goat, guinea pig, rabbit and rat *(2)* are consistent as to the effects of hydrogen sulfide: at 150–225 mg/m^3 signs of local irritation of eyes and throat after many hours of exposure; at 300–400 mg/m^3 eye and mucous membrane irritation in 1 hour and slight general effects with longer exposure; at 750–1000 mg/m^3 slight systemic symptoms in less than 1 hour and possible death after several hours; at 1350 mg/m^3 grave systemic effects within 30 minutes and death in less than 1 hour; at 2250 mg/m^3 collapse and death within 15–30 minutes, and at 2700 mg/m^3 immediate collapse, respiratory paralysis and death.

Injection of sodium nitrite, inducing methaemoglobinaemia, had protective and antidotal effects against hydrogen sulfide poisoning in mice, armadillos, rabbits, and dogs *(2)*. There is very limited information on chronic hydrogen sulfide intoxication in experimental animals.

No information is available on the mutagenic, carcinogenic or teratogenic effects of hydrogen sulfide in experimental animals.

Effects on humans

In its acute form, hydrogen sulfide intoxication is mainly the result of action on the nervous system. At concentrations of 15 mg/m^3 and above, hydrogen sulfide causes conjunctival irritation, because sulfide and hydrogen sulfide anions are strong bases *(11)*. Hydrogen sulfide affects the sensory nerves in the conjunctivae, so that pain is diminished rapidly and the tissue damage is greater *(11)*. Serious eye damage is caused by a concentration of 70 mg/m^3. At higher concentrations (above 225 mg/m^3, or 150 ppm), hydrogen sulfide has a paralysing effect on the olfactory perception *(2)*, so that the odour can no longer be recognized as a warning signal. At higher concentrations, respiratory irritation is the predominant symptom, and at a concentration of around 400 mg/m^3 there is a risk of pulmonary oedema. At even higher concentrations there is strong stimulation of the CNS *(2)*, with hyperpnoea leading to apnoea, convulsions, unconsciousness, and death. At concentrations of over 1400 mg/m^3 there is immediate collapse. In fatal human intoxication cases, brain oedema, degeneration and necrosis of the cerebral cortex and the basal ganglia have been observed *(11)*.

If respiration can be maintained, the prognosis in a case of acute hydrogen sulfide intoxication, even a severe one, is fairly good. There are reports of neurasthenic symptoms after severe acute intoxication, such as amnesia, fatigue, dizziness, headache, irritability, and lack of initiative *(11)*. A decrease of delta-aminolaevulinic acid dehydrase (ALAD) synthase and haem synthase activity in reticulocytes one week after hydrogen sulfide intoxication has been reported *(12)*, together with low levels of erythrocyte protoporphyrin. The ALAD and haem synthase activities returned to normal two months after the accident, erythrocyte protoporphyrin remaining low. Changes in the electrocardiogram have been reported after acute hydrogen sulfide intoxication, these changes being reversible *(11)*. No tolerance to the acute effects of hydrogen sulfide has been reported to develop *(11)*.

The mortality in acute hydrogen sulfide intoxications seems to be lower than that reported in 1977; according to a recent Canadian report it is now 2.8% *(13)*, whereas formerly it was 6% *(2)*. This may be a result of improved first-aid procedures and increased awareness of the dangers of hydrogen sulfide.

Information about longer-term exposures to hydrogen sulfide is scanty. Eighty-one Finnish pulp mill workers who were exposed to hydrogen sulfide concentrations of less than 30 mg/m^3 (20 ppm) and to methyl mercaptan concentrations of less than 29.6 mg/m^3 (15 ppm), displayed loss of concentration capacity and chronic or recurrent headache more often than a nonexposed control group of 81 workers. Restlessness and lack of vigour also appeared more often, but the findings were not statistically significant. There was also a tendency towards more frequent sick leave among the exposed group *(6)*. One report cites decreased activity of haem synthesizing enzymes in reticulocytes of pulp mill workers exposed for years to organic and inorganic sulfides, with hydrogen sulfide concentrations of between 0.075 mg/m^3 and 7.8 mg/m^3 *(12)*. No information is available as to whether the observed effect was related to peak concentrations or average concentrations. It can, however, be assumed that average exposure was considerably higher than 0.075 mg/m^3 (around 1.5–3 mg/m^3). Furthermore, there is no firm proof that hydrogen sulfide was the causative agent, as there may be confounding factors (other substances).

Epidemiological data concerning longer-term exposures are limited. Seventy per cent of workers exposed to hydrogen sulfide daily, often at 30 mg/m^3 or more, complained of such symptoms as fatigue, somnolence, headache, irritability, poor memory, anxiety, dizziness, and eye irritation *(14)*. In a Finnish mortality study workers in a sulfate pulp mill showed excess mortality from cardiovascular diseases (standardized mortality rate 140), and especially from heart infarction (standardized mortality rate 142). The findings were statistically significant. In the same study population, cancer incidence was not significantly different from that of the general Finnish population *(15)*.

Sensory effects

Hydrogen sulfide is an odorant, which in pure form has an odour detection threshold of 0.2–2.0 μg/m^3 depending on the purity *(16,17)*. Its characteristic smell of rotten eggs appears at concentrations 3–4 times higher than the odour threshold *(18)*. In practical situations, such as in the effluents of kraft pulp mills, hydrogen sulfide is accompanied by other odorous substances, such as methyl mercaptan, dimethyl disulfide and dimethyl monosulfide, and, in the case of effluents of the viscose industry, by carbon disulfide. The odour quality of these emissions changes with the specific composition of the mixtures *(19)*.

Hydrogen sulfide causes odour nuisance at concentrations far below those that cause health hazards. On the basis of the scientific literature, it is not possible to state a specific concentration of hydrogen sulfide at which odour nuisance starts to appear. Half-hour average concentrations exceeding 7 $\mu g/m^3$ are likely to produce substantial complaints among persons exposed *(19,20)*. A reduction in the concentration of hydrogen sulfide does not guarantee a substantial reduction of the odour nuisance, since hydrogen sulfide in many effluents provides only a small contribution to the odour strength of the total effluent *(21)*. Moreover, the interaction between hydrogen sulfide and other odorous components in the effluent cannot explain the odour strength of the total effluents from pulp mills *(21)*. Better short-term studies are required to elucidate the relationship between actual concentrations and reports of odour nuisance.

Evaluation of Human Health Risks

Exposure

Typical symptoms and signs of hydrogen sulfide intoxication are most often caused by relatively high concentrations in occupational exposures. There are many occupations where there is a potential risk of hydrogen sulfide intoxication and, according to the US National Institute for Occupational Safety and Health *(14)*, in the United States alone approximately 125 000 employees are potentially exposed to hydrogen sulfide. Low-level concentrations can occur more or less continuously in certain industries, such as in viscose rayon and pulp production, at oil refineries and in geothermal energy installations.

In geothermal areas there is a risk of exposure to hydrogen sulfide for the general population *(2)*. The biodegradation of industrial wastes has been reported to cause ill effects in the general population *(2)*. An accidental release of hydrogen sulfide into the air surrounding industrial facilities can cause very severe effects, as at Poza Rica, Mexico, where 320 people were hospitalized and 22 died *(2)*. The occurrence of low-level concentrations of hydrogen sulfide around certain industrial installations is a well known fact (see pp. 233–234).

Dose–effect and dose–response relationship

The first noticeable effect of hydrogen sulfide at low concentrations is its unpleasant odour. Conjunctival irritation is the next subjective symptom and can cause so-called gas eye at hydrogen sulfide concentrations of 70–140 mg/m^3. Table 1 shows the established dose–effect relationships for hydrogen sulfide.

Health risk evaluation

The hazards caused by high concentrations of hydrogen sulfide are relatively well known, but information on human exposure to very low concentrations

Table 1. Hydrogen sulfide: established dose- effect relationships

Hydrogen sulfide concentration		Effect	Reference
mg/m³	ppm		
1400-2800	1000-2000	Immediate collapse with paralysis of respiration	*(2)*
750-1400	530-1000	Strong CNS stimulation, hyperpnoea followed by respiratory arrest	*(2)*
450-750	320-530	Pulmonary oedema with risk of death	*(2)*
210-350	150-250	Loss of olfactory sense	*(11)*
70-140	50-100	Serious eye damage	*(11)*
15-30	10-20	Threshold for eye irritation	*(11)*

is scanty. Workers exposed to hydrogen sulfide concentrations of less than 30 mg/m³ are reported to have rather diffuse neurological and mental symptoms *(6)* and to show no statistically significant differences when compared with a control group. On the other hand, changes of haem synthesis have been reported at hydrogen sulfide concentrations of less than 7.8 mg/m³ (1.5–3 mg/m³ average) *(12)*. It is not known whether the inhibition is caused by the low concentration levels or by the cumulative effects of occasional peak concentrations. Most probably, at concentrations below 1.5 mg/m³ (1 ppm), even in exposure for longer periods, there are very few detectable health hazards in the toxicological sense. The malodorous property of hydrogen sulfide is a source of annoyance for a large proportion of the general population at concentrations below 1.5 mg/m³, but from the existing data it cannot be concluded whether any health effects result. The need for epidemiological studies on possible effects of long-term, low-level hydrogen sulfide exposure is obvious. A satisfactory biological exposure indicator is also needed.

Guidelines

The lowest-adverse-effect level of hydrogen sulfide is 15 mg/m³, when eye irritation is caused. In view of the steep rise in the dose–effect curve implied by reports of serious eye damage at 70 mg/m³, a relatively high protection (safety) factor of 100 is recommended, leading to a guideline value of 0.15 mg/m³ with an averaging time of 24 hours. A single report of changes in haem synthesis at a hydrogen sulfide concentration of 1.5 mg/m³ should be borne in mind.

In order to avoid substantial complaints about odour annoyance among the exposed population, hydrogen sulfide concentrations should not be allowed to exceed 7 $\mu g/m^3$, with a 30-minute averaging period.

When setting concentration limits in ambient air, it should be remembered that hydrogen sulfide is emitted from natural sources in many places.

References

1. **Steudler, P.A. & Peterson, B.J.** Contribution of gaseous sulphur from salt marshes to the global sulphur cycle. *Nature,* **311**: 455–457 (1984).
2. *Hydrogen sulfide.* Geneva, World Health Organization, 1981 (Environmental Health Criteria, No. 19).
3. *Hydrogen sulfide in the atmospheric environment: scientific criteria for assessing its effects on environmental quality.* Ottawa, National Research Council Canada, 1981 (Publication No. 18467).
4. **Häkkinen, A.J. et al.** *Imatran ilman rikkidioksidin ja keskeisten hajurikkiyhdisteiden pitoisuustasot sekä alueen havupuuvauriot. Ilmatieteen laitos* [Concentration levels of sulfur dioxide and main odorous sulfur compounds in Imatra, and damage to the coniferous trees of the area]. Helsinki, Meteorological Institute, 1985.
5. *Valkeankosken ilma ja terveys. Epidemiologinen tutkimus yhdyskuntailman ja terveyden välisistä suhteista rekistereistä saatavien tietojen valossa* [Ambient air and health status in Valkeankoski]. Helsinki, Government Printing Centre, 1982 (National Board of Health, Working Group Report, No. 3).
6. **Kangås, J. et al.** Exposure to hydrogen sulfide, mercaptans and sulfur dioxide in pulp industry. *American Industrial Hygiene Association journal,* **45**: 787–790 (1984).
7. **Higashi, T. et al.** Cross sectional study of respiratory symptoms and pulmonary functions in rayon textile workers with special reference to hydrogen sulfide exposure. *Industrial health,* **21**: 281–292 (1983).
8. **Donham, K.J. et al.** Acute toxic exposure to gases from liquid manure. *Journal of occupational medicine,* **24**: 142–145 (1982).
9. **Ronk, R. & White, M.** Hydrogen sulfide and the probabilities of "inhalation" through a tympanic membrane defect. *Journal of occupational medicine,* **27**: 337–340 (1985).
10. *Guidelines for drinking-water quality. Vol. 2. Health criteria and other supporting information.* Geneva, World Health Organization, 1984.
11. **Savolainen, H.** Nordiska expertgruppen för gränsvärdesdokumentation. 40. Dihydrogensulfid [Nordic expert group for TLV evaluation. 40. Hydrogen sulfide]. *Arbeta och hälsa,* **31**: 1–27 (1982).
12. **Tenhunen, R. et al.** Changes in haem synthesis associated with occupational exposure to organic and inorganic sulphides. *Clinical science,* **64**: 187–191 (1983).
13. **Arnold, I.M.F. et al.** Health implication of occupational exposures to hydrogen sulfide. *Journal of occupational medicine,* **27**: 373–376 (1985).

14. *Occupational exposure to hydrogen sulfide.* Cincinnati, OH, US Department of Health, Education, and Welfare, 1977 (DHEW Publication (NIOSH) No. 77-158).
15. **Jäppinen, P. et al.** Cancer incidence of workers in the Finnish pulp and paper industry. *Scandinavian journal of work, environment & health,* **13**: 197–202 (1987).
16. **Winneke, G. et al.** Zur Wahrnehmung von Schwefelwasserstoff unter Labor- und Feldbedingungen [Determination of hydrogen sulfide in laboratory and field conditions]. *Staub, Reinhaltung der Luft,* **39**: 156–159 (1979).
17. **van Gemert, L.J. & Nettenbreijer, A.H., ed.** *Compilation of odour threshold values in air and water.* Zeist, Central Institute for Nutrition and Food Research, 1977 and Supplement V, 1984.
18. **Leonardos, G. et al.** Odour threshold determinations of 53 odorant chemicals. *Journal of the Air Pollution Control Association,* **19**: 91–95 (1969).
19. **National Research Council.** *Odors from stationary and mobile sources.* Washington, DC, National Academy of Sciences, 1979, p. 491.
20. **Lindvall, T.** On sensory evaluation of odorous air pollutant intensities. Measurements of odor intensity in the laboratory and in the field, with special reference to effluents of sulfate pulp factories. *Nordisk hygienisk tidskrift,* **51**(Suppl. 2): 36–39 (1970).
21. **Berglund, B. et al.** Perceptual interaction of odors from a pulp mill. *In: Proceedings of the Third International Clean Air Congress, Düsseldorf, 1973.* Düsseldorf, VDI, 1973, pp. A40–A43.

23

Lead

General Description

Lead is a bluish or silvery-grey soft metal with a melting-point of 327.5 °C and a boiling-point at atmospheric pressure of 1740 °C. It has four naturally occurring isotopes (atomic weights: 208, 206, 207 and 204 in order of abundance). The isotopic ratios differ for various mineral sources. This property has been used in nonradioactive-tracer studies to investigate the environmental and metabolic pathways of lead.

The usual oxidation state of lead in inorganic compounds is +2. Apart from nitrate, chlorate and, to a much lesser degree, chloride, most of the inorganic salts of lead (II) have a poor solubility in water.

Organic lead compounds such as tetraethyl lead and tetramethyl lead are of great importance due to their extensive use as fuel additives. Tetraethyl lead and tetramethyl lead are colourless liquids with boiling-points of 110 °C and 200 °C respectively. Since their volatility is lower than that of most petrol components, the evaporation of petrol tends to concentrate tetraethyl lead and tetramethyl lead. Both compounds are decomposed at boiling-point as well as by ultraviolet light and trace chemicals in air, such as halogens, acids, or oxidizing agents.

Sources

The combustion of alkyl lead additives in motor fuels accounts for the major part of all lead emissions into the atmosphere. An estimated 80–90% of lead in ambient air derives from the combustion of leaded petrol. The degree of pollution from this source differs from country to country, depending on motor vehicle density and the efficiency of efforts to reduce the lead content of petrol.

The mining and smelting of lead ores create pollution problems in some areas. The level of contamination of the surrounding air and soil depends on the amount of lead emitted, the height of the stack, topography, and other local features. Secondary lead smelters, the refining and manufacture of compounds and goods containing lead, and refuse incineration also give rise to lead emissions.

Since coal, like many minerals, rocks and sediments, usually contains low concentrations of lead, a number of other industrial activities such as iron and steel production, copper smelting and coal combustion must be regarded as additional sources of lead emissions into the atmosphere.

The presence of lead water-pipes in old houses can be an important source of lead exposure for humans, particularly in soft-water areas. In certain areas lead-containing paint in old houses can be an additional source of exposure.

Occurrence in air

Current "baseline" levels of lead in the atmosphere are estimated to be in the range of $5 \times 10^{-5} \mu g/m^3$ *(1)*. Whether or not this level is wholly natural or a composite of natural and anthropogenic sources can be determined by analysing the isotopic composition. Even in the remotest sites, human activity has raised the lead concentration in the air considerably higher than natural levels. In nonurban sites located near urban areas, air lead levels average around $0.5 \mu g/m^3$, while in rural areas, levels in the range of 0.1 to $0.3 \mu g/m^3$ are found.

High concentrations of lead in ambient air are found in urban areas with high traffic density. At present, urban air lead levels are in the range of 0.5 to $3.0 \mu g/m^3$ (annual means) in most European cities. However, owing to decreases in the lead content of petrol, there is a trend towards lower air lead values *(2)*. High air lead levels are found in the vicinity of lead smelters.

Most of the lead in the air is in the form of fine particles with a mass median equivalent diameter of less than $1 \mu m$. The fraction of organic lead (predominantly lead alkyls that escaped combustion) is generally below 10% of the total atmospheric lead, the majority (>90%) of lead from leaded petrol emission being emitted as inorganic particles (e.g. PbBrCl). In the immediate vicinity of smelters, the particle size distribution usually shows a predominance of larger particles. However, these particles settle at distances of a few hundred metres or 1–2 km, so that further away the particle size distribution is indistinguishable from that of other urban sites.

Since people spend much of their time indoors, ambient air data may not accurately indicate actual exposure to airborne lead. Studies on indoor/outdoor air lead levels show that the indoor concentrations, in general, are lower than the corresponding outdoor values. Overall, the data suggest that indoor/outdoor ratios in the range of 0.6–0.8 are typical for airborne lead in houses without air-conditioning. Lower indoor/outdoor ratios have been observed during winter, when windows and doors are tightly closed *(3)*.

Lead is removed from the atmosphere by dry or wet deposition. The residence time of lead-containing particles in the atmosphere varies according to a number of factors, such as particle size, wind currents, rainfall and height of emission. Soil and water pollution from car emission fallout is predominantly limited to the immediate urban area. Fallout from the emission of industrial sources, such as smelters, is likewise limited mainly to the immediate vicinity. However, strong evidence indicates that a fraction of airborne lead is transported over long distances. As a result, a long-term global accumulation of lead has occurred in recent decades. This has been demonstrated convincingly by analyses of glacial ice and snow deposits in remote areas *(1)*.

Routes of Exposure

Air

Most of the lead in ambient air is in the form of submicron-sized particles. Some 30–50% of these inhaled particles are retained in the respiratory system. Virtually all of this retained lead is absorbed into the body. Particles in the size range of 1–3 μm are also efficiently deposited in the lungs. Larger particles are deposited with variable efficiency, mainly in the upper respiratory tract with incomplete absorption. All lead particles that are cleared by the lung can be swallowed and result in further lead absorption in the gastrointestinal tract.

Drinking-water

Lead concentrations in drinking-water and groundwater vary from 1 to 60 μg/litre. In most European countries, the levels of lead in domestic tap-water are relatively low, i.e. normally below 20 μg/litre. Consequently, man's exposure to lead through water is generally low in comparison with exposure through food *(4,5)*. However, in some places (areas with soft water, where lead water-pipes and lead plumbing are common), lead in drinking-water can contribute significantly to the total lead intake *(6)*.

Food

Most people receive the largest portion of their daily lead intake via food. Most lead enters food during storage and manufacture, e.g. in canned food and in alcoholic drinks. The most important pathway whereby atmospheric lead enters the food chain is thought to be direct foliar contamination of plants. This contamination depends on the rate of fallout of lead in the districts where food is grown; it tends to be higher in heavily industrialized areas. Additionally, air deposits raise the level of lead in soil, which, in the course of decades and centuries, may result in an increased uptake of lead through the roots.

The concentrations of lead in various food items are highly variable. Several studies have reported average lead intakes in the range of 100–500 μg/day for adults, with individual diets covering a much greater range. More recent data indicate total daily intakes of about 100 μg or less *(2,7)*. For young children, estimates of total daily intakes are about one half the figures for adults. Recent data suggest that levels of lead in the diet appear to have been falling in the last few years.

Additional exposures

An individual's lead exposure may be increased by choice, habit or unavoidable circumstances, in addition to the "normal" environmental exposure through food, drinking-water and air. These additional exposures can be categorized as being either due to lead in the ambient air or independent of lead in the ambient air.

The former category includes high lead levels in dustfall and soil in residential areas near smelters or refineries, high-density traffic, and the consumption of vegetables and fruit grown on high-lead soils or near sources of lead emissions (smelters, roadways with high traffic density).

The latter category includes occupational exposures; secondary occupational exposures of members of the families of lead workers; contamination of house dust in houses with interior lead paint; contamination of tap-water in houses with lead water-pipes or lead plumbing; use of improperly glazed earthenware vessels; tobacco smoke *(8)* and alcohol consumption (in particular, wine) *(8,9)*.

Lead in dust, indoors as well as outdoors, is an important potential source of intake by ingestion, particularly for young children living in contaminated areas, e.g. lead smelter areas and central urban areas *(10–14)*.

Relative significance of different routes of exposure

Exposure to lead from water, food, air and other sources can vary significantly for different individuals and population groups. Since the relative contribution of each of these sources can also vary substantially, comprehensive information covering a wide range of circumstances cannot be provided. To give some idea of possible situations, a few simplified calculations are presented in Table 1.

Contributions from occupational exposures, cigarette-smoking and various other sources were not taken into account. Regarding the absorbed dose of lead, the contribution of inhaled airborne lead is in the range of 15–70% in adults and 2–17% in children. As far as adults are concerned, these figures are consistent with the results of isotopic tracing studies, which indicate that the percentage of petrol lead contributing to total human blood lead can be in the order of 25% or somewhat higher *(15)*.

An important group omitted in Table 1 is that of infants (up to 1 year old). At present, insufficient information is available on the lead content of their diet and its absorption for reasonable estimates to be made. However, the contribution of drinking-water in this group is likely to be high, probably higher than that for the children aged 1–5 years referred to in Table 1.

It should be emphasized that preschool children represent the most important risk group. For this group the contribution of air lead to blood lead by way of inhalation alone, as estimated in Table 1, clearly underestimates the contribution of environmental lead to blood lead, as air lead can only be taken as an indicator of general multimedia lead pollution. Because of the breathing behaviour of preschool children, outdoor lead deposition is the most important single explanation of differences between inner city and suburban areas in the blood lead of children *(13)*. Table 2 therefore illustrates some possible situations based on various assumptions of dust intake by children along with intake of lead in air, food and water. It is obvious that lead in dust can make a substantial contribution to absorbed lead, sometimes up to 80% of the total amount.

Table 1. Estimates of lead (μg/day) absorbed by adults and by children aged 1–5 years from air, food and drinking-water at different air lead levels

Mean air lead level (μg/m³)	Source			Total	Air/total (%)
	Air	Food	Water		
Adults					
0.3	2.4	10	2	14.4	17
0.5	4.0	10	2	16	25
1.0	8.0	10	2	20	40
2.0	16.0	10	2	28	57
3.0	24.0	10	2	36	67
Children (1–5 years old)					
0.3	0.6	25	5	30.6	2.0
0.5	1.0	25	5	31	3.2
1.0	2.0	25	5	32	6.3
2.0	4.0	25	5	34	11.8
3.0	6.0	25	5	36	16.7

Note. The estimates are based on the following assumptions:

Air: respiratory volume:
- adults 20 m³/day
- children 5 m³/day

respiratory absorption: 40%

Food: intake:
- adults 100 μg/day; absorption 10%
- children 50 μg/day; absorption 50%

Water: concentration 20 μg/litre:
- adults 1 litre/day; absorption 10%
- children 0.5 litre/day; absorption 50%

Kinetics and Metabolism

Absorption

Absorption through the respiratory tract is influenced by the particle size distribution and the ventilation rate. For adults the retention rates of airborne particulates range from 20% to 60%. Although lead salts differ widely in terms of water solubility, the chemical form of lead is not considered an important factor for respiratory absorption.

The proportion of lead absorbed from the gastrointestinal tract is about 10% in adults, whereas levels of 40–50% have been reported in children *(16,17)*. Gastrointestinal absorption is highly dependent on dietary or

Table 2. Estimates of lead (μg/day) absorbed by children from ingested dust, inhaled air, food and drinking-water, assuming different amounts of dust intake

Source				Total	Dust/total (%)
Dust	Air	Food	Water		
0	2	25	5	32	0
12.5	2	25	5	44.5	28.1
25	2	25	5	57	43.9
50	2	25	5	82	61.0
100	2	25	5	132	75.8

Note. The estimates are based on the following assumptions:

Dust: lead concentration: 1 mg/g

Dust intake: 0, 25, 50, 100, 200 mg/day; absorption 50%

Air concentration: 1 μg/m³; respiratory volume 5 m³/day; respiratory absorption 40%

Food intake: 50 μg/day; absorption 50%

Water concentration: 20 μg/litre; water intake 0.5 litre/day; absorption 50%

nutritional factors *(5)*: both milk and fasting enhance absorption. Diets with low levels of calcium, vitamin D and iron have been shown to increase lead absorption in laboratory animals.

Distribution

The nonexcreted fraction of absorbed lead is distributed among three compartments: blood, soft tissues and the mineralizing tissues (bones, teeth). About 95% of the lead body-burden in adults is located in the bones, compared with about 70% in children *(18)*. Ninety-nine per cent of the lead in the bloodstream is bound to erythrocytes. The biological half-time of lead in blood can be as short as 20–40 days (isotopic tracer data), although longer half-time values have been reported in lead workers, and these may depend on the lead body-burden *(19,20)*.

Lead concentration in bones increases with age and this increase is more noticeable in males in the more dense tibial bones *(21)*. Lead stored in bones was shown to have a biological half-time of some years *(22)*.

Lead may be released from the bones in decalcification processes related to elderly people, pregnancy, acidosis, thyrotoxicosis or active remodelling processes in the bones of children. Animal experiments have shown mobilization of lead in pregnancy *(23)*. The evidence for the release of lead from bone in disease states involving febrile illness or altered metabolic activity in humans is at present speculative and more information is required.

Elimination

Nonabsorbed lead passes through the gastrointestinal tract and is excreted in the faeces. Of the absorbed fraction, 50–60% is removed by renal and biliary excretions. Intestinal clearance is about 50% of the renal clearance. (These figures relate to adult subjects.) Surprisingly little information exists about the age dependency of lead retention and excretion. Some data indicate that children, particularly infants, retain a higher amount of lead *(16,17)*.

Health Effects

The toxicity of lead may to some extent be explained by its interference with different enzyme systems: lead inactivates these enzymes by binding to SH-groups of its proteins or by displacing other essential metal ions. For this reason, almost all organs or organ systems may be considered potential targets for lead, and a wide range of biological effects of lead has been documented. These include effects on haem biosynthesis, the nervous system (neurotoxic effects), the kidneys, reproduction, the immune system, and also cardiovascular, hepatic, endocrinal and gastrointestinal effects. In conditions of low-level and long-term lead exposure such as are found in the general population, the most critical effects are those on haem biosynthesis, erythropoiesis, the nervous system and blood pressure.

Effects on experimental animals and *in vitro* test systems

Toxicological effects

Effects on both haem biosynthesis and the nervous system have been studied in laboratory animals. Inhibition of the activity of delta-aminolaevulinic acid dehydrase (ALAD), an enzyme involved in haem biosynthesis, is among the earliest biological effects of lead. Delta-aminolaevulinic acid dehydrase inhibition has also been observed in brain tissue. Comparative studies in rodents suggest that the brain of suckling rodents is more vulnerable to lead-induced ALAD inhibition than the adult brain *(24,25)*.

Neurobehavioural models of learning and memory have been used to study the effects of low-level lead exposure on the nervous system of rodents and monkeys. The more recent work in this field has been selectively reviewed *(26)*: learning and memory deficit has been found in rats after prenatal and postnatal lead exposure at blood lead levels below 0.2 μg/ml. Similar effects have been observed in monkeys at blood lead levels in excess of 0.4 μg/ml. There is some evidence that, if exposure occurred during the early stages of brain maturation, learning and memory deficit persists into adulthood long after the cessation of lead exposure.

Mutagenic and carcinogenic effects

There is no evidence that lead acetate or lead chloride induce mutations in bacteria. However, some chromosomal aberration tests in mammalian systems (either *in vitro* or *in vivo*) have given positive results.

According to IARC *(27)*, there is sufficient evidence that lead acetate, lead subacetate and lead phosphate are carcinogenic in rats and that lead subacetate is carcinogenic in mice. These compounds induce benign and malignant tumours of the kidney following oral or parenteral administration. Moreover, gliomas were observed in rats given lead subacetate by the oral route.

There are insufficient data to evaluate the carcinogenicity and mutagenicity of organometallic lead compounds.

Effects on humans

Toxicological effects

As far as long-term and low-level lead exposure is concerned, the following effects have to be considered in relation to the general population:

(*a*) effects on haem biosynthesis;

(*b*) effects on the nervous system; and

(*c*) effects on blood pressure.

The present discussion is, therefore, limited to these aspects of lead toxicity.

Effects on haem biosynthesis and erythropoiesis. The normal process of haem biosynthesis and its disturbance by lead is well understood. On the cellular level, the initial and final steps of haem formation are mitochondrial, whereas the intermediate steps take place in the cytoplasm.

Essentially, lead interferes with the activity of three enzymes:

(*a*) it indirectly stimulates the mitochondrial enzyme delta-aminolaevulinic acid synthetase (ALAS);

(*b*) it directly inhibits the activity of the cytoplasmatic enzyme delta-aminolaevulinic acid dehydrase (ALAD);

(*c*) it interferes with the normal functioning of intramitochondrial ferrochelatase, which is responsible for the insertion of iron (II) into the protoporphyrin ring.

Delta-aminolaevulinic acid synthetase stimulation has been found in lead workers at blood lead levels of about 0.4μg/ml *(28)*. In contrast, ALAD in the erythrocytes is inhibited at very low blood lead levels. According to Hernberg & Nikkanen *(29)*, activity inhibition in urban adults was 50% at a blood lead level of 0.16μg/ml. Roels et al. *(30)* were unable to determine a threshold for ALAD inhibition in children.

Delta-aminolaevulinic dehydrase inhibition results in an accumulation of its substrate, ALA, in blood, plasma and urine. Although the threshold for urinary ALA elevation is widely accepted as being 0.4μg/ml, some studies demonstrated blood lead/urinary ALA correlations at an even lower blood lead value *(28)*.

Lead's interference with the formation of haem from protoporphyrin is apparent from increased levels of erythrocyte protoporphyrin or zinc protoporphyrin in blood. Studies based on pooled data from various authors point to a threshold of about 0.2μg/ml in adults *(31)*. In children the threshold at which an increase of erythrocyte protoporphyrin occurs is in the range of 0.1–0.2μg/ml *(32,33)*. There are some data suggesting that the no-response level in children is even lower than 0.1μg/ml *(33)*.

Effects of lead on erythropoiesis and erythrocyte physiology represent more direct signs of damage to the haemopoietic system than haem precursors in blood or urine. Anaemia is a frequent outcome of chronic lead intoxication. Calculations by Piotrowski & O'Brien *(31)* using pooled data of various authors suggest that the threshold in children is about 0.25μg/ml. A corresponding level of 0.5μg/ml has been found in adult lead workers *(34)*.

In young children, lead exposure is associated with a decrease in the biosynthesis of the important hormonal metabolite of vitamin D: 1,25-dihydroxy-vitamin D *(35)*. This effect is correlated with blood lead in a group of 177 children over the blood lead range of 0.12–1.2μg/ml. While this finding is based on limited information, its potential significance may be considerable.

Effects on the nervous system. Encephalopathy has been observed in adults at blood lead levels exceeding 1.2μg/ml, and in children at levels of 0.8–1.0μg/ml. The outcome is frequently fatal in children, and those who survive often present irreversible neurological and neuropsychological sequelae *(36,37)*. This aspect of lead toxicity is noncontroversial. Controversy exists as to whether lead exposure associated with blood lead levels below 0.7μg/ml may be detrimental to the structure and function of the peripheral nervous system and/or CNS.

With regard to the peripheral nervous system, a decreased sensory and motor nerve conduction velocity has been found in lead-exposed workers at blood lead levels between 0.3 and 0.8μg/ml *(38,39)*. The effect is reversible after discontinuation of exposure *(40)*. However, the effect is usually small, i.e. between 4 and 9 m/second for blood lead levels ranging from 0.05 to 0.6μg/ml. Some negative findings *(41–43)* do exist, however, and the clinical significance has been questioned *(44)*.

Children as a risk group for CNS effects have received particular attention in studies dealing with lead-induced neuropsychological deficits at blood lead levels between 0.15 and 0.7μg/ml. The earlier work describing cognitive dysfunctions, namely IQ deficit, impairment of eye–hand coordination and attention deficits, as well as behavioural abnormalities such as hyperactivity, has been critically reviewed *(45,46)*. The reviewers concluded that these earlier studies presented no convincing evidence of cognitive deficits associated with blood lead levels below 0.4μg/ml in children; at higher levels, concern is warranted.

More recent work extends the range of concern to lower blood lead levels of about 0.2μg/ml. This work also has recently been reviewed *(47)*. In the absence of prospective studies, final conclusions cannot be drawn from these largely retrospective approaches, but extending the area of concern for lead-related neuropsychological deficits to blood lead levels below 0.3 μg/ml seems justified.

In addition to neuropsychological data, recent electrophysiological findings in children exposed to blood lead must be considered. A significant linear relationship between slow-wave voltage and blood lead levels ranging from 0.06 to 0.59 μg/ml has been found *(48)*. A 2-year follow-up *(49)* essentially confirmed these findings, thus demonstrating a persistence of the effect over at least this time-span. Although the clinical significance of these findings is not established, they clearly indicate altered CNS function at blood lead levels below 0.3 μg/ml.

Effects on blood pressure. Epidemiological and animal studies indicate that lead increases blood pressure. Data obtained from the National Health Assessment and Nutritional Evaluation Survey II in the USA showed that, after adjustment for age, body mass index, nutritional factors and blood biochemistries in a multiple regression model, the relationship of systolic and diastolic blood pressures to blood lead levels was statistically significant ($P < 0.01$) in white males aged 20–74 years *(50,51)*. A threshold blood lead level for this relationship was not evident. These findings were essentially confirmed in an epidemiological study on 7735 men aged 40–59 years in the United Kingdom *(52)*; however, the dose–response relationship was much weaker than in the United States survey. Although these data alone do not prove a causal relationship between low blood lead levels and blood pressure, the findings are consistent with results from animal studies, indicating that a causal relationship is probable. Considering the fact that high blood pressure is a major health problem in many industrialized countries and that a reduction of blood lead would decrease the number of hypertensive subjects and the cardiovascular and cerebrovascular risk implications associated with high blood pressure, it is highly desirable to keep the blood lead level of the general population as low as possible.

Carcinogenic effects

According to IARC *(53)*, evidence of the carcinogenicity of lead and lead compounds in humans is inadequate.

Evaluation of Human Health Risks

Exposure

Average lead levels are usually below 0.5 μg/m^3 at nonurban sites. Urban air lead levels at sites close to streets are between 0.5 and 3 μg/m^3 (annual means) in most European cities. Additional routes of exposure must not be neglected, e.g. lead in dust, a cause of special concern for children.

Quantitative relationships of multimedia exposure to blood lead

As shown in Fig. 1, air lead not only enters the body directly through the lungs but will also pass into other media which, in turn, are routes of human exposure.

Fig. 1. Air lead pathways contributing to blood lead levels

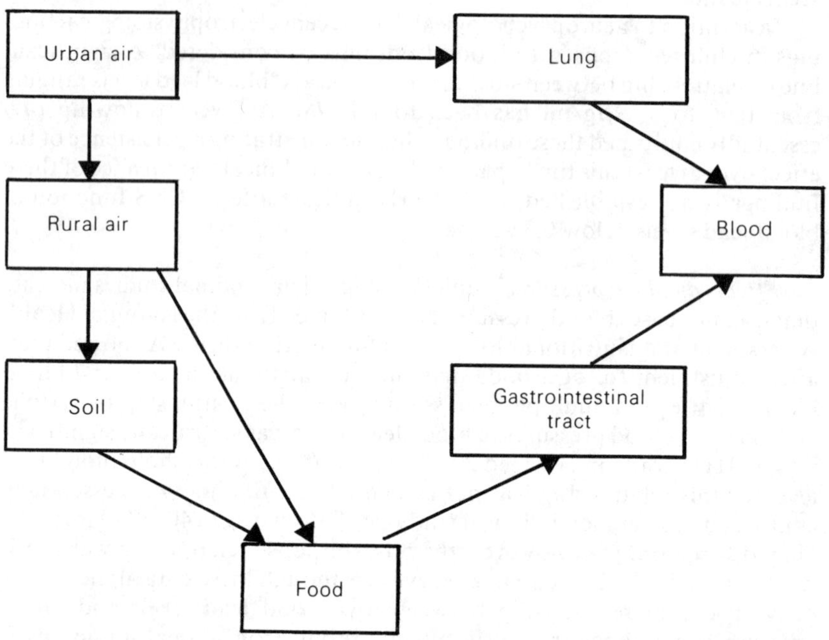

The relationships between air lead exposure and blood lead have been shown to exhibit downward curvilinearity if the range of exposures is sufficiently large (i.e. if it includes high levels). At lower levels of exposure the deviation from linearity is negligible and linear models of the relationship between intake and blood lead are as good as nonlinear model fits. For these lower ranges of the relationship (lead below $3 \mu g/m^3$ in air and below $0.3 \mu g/ml$ in the blood) various data show that a change of $1 \mu g/m^3$ in air lead is associated with a blood lead change of $0.01–0.02 \mu g/ml$ *(1)*. However, this relationship is only valid for adults. On the basis of published studies it is not possible to make a reliable quantitative estimate of the relationship between air lead and children's blood lead.

Quantification of the relationship between lead in other media affected by air lead and blood lead is not well understood, but a multimedia modelling approach may be employed to obtain rough estimates of such inputs.

Dose–effect relationships

It is generally accepted that the level of lead in blood is the best indicator of current exposure and a reasonably good indicator of the lead body-burden. It is therefore useful to relate the biological effects of lead to blood lead as an indicator of internal exposure.

Table 3 summarizes the threshold effect levels for haematological and neurological parameters in adults. Whereas clinically defined and severe anaemia does not occur at blood lead levels below 0.8μg/ml, elevated erythrocyte protoporphyrin levels have been observed at blood lead levels of above 0.2μg/ml in men and of 0.15–0.2μg/ml in women. Neurological and neuropsychological effects of questionable validity have been found in lead workers at blood lead levels of 0.4–0.6μg/ml, whereas in one study a decrease of nerve conduction velocity has been found to start at blood lead levels of about 0.3μg/ml.

Table 4 summarizes threshold effect levels for haematological and neurological parameters in children. Whereas anaemia has not been observed at

Table 3. Summary of lowest-observed-effect levels for lead-induced health effects in adults

Lowest-observed-effect levels of blood lead	Haem synthesis and haematological effects	Effects on nervous system
1.0 -1.2μg/ml		Encephalopathic signs and symptoms
0.8μg/ml	Frank anaemia	
0.6μg/ml		------------- ↑
0.5μg/ml	Reduced haemoglobin production	Overt subencephalopathic neurological symptoms ↓
0.4μg/ml	Increased urinary ALA and elevated coproporphyrin	------------- ↑
0.3μg/ml		Peripherical nerve dysfunction (slowed nerve conduction velocities) ↓
0.2 -0.3μg/ml	Erythrocyte protoporphyrin elevation in males	?
0.15–0.2μg/ml	Erythrocyte protoporphyrin elevation in females	
0.1μg/ml	ALAD inhibition	

Table 4. Summary of lowest-observed-effect levels for lead-induced health effects in children

Lowest-observed-effect levels of blood lead	Haem synthesis and haematological effects	Effects on nervous system
0.8 -1.0 µg/ml		Encephalopathic signs and symptoms
0.7 µg/ml	Frank anaemia	
0.6 µg/ml		
0.5 µg/ml		——— ↑
0.4 µg/ml	Increased urinary ALA and elevated coproporphyrin	
0.25-0.3 µg/ml	Reduced haemoglobin synthesis	Cognitive CNS deficit Peripheral nerve dysfunction (slowed nerve conduction velocities) ↓ ↑
0.1 -0.2 µg/ml	Erythrocyte protoporphyrin elevation	CNS electrophysiological changes ↓
0.1 µg/ml	ALAD inhibition	----- ? -----

blood lead levels below 0.7 µg/ml, haemoglobin reduction was found at blood lead levels of about 0.25 µg/ml, and erythrocyte protoporphyrin elevation at blood lead levels between 0.15 and 0.2 µg/ml. Neuropsychological effects, although not universally agreed upon, have been observed at blood lead levels between 0.2 and 0.5 µg/ml, and electrophysiological findings of unknown clinical significance as well as attentional deficits have been found at blood lead levels below 0.3 µg/ml. Dose–response relationships have been established for erythrocyte protoporphyrin elevation in children *(32)*, elevation of free erythrocyte protoporphyrin in children and adults (females and males) *(30)*, urinary ALA increase and ALAD inhibition in children *(31)*.

No comparable dose–response relationships can as yet be given for measures of neurological and/or neuropsychological outcome.

Health risk evaluation

The decisive parameter upon which the guidelines for lead in air should be based is the concentration of lead in blood. After a review of the array of toxic effects of lead, weight was given to the elevation of erythrocyte protoporphyrin. In selecting the appropriate limit value for lead in blood, a value of 0.2μg/ml may be regarded as the borderline dividing the no-adverse-effect level from the lowest-adverse-effect levels. At this level, the increase of free erythrocyte protoporphyrin starts in a small percentage of adult subjects. At concentrations slightly higher, the whole chain of effects may become apparent, including haemoglobin decrease (in children only), subtle neurological changes, and disturbances in vitamin D levels.

On the other hand, at levels lower than 0.2μg/ml the most sensitive effect is manifested (inhibition of ALAD activity); this, however, was not accepted as the basis, since it was judged to be a subcritical effect of no clear biological significance.

Certain lead compounds have been found to be carcinogenic in rats and mice after oral exposure. There is inadequate evidence of lead being carcinogenic in humans. Accordingly, IARC has classified lead in Group 3.

The following assumptions are made in constructing the guideline value.

Adult urban population

1. Ninety-eight per cent of the adult population should have blood lead levels below 0.2μg/ml.

2. From the frequency distribution of blood lead levels in the general population it can be calculated that the median value, corresponding to a 98th percentile of 0.2μg/ml, is approximately 0.1μg/ml.

3. There is a "baseline" level of blood lead, which can be taken as either nonanthropogenic in origin or as resulting from relatively minimal anthropogenic lead exposure. Based on the population data of three large cities in which low values of blood lead were reported *(54)* and data from rural areas in Japan *(55)*, the "baseline" level of blood lead seems to be, on average, around 0.04–0.06μg/ml. It is somewhat lower in women than in men. The distance between the upper limit of tolerable median (0.1μg/ml) and the average "baseline" level of blood lead is thus 0.04–0.06μg/ml.

4. As shown in Table 1, it can be assumed that, at a typical urban concentration of lead in air of the order of 1μg/m^3, the contribution of direct inhalation to the total lead absorption of adults is approximately 40%. This percentage indicates that inhalation of airborne lead is an important route of exposure in adult humans.

5. The WHO Environmental Health Criteria document on lead *(5)* explains that in experimental conditions, where one deals with direct inhalation only, 1μg per m^3 air contributes to 0.01–0.02μg per ml blood. The upper figure is used hereafter as a more conservative one. By simple conversion 0.04–0.06μg/ml corresponds to 2–3μg/m^3.

6. It must then be taken into account that in real-life situations an increase of lead in air also contributes to increased lead uptake by indirect

ways, through other environmental pathways. To allow for uptake by other routes also, a recalculation factor of 5 : 1 is introduced (i.e. $1 \mu g/m^3$ would contribute to $0.05 \mu g/ml$). This relationship, then, already includes a certain protection factor.

7. In order to ensure that any anthropogenic input of lead into the blood should be limited so as not to exceed a value of $0.2 \mu g/ml$ in 98% of the adult population and the corresponding median value of $0.1 \mu g/ml$, it would be reasonable that air lead, measured as a long-term average value, should not exceed $1 \mu g/m^3$.

8. Taking into account some uncertainties that are inherent in the above evaluation, a guideline in a range of values $0.5–1.0 \mu g/m^3$, measured as long-term average values in urban areas, is recommended.

Children of preschool age

Children up to 6 years of age are a population at increased risk for lead exposure as well as for adverse health effects, for the following reasons.

1. Children have behavioural characteristics (outdoor activity, less concern for hygienic conditions, mouth activities or even pica), which increase the risk of lead exposure.

2. Children eat and drink more per unit of body weight than adults, so that their relative lead intake is increased.

3. Lead absorption in the gastrointestinal tract is substantially higher in children (about 50%, compared with about 10% in adults) *(9,10)*.

4. Among children there is a greater prevalence of nutritional deficiencies (e.g. iron and vitamin D), which enhance absorption of lead from the gastrointestinal tract.

5. The blood–brain barrier is not yet fully developed in young children.

6. Haematological and neurological effects of lead occur at lower thresholds in children than in adults.

Since the placenta is no effective biological barrier, pregnant women represent a second group at increased risk because of exposure of the fetus to lead.

In establishing guidelines for the critical group of children of preschool age, the following reasoning can be applied.

1. As seen from Table 1, the contribution of direct inhalation to total lead absorption, at an air concentration of $1 \mu g/m^3$, is smaller than in adults (approximately 6% only). Most of the absorbed dose can be accounted for by other pathways, mainly food. In addition, the specific behaviour of children entails an additional input in the form of dust (dirt) passing by hand to the mouth.

2. Uncertainties connected with numerical values of transfer coefficients are too great to calculate alternative pathways of lead uptake from other sources and the relationships between lead in other media affected by air lead on the one hand and blood lead on the other.

3. Since in children the estimated contribution of direct inhalation is smaller compared with adults, decreases in air lead will have correspondingly little direct effect. Therefore, guideline values for adults should be used for the protection of the whole population, including children.

Because of the limited direct influence of atmospheric lead on blood lead levels in children, the air guidelines for lead should not be regarded as an adequate way of protecting this population. Other measures may need to be taken in parallel. These should specifically take the form of monitoring the lead content of dust and soils arising from air lead fallout. The normal behaviour of children with regard to dust and soil defines these media as potentially serious exposure sources. An additional source may be lead from paint. A specific monitoring value is not recommended. It should be mentioned, however, that the Greater London Council has selected an external dust lead value of 500 μg per gram dust as a basis for further evaluation and a value of 5000 μg per gram dust as the basis for implementing control strategies. In the Federal Republic of Germany, the lead fallout has been limited to 250 $\mu g/m^2$ per day. According to a recent study *(14)*, this limit is sufficient to prevent excessive intake of lead from environmental dust by children.

Guidelines

A guideline in the range of 0.5–1.0 μg lead per m^3 (long-term average, e.g. annual mean) is recommended; this incorporates a protection (safety) factor close to 2.

This guideline value is based on the assumption that 98% of the general population will be maintained below a blood lead level of 0.2 μg/ml, predicated on the selection of elevated erythrocyte protoporphyrin as the effect. This particular blood level was carefully considered, but it was recognized that other starting-points in terms of both effects and statistical limits might be chosen. In that case, another air lead guideline might result.

Because of the limited influence of atmospheric lead on the overall exposure of children, air quality guidelines are not sufficient to protect this population, and other initiatives may be necessary.

The monitoring of lead deposition in dust and soil in areas near point emissions and in high-density traffic areas is advisable in order to protect young children from excessive exposure and consequent adverse health effects.

References

1. **Nriagu, J.O.** Lead in the atmosphere. *In:* Nriagu, J.O., ed. *The biogeochemistry of lead in the environment.* Amsterdam, Elsevier-North Holland, 1978, Part A, pp. 137–184.

2. **Gezondheidsraad.** *Advies inzake lood. Advieswaarde voor de kwaliteit van de buitenlucht* [Guidelines for lead. Recommended standards for outdoor air quality]. The Hague, Ministry of Health and Environmental Hygiene, 1984.
3. **Yocum, J.E.** Indoor–outdoor air quality relationships: a critical review. *Journal of the Air Pollution Control Association,* **32**: 500–520 (1982).
4. **Mahaffey, K.R.** Environmental exposure to lead. *In:* Nriagu, J.O., ed. *The biogeochemistry of lead in the environment.* Amsterdam, Elsevier-North Holland, 1978, Part B, pp. 1–36.
5. *Lead.* Geneva, World Health Organization, 1977 (Environmental Health Criteria, No. 3).
6. **Elwood, P.C. et al.** Greater contribution to blood lead from water than from air. *Nature,* **310**: 138–140 (1984).
7. *Lead in the environment:* ninth report of the Royal Commission on Environmental Pollution. London, H.M. Stationery Office, 1983.
8. **Grandjean, P. et al.** Influence of smoking and alcohol consumption on blood lead levels. *International archives of occupational and environmental health,* **48**: 391–397 (1981).
9. **Elinder, C.G. et al.** Lead and cadmium levels in blood samples from the general population of Sweden. *Environmental research,* **30**: 233–253 (1983).
10. **Roels, H.A.** Exposure to lead by the oral and the pulmonary routes of children living in the vicinity of a primary lead smelter. *Environmental research,* **22**: 81–94 (1980).
11. **Duggan, M.J.** Contribution of lead in dust to children's blood lead. *Environmental health perspectives,* **50**: 371–381 (1983).
12. **Brunekreef, B.** The relationship between air lead and blood lead in children: a critical review. *Science of the total environment,* **38**: 79–123 (1984).
13. **Brunekreef, B.** *The relationship between environmental lead and blood lead in children: a study in environmental epidemiology.* Department of Environmental and Tropical Health, Agricultural University of Wageningen, Netherlands, 1985 (Report 1985-211).
14. **Ewers, U. et al.** Contribution of lead and cadmium in dustfall to blood lead and blood cadmium in children and adults living in two non-ferrous smelter areas of West Germany. *In:* Lekkas, T.D., ed. *International Conference on Heavy Metals in the Environment, Athens, September 1985.* Edinburgh, CEP Consultants, 1985.
15. **Facchetti, S. & Geiss, F.** *Isotopic lead experiment: status report.* Luxembourg, Commission of the European Communities, 1982 (Publication No. EUR 8352 EN).
16. **Alexander, F.W. et al.** The uptake and excretion by children of lead and other contaminants. *In: Environmental health aspects of lead.* Luxembourg, Commission of the European Communities, 1973, pp. 319–330.
17. **Ziegler, E.E. et al.** Absorption and retention of lead by infants. *Pediatric research,* **12**: 29–34 (1978).
18. **Barry, P.S.I.** Concentrations of lead in the tissues of children. *British journal of industrial medicine,* **38**: 61–71 (1981).

19. **Kang, H.K. et al.** Determination of blood-lead elimination patterns of primary lead smelter workers. *Journal of toxicology and environmental health,* **11**: 199–210 (1983).
20. **O'Flaherty, E.J. et al.** Dependence of apparent blood lead half-life on the length of previous lead exposure in humans. *Fundamental and applied toxicology,* **2**: 49–54 (1982).
21. **Barry, P.S.I. & Mossman, D.B.** Lead concentrations in human tissues. *British journal of industrial medicine,* **27**: 339–351 (1970).
22. **Rabinowitz, M.B. et al.** Kinetic analysis of lead metabolism in healthy humans. *Journal of clinical investigation,* **58**: 260–270 (1976).
23. **Buchet, J.P. et al.** Mobilization of lead during pregnancy in rats. *International archives of occupational and environmental health,* **40**: 33–36 (1977).
24. **Gerber, G. et al.** Effect of dietary lead on placental blood flow and on fetal uptake of α-amino isobutyrate. *Archives of toxicology,* **41**: 125–131 (1978).
25. **Millar, J.A. et al.** Lead and delta-aminolevulinic acid dehydratase levels in mentally retarded children and in lead-poisoned suckling rats. *Lancet,* **2**: 695–698 (1970).
26. **Winneke, G.** Animal studies. *In:* Lansdown, R. & Yule, W., ed. *The lead debate: the environment, toxicology and child health.* London & Sydney, Croom Helm, 1986, pp. 217–234.
27. *Some metals and metallic compounds.* Lyon, International Agency for Research on Cancer, 1980 (IARC Monographs on the Evaluation of the Carcinogenic Risk of Chemicals to Humans, Vol. 23).
28. **Meredith, P.A. et al.** Delta-aminolevulinic acid metabolism in normal and lead exposed humans. *Toxicology,* **9**: 1–9 (1978).
29. **Hernberg, S. & Nikkanen, J.** Enzyme inhibition by lead under normal urban conditions. *Lancet,* **1**: 63–64 (1970).
30. **Roels, H.A. et al.** Response of free erythrocyte porphyrin and urinary delta-aminolevulinic acid in men and women moderately exposed to lead. *Internationales Archiv für Arbeitsmedizin,* **34**: 97–108 (1975).
31. **Piotrowski, J.K. & O'Brien, B.J.** *Analysis of the effects of lead in tissue upon human health, using dose–response relationships.* London, University of London Monitoring and Assessment Research Centre, 1980 (MARC Report No. 17).
32. **Piomelli, S. et al.** Threshold for lead damage to heme synthesis in urban children. *Proceedings of the National Academy of Sciences of the USA,* **79**: 3335–3339 (1982).
33. **Cavallieri, A. et al.** Biological response of children to low levels of inorganic lead. *Environmental research,* **25**: 415–423 (1981).
34. **Tola, S. et al.** Parameters indicative of absorption and biological effect in new lead exposure: a prospective study. *British journal of industrial medicine,* **30**: 134–141 (1973).
35. **Mahaffey, K.R. et al.** Association between age, blood lead concentration, and serum 1,25-dihydroxycholecalciferol levels in children. *American journal of clinical nutrition,* **35**: 1327–1331 (1982).

36. **Perlstein, M.A. & Attala, R.** Neurologic sequelae of plumbism in children. *Clinical pediatrics,* **5**: 292–298 (1966).
37. **Smith, H.D. et al.** The sequelae of pica with and without lead poisoning. A comparison of the sequelae five or more years later. I. Clinical and laboratory investigations. *American journal of diseases of children,* **105**: 609–616 (1963).
38. **Seppäläinen, A.M. et al.** Subclinical neuropathy at "safe" levels of lead exposure. *Archives of environmental health,* **30**: 180–183 (1975).
39. **Seppäläinen, A.M. & Hernberg, S.** A follow-up study of nerve conduction velocities in lead-exposed workers. *Neurobehavioural toxicology and teratology,* **4**: 721–723 (1982).
40. **Araki, S. et al.** Recovery of slowed nerve conduction velocity in lead-exposed workers. *International archives of occupational and environmental health,* **46**: 151–157 (1980).
41. **Lilis, R. et al.** Prevalence of lead disease among secondary lead smelter workers and biological indicators of lead exposure. *Environmental research,* **14**: 255–285 (1977).
42. **Paulev, P.-E. et al.** Motor nerve conduction velocity in asymptomatic lead workers. *International archives of occupational and environmental health,* **43**: 37–43 (1979).
43. **Englert, N.** Periphere motorische Nervenleitgeschwindigkeit bei Probanden mit beruflicher Blei-Exposition [Peripheral motor nerve conduction velocity in test subjects occupationally exposed to lead]. *Arbeitsmedizin, Sozialmedizin, Präventivmedizin,* **15**: 254–260 (1980).
44. **Buchtal, F. & Behse, F.** Electrophysiology and nerve biopsy in men exposed to lead. *British journal of industrial medicine,* **36**: 135–147 (1979).
45. **Bornschein, R. et al.** Behavioral effects of moderate lead exposure in children and animal models. Part 2: animal studies. *CRC critical reviews in toxicology,* **8**: 101–152 (1980).
46. **Rutter, M.** Raised lead levels and impaired cognitive/behavioral functioning. *Developmental medicine and child neurology,* **22**(Suppl. 42): 1–26 (1980).
47. **Lansdown, R.** Lead, intelligence and behaviour. *In:* Lansdown, R. & Yule, W. ed. *The lead debate: the environment, toxicology and child health.* London & Sydney, Croom Helm, 1986, pp. 235–270.
48. **Otto, D. et al.** Effects of age and body lead burden on CNS function in young children. I. Slow cortical potentials. *Electroencephalography and clinical neurophysiology,* **52**: 229–239 (1981).
49. **Otto, D. et al.** Effects of low to moderate lead exposure on slow cortical potentials in young children: two year follow-up study. *Neurobehavioral toxicology and teratology,* **4**: 733–737 (1982).
50. **Pirkle, J.L. et al.** The relationship between blood lead levels and blood pressure and its cardiovascular risk implications. *American journal of epidemiology,* **121**: 246–258 (1985).
51. **Schwartz, J.** The relationship between blood lead and blood pressure. *In:* Lekkas, T.D., ed. *International Conference on Heavy Metals in the Environment, Athens, September 1985.* Edinburgh, CEP Consultants, 1985.

52. **Pocock, S.J. et al.** Blood lead and blood pressure in middle-aged men. *In:* Lekkas, T.D., ed. *International Conference on Heavy Metals in the Environment, Athens, September 1985.* Edinburgh, CEP Consultants, 1985.
53. *Chemicals, industrial processes and industries associated with cancer in humans. IARC Monographs, Volumes 1 to 29.* Lyon, International Agency for Research on Cancer, 1982 (IARC Monographs on the Evaluation of the Carcinogenic Risk of Chemicals to Humans, Supplement 4).
54. **Friberg, L. & Vahter, M.** Assessment of exposure to lead and cadmium through biological monitoring: results of a UNEP/WHO Global Study. *Environmental research,* **30**: 95–128 (1983).
55. **Watanabe, T. et al.** Baseline level of blood lead concentration among Japanese farmers. *Archives of environmental health,* **40**: 170–176 (1985).

24

Manganese

General Description

Sources

Manganese (Mn) is an element widely distributed in the earth's crust. It is considered to be the twelfth most abundant element and the fifth most abundant metal. Manganese does not occur naturally in its native state. Oxides, carbonates and silicates are the most important among manganese-containing minerals. The most common manganese mineral is pyrolusite (MnO_2), usually mined in sedimentary deposits by opencast techniques. Manganese occurs in most iron ores. Its content in coal is in the range 6–100 μg/g; it is also present in crude oil, but at substantially lower concentrations *(1)*.

Manganese is mainly used in metallurgical processes, as a deoxidizing and desulfurizing additive and as an alloying constituent. It has other uses, e.g. in the production of dry-cell batteries, the production of manganese chemicals and in some other chemical processes, as well as in the manufacture of glass, in the leather and textile industries, and as a fertilizer. Organic carbonyl compounds are used as fuel-oil additives, smoke inhibitors, and as anti-knock additives in petrol *(2)*.

Crustal manganese enters the atmosphere by a number of natural and anthropogenic processes, which include the suspension of road dusts by vehicles, wind erosion and suspension of soils, particularly in agricultural and building activities and quarrying processes. The resulting mechanically generated aerosols consist primarily of coarse particles ($\geqq 2.5 \mu$m mass median aerodynamic diameter (MMAD)). The smelting of natural ores and the combustion of fossil fuels also result in the ejection of crustal manganese to the atmosphere in the form of fume or ash in the fine-particle range ($\leqq 2.5 \mu$m MMAD). Manganese is released to the atmosphere during the manufacture of ferroalloys and other industrial processes. Nearly one half of all industrial and combustive emissions of manganese are from ferroalloy manufacture and about one tenth from fossil-fuel combustion. Minor amounts are generated by other processes. The use of manganese fuel additives constitutes an additional source *(2)*.

As an element of low volatility, manganese tends to settle out near sources of pollution, but fine particles containing manganese can be distributed very widely. The most common forms of manganese compounds in the coarse particulates of crustal origin are oxides or hydroxides of oxidation state +2, +3, +4, and manganese carbonate. The manganese emitted by metallurgical processes consists of oxides. The manganese from combusted methylcyclopentadienyl manganese tricarbonyl (MMT), used in some countries as a fuel additive, is emitted primarily as Mn_3O_4. Minute amounts of organic manganese compounds may be present in ambient air under certain conditions. However, MMT is rapidly photodegraded to inorganic manganese in sunlight. The estimated half-time is 10–15 seconds *(2)*.

Manganese dioxide can react with sulfur dioxide or nitrogen dioxide to form manganous sulfate and dithionate, or manganese nitrate, respectively. It has been shown that aerosols of manganous sulfate can catalyse the oxidation of atmospheric sulfur dioxide to sulfur trioxide, thus promoting the formation of sulfuric acid *(2)*.

Occurrence in air

The natural level of manganese in air is low. Background concentrations of 0.05 to 5.4 ng/m^3 over the Atlantic Ocean *(3)*, and 0.01 ng/m^3 at the South Pole *(4)* have been reported. A concentration of 0.006 μg/m^3 in air at a height of 2500 m and an annual average concentration of 0.027 μg/m^3 at 823 m were reported *(5)*. The national air surveillance network of urban areas in the USA indicated an annual average manganese concentration of 0.033 μg/m^3 in 1982 *(2)*. In two cities of the Federal Republic of Germany (Frankfurt am Main and Munich), annual mean concentrations of manganese ranged between 0.03 and 0.16 μg/m^3 *(6)*, and in Belgium over the period 1972–1977 concentrations of manganese, expressed also as annual means, between 0.042 and 0.456 μg/m^3 were reported *(7)*. The Environmental Agency Japan *(8)* reported an annual mean manganese concentration in the air of Japanese cities of about 0.02–0.80 μg/m^3, with maximum 24-hour concentrations of 2–3 μg/m^3.

From these and other data it can be concluded that annual average levels of manganese in ambient air in nonpolluted areas range from approximately 0.01 to 0.03 μg/m^3, while in urban and rural areas without significant manganese pollution annual averages are mainly in the range of 0.01–0.07 μg/m^3. With local pollution near foundries, this level can rise to an annual average of 0.2–0.3 μg/m^3 and, in the presence of ferro- and silicomanganese industries, to over 0.5 μg/m^3 *(9)*. In such places the average 24-hour concentrations may exceed 10 μg/m^3.

About 80% of manganese emitted into the atmosphere is associated with particles with an MMAD of less than 5 μm, and about 50% with particles with an MMAD of less than 2 μm. More recent data, however, tend to indicate that less than 50% of the total measured manganese in ambient air is found in fine particles *(9)*.

Atmospheric particulate matter, including manganese, is transported by air currents until it is lost from the atmosphere by either dry or wet deposition. Manganese deposit in dustfall is more than twice that in rainfall *(2)*.

The highest values of manganese concentrations in the working environment have been reported from manganese mines, ore-processing plants, dry-cell battery plants, and ferromanganese plants. In mining operations, manganese concentrations up to 250 mg/m^3 or even higher have sometimes been found. In dry-cell battery plants and ferromanganese plants the concentrations of manganese in air are lower. Values up to 5–8 mg/m^3, but occasionally also higher — up to 20 mg/m^3 or even more — have been reported *(10)*. An important point is that in ferromanganese plants, but also in dry-cell battery plants, the size distribution of manganese aerosols is such that small particles prevail absolutely, compared with mining operations, where a smaller proportion of respirable particles is usually encountered. There is also some evidence that aerosols formed by condensation may be more harmful than those formed by disintegration. Whether or not this is caused by differences in the distribution of particle size remains to be clarified *(10)*.

Routes of Exposure

Air

Because of the low solubility of manganese oxides, only inhaled manganese particles small enough to reach the alveoli ($\leqq 2.5\,\mu$m MMAD) are likely to enter the bloodstream. Alveolar deposition of manganese may be estimated to be in the range of 0.07–0.5 μg/day as an average and 6–7 μg/day under high-exposure conditions *(2)*.

Drinking-water

Concentrations of manganese in fresh water may vary from less than one to several thousand μg/litre *(11)*. Drinking-water generally contains less than 100 μg manganese per litre. In 100 of the largest cities in the USA, 97% of the surveyed public water supplies contained concentrations below 100 μg/litre.

Food

Manganese concentration in foodstuffs varies markedly, but on the whole food constitutes a major source of manganese intake for humans. The highest concentrations are found in some foods of plant origin, especially wheat and rice, with concentrations between 10 and 100 mg/kg. Polished rice and wheat flour contain less manganese, since most of it is in the bran. High concentrations of manganese have been found in tea leaves. Eggs, milk, fruits and meat generally contain less than 1 mg manganese per kg *(2)*. In a study performed in Canada *(12)* it was estimated that, of people's total manganese intake via food, 54% came from cereals and 14% from potatoes, whereas meat, fish and poultry provided only 2% of manganese intake. However, there is a difference in manganese concentrations for the same items in different countries and areas. Thus, daily manganese intake by adults from food may vary over a rather wide range (2–12 mg). The daily intake for children aged 3–5 years averages 1.4 mg/day and for children aged 9–13 years 2.18 mg/day *(13)*. The daily intake of manganese by bottle-fed and breastfed infants is very low because of the low concentrations of manganese in both breast-milk and cow's milk *(14)*.

Relative significance of different routes of exposure
Total human exposure to manganese may be estimated from information on the levels in air, water and diet. The degree of uptake of manganese by inhalation is dependent upon particle size because of the low solubility of manganese oxides. Deposition of manganese in the alveoli can be calculated from the ambient concentration and the fraction present in fine particles. Alveolar deposition of manganese at current ambient levels may be estimated at 0.07 μg/day as an average and 6–7 μg/day under high-exposure conditions *(2)*. Thus, daily intake through inhalation generally constitutes less than 0.1% of the total daily intake by the general population and rarely exceeds 1%, even in heavily polluted areas *(1)*. Estimates of total thoracic deposition, including particles deposited in the tracheobronchial region, range from about 0.26 μg/day (average) to 10 μg/day (high) *(2)*.

Ingested manganese in diet is estimated to be 0.002–0.004 mg/kg body weight per day in infants and 0.06–0.08 mg/kg body weight per day in adults *(1)*. The daily intake of manganese from food by adults appears to be 2–9 mg. In Europe and the USA studies suggest a likely range of 2–5 mg, while in countries where grain and rice make up a major portion of the diet, the intake is more likely to be in the range of 5–9 mg. The consumption of tea may substantially add to the daily intake *(1)*. The median intake via drinking-water is about 0.008 mg/day, but can be as high as 2.0 mg/day for some water supplies. However, the average daily intake of manganese with water is unlikely to constitute more than 1–2% of the total intake of manganese. The ingestion of particles cleared from the respiratory tract probably constitutes no more than 0.01 mg/day under the highest ambient exposure conditions *(2)*.

Kinetics and Metabolism

There are no quantitative animal data on absorption rates for inhaled manganese compounds. Mena et al. *(15)* found that, in 17 humans exposed to a nebulized solution of manganese chloride and in 4 humans exposed to manganese oxide in a similar fashion, 40–70% of the deposited amount was recovered in the faeces. Both compounds were labelled with ^{54}Mn.

Manganese absorption is controlled by homeostatic mechanisms. The absorption rate will depend on the amount ingested and on tissue levels of manganese. In mice and rats, absorption of ingested manganese varies between 1 and 3.5%. It seems that manganese is absorbed equally well throughout the small intestine.

An increase in the iron content of milk decreases the whole-body retention of orally administered ^{54}Mn in rats by a factor of 10 *(16)*. Manganese interactions with other elements (cadmium, nickel, indium, rhodium, selenium) and ethanol at the level of gastrointestinal absorption were also observed.

Mena et al. *(15)* found that 11 normal individuals absorbed an average of 3% of a dose of 200 μg manganese chloride labelled with ^{54}Mn. Another human study *(17)* indicates that manganese absorption takes place by diffusion in iron overload states and by active transportation in the duodenum and jejunum in iron deficiency states.

Absorbed manganese is rapidly eliminated from the blood and at first concentrates mainly in the liver. Excess metal may be distributed to other tissues. Concentrations of manganese are characteristic of the individual tissues, and almost independent of the species. Manganese preferentially accumulates in tissues rich in mitochondria. It penetrates the placental barrier in all species and is secreted in milk *(2)*.

The highest concentrations of manganese in man have been found in the liver, kidneys, endocrine glands, and in the small and large intestines. The total body-burden for manganese has been estimated at 8–10 mg; about 35% of this amount is found in muscle tissue and 20% in the liver *(18)*.

In blood, manganese is bound to proteins. In the trivalent state it can bind to transferrin and in the divalent state to an α-macroglobulin. The organic compound MMT is rapidly metabolized. The distribution in general is similar to that seen after exposure to inorganic manganese *(1)*.

The apparent absence of cases of chronic oral manganese toxicity could be attributed to the extremely efficient homeostatic mechanism, well documented in animal and human studies, which prevents the accumulation of manganese in the body and maintains systemic plasma manganese concentrations at constant levels *(2,19)*. Manganese is excreted primarily via biliary clearance *(2)*. However, it is possible that the capacity of the liver to excrete higher amounts of inhaled or injected manganese is limited, resulting in an increase in the tissue retention of manganese and, consequently, in the manifestation of toxic effects *(19,20)*. In excessive exposure other gastrointestinal routes may participate. Urinary excretion is small; in humans it is less than 1 μg/day. This means that only about 0.01% of the body-burden is excreted daily via that route. Manganese is also excreted via sweat, hair, placenta and milk.

The biological half-time of manganese depends on the body-burden of manganese. Experiments have shown that manganese elimination from the brain is slower than that from the whole body. In humans without occupational exposure to manganese the half-time is around 35 days *(21)*.

Health Effects

Manganese is an essential element. It is a constituent of several enzymes and can also activate many enzymes. Manganese deficiency was described only once in a man given a synthetic diet in which manganese had been omitted by mistake. Among the symptoms and signs were dermatitis, pigment changes of hair, retarded hair growth and hypocholesterolaemia *(1)*. Excessive exposure to manganese has been shown to cause toxic effects in animals and humans. As manganese is regarded as a metal with a relatively low toxicity, acute poisoning by manganese in humans is very rare *(2)*.

Effects on experimental animals

A large number of studies on laboratory animals were performed in order to explain the mechanism underlying the neurotoxicity of manganese seen in exposed workers *(2)*. Although animal experiments show that in chronic manganese intoxication the central dopaminergic system is disturbed, the mechanism of manganese neurotoxicity has not been elucidated *(2,22)*.

Recent studies indicate that age may play a role in the extrapyramidal disturbances seen in human manganism, suggesting that the aging brain shows differing susceptibility vis-à-vis the toxic effects of manganese *(2,23)*. However, some new studies raise the intriguing possibility that lifetime treatment with manganese has some beneficial effects on the aging brain *(23)*. An accurate dose–response relationship for inhalation exposure and neurotoxicity is unobtainable at present from the available animal studies *(2)*.

Inhalation studies of the pulmonary effects show the occurrence of acute respiratory effects (oedema and leukocyte infiltration) when the level of exposure exceeds 20 mg MnO_2 per m^3 *(2)*. Mice and monkeys exposed to MnO_2 by inhalation showed pathological effects after chronic (10 months) exposure to 0.7 mg/m^3 *(24)*. Studies in which animals were exposed for a long period of time (66 and 40 weeks) to about 0.1 mg MnO per m^3 as Mn_3O_4 particles or aerosols of respirable particle size showed no respiratory effects, but the studies had several methodological deficiencies which reduce confidence in the negative results *(25,26)*.

It is plausible that exposure to manganese may increase susceptibility to pulmonary infections by disturbing the normal mechanism of lung clearance. However, it seems that a primary inflammatory reaction can occur in the lung after exposure to MnO_2, if the concentration is high enough, without the presence of pathogenic bacteria *(27)*. Studies on the influence of manganese on susceptibility to bacterial infections showed increased morbidity and mortality rates in animals infected before, during or after exposure to manganese dioxide *(28,29)*.

Effects on humans

The neurological disorder known as manganism may result from occupational exposure to manganese dusts and fumes. Symptoms and signs of manganism have often been compared with Parkinson's disease, but certain differences should be noted. Patients with Parkinson's disease show pronounced disturbances of motor behaviour, which include tremor observed at rest rather than during intentional motor activity as in manganism. Fully developed manganism causes severe rigidity, with the extremities showing the "cog-wheel" phenomenon *(2,30)*. Manganism usually appears after a prolonged exposure of two or more years, but it may result from exposure to high concentrations of manganese for only a few months. The full clinical picture of chronic manganese poisoning has been reported mainly in manganese miners, but also in other occupations where exposure to manganese is high *(2,30)*. The disease has been found less frequently at exposure levels below 5 mg manganese per m^3. However, there are reports of signs and symptoms which may be connected with subclinical or early clinical stages of chronic manganese poisoning in workers exposed to 0.3–5 mg manganese per m^3. In connection with the toxic effects of manganese, a marked individual susceptibility has been observed *(10,30)*. The data available for identifying effects below 1 mg manganese particles per m^3 are equivocal and inadequate. Furthermore, no good biological indicators of manganese exposure are available at present.

The toxic effects of manganese on the pulmonary system vary in type and severity. There are reports of humans developing pneumonia in occupational but also in ambient exposure to manganese. Increased incidence of pneumonia was observed in workers, mainly miners, exposed to manganese at levels higher than 5 mg/m^3 *(2)*. In a more recent study *(31)*, an increased incidence of pneumonia and bronchitis was found in workers exposed to manganese concentrations of 0.4–16 mg/m^3 in a factory producing manganese alloys.

Elstad *(32)* noted an 8-fold increase in mortality from pneumonia and a 4-fold increase in pneumonia morbidity in the general population living near a ferromanganese plant in Norway. The air concentrations of manganese were measured only once and were reported to be 46 μg/m^3 at a distance of 3 km from the plant. It was also reported that the incidence of pneumonia followed the rate of production of manganese alloys. Analyses of lung tissues from 11 persons who died from pneumonia showed manganese concentrations of 0.35–1.63 μg/kg wet weight. A higher prevalence of nose and throat symptoms, and lower values of lung function tests compared with controls, were observed in children attending a school situated near a ferromanganese plant *(33)*, where the average manganese exposure was about 7 μg/m^3 (range 3–11 μg/m^3). The study involved several hundred children, had a participation rate of over 97% and documented monitored levels of settled manganese dust for several years. Effects of still lower levels of airborne manganese (about 1 μg/m^3) have been claimed to occur in a population living near a manganese alloy plant, with an increased incidence of acute bronchitis over an observation period of 4 years. The incidence of pneumonia did not seem to exceed the expected values *(34)*. Although sulfur dioxide concentrations were measured (annual means: 10–30 μg/m^3), other factors, including the socioeconomic factor, which had not been considered, may have influenced the results. Chronic bronchitis has been reported to be more prevalent in workers exposed to manganese if they are smokers *(35)*.

A number of effects of manganese in other organs and systems have been claimed on the basis of results obtained from animal experiments and from epidemiological and clinical studies. These include a decrease in systolic blood pressure values, an increased rate of spontaneous abortions, changes in erythropoiesis and granulocyte formation, disturbed excretion of 17-ketosteroids, and changes in the activity of some enzymes *(2)*. Reports of impotence in a number of patients with chronic manganese poisoning are also common. However, these effects have not been observed consistently. In some cases the implications of the results of animal experiments for human health are uncertain and therefore they cannot be regarded as relevant in the assessment of the potential health risk of ambient exposure to manganese.

Evaluation of Human Health Risks

Exposure

In urban and rural areas without significant manganese pollution, annual averages are mainly in the range of 0.01–0.07 μg/m^3; near foundries the level can rise to an annual average of 0.2–0.3 $\mu g/m^3$ and, where ferro- and silicomanganese industries are present, to more than 0.5 $\mu g/m^3$, with individual 24-hour concentrations exceeding 10 $\mu g/m^3$.

Health risk evaluation

Manganese is both an essential and, at higher levels, a toxic element. In assessing the health impact of manganese exposure, the effect on the CNS and the lungs should be regarded as the most significant. The neurological disorder known as manganism has been reported in the context of occupational exposure to manganese, seldom at levels below 5 mg/m^3. Studies related to signs and symptoms which may be connected with a subclinical or early stage of manganese poisoning in workers exposed to concentrations below 1 mg/m^3 are equivocal or negative.

Respiratory symptoms seem to occur at lower levels of exposure to manganese, e.g. below 1 mg/m^3, than do neurological symptoms and signs. Therefore, respiratory effects may be considered to be critical in ambient exposure to manganese. Schoolchildren exposed to about 7 μg manganese per m^3 (range 3–11 $\mu g/m^3$) emitted from a ferromanganese plant had an increased prevalence of respiratory symptoms. This level of 7 $\mu g/m^3$ may therefore be considered the lowest-observed-effect level *(33)*. This conclusion is supported by a report of increased incidence of acute bronchitis at levels of about 1 $\mu g/m^3$ in a population living near a manganese alloy plant *(34)*.

The available evidence indicates that the current manganese levels generally found in industrialized countries are not in the concentration range associated with potentially harmful effects.

Guidelines

Available data from epidemiological studies suggest that the lowest-observed-adverse-effect concentration of manganese is about 7 $\mu g/m^3$. It is assumed that below 1 $\mu g/m^3$ (annual average), adverse health effects of environmental exposure to manganese are not likely to occur and therefore an annual average of 1 $\mu g/m^3$ is recommended as a guideline value. This value incorporates a sufficient margin of protection for the most sensitive population group. As the critical effect is one of respiratory irritancy it is desirable to have a short-term guideline value, but the present data base does not allow such estimations.

References

1. *Manganese.* Geneva, World Health Organization, 1981 (Environmental Health Criteria, No. 17).
2. *Health assessment document for manganese.* Cincinnati, OH, US Environmental Protection Agency, 1984 (Report No. EPA-600/8-83-013F).
3. **Duce, R.A. et al.** Atmospheric trace metals at remote northern and southern hemisphere sites: pollution or natural? *Science,* **187**: 59–61 (1975).

4. **Zoller, W.H. et al.** Atmospheric concentrations and sources of trace metals at the South Pole. *Science,* **183**: 198–200 (1974).
5. **Georgii, H.W. et al.** Die Verteilung von Schwermetallen in reiner und verunreinigter Atmosphäre [Distribution of heavy metals in clean and polluted atmospheres]. *Staub, Reinhaltung der Luft,* **34**: 15–17 (1974).
6. **Georgii, H.W. & Müller, J.** Schwermetallaerosole in der Großstadtluft [Heavy metal aerosols in the air of large cities]. *Schriftenreihe des Vereins für Wasser-, Boden- und Lufthygiene,* **42**: 39–50 (1974).
7. **Kretzschmar, J.G. et al.** Heavy metal levels in Belgium: a five-year survey. *Science of the total environment,* **14**: 85–97 (1980).
8. [*Results of measurements of national air sampling stations during 1973 (fiscal year)*]. Tokyo, Environmental Agency Japan, 1975 (in Japanese).
9. **Pace, T.G. & Frank, N.H.** *Procedures for estimating probability of non-attainment of a PM_{10} NAAQS using total suspended particulate or inhalable particulate data.* Research Triangle Park, NC, US Environmental Protection Agency, 1983.
10. WHO Technical Report Series, No. 647, 1980 (*WHO recommended health-based limits in occupational exposure to heavy metals*).
11. *WHO guidelines for drinking-water quality. Vol. 2. Health criteria and other supporting information.* Geneva, World Health Organization, 1984, pp. 275–278.
12. **Méranger, J.C. & Smith, D.C.** The heavy metal content of a typical Canadian diet. *Canadian journal of public health,* **63**: 53–57 (1972).
13. **Schlage, C. & Wortberg, B.** Manganese in the diet of healthy preschool and schoolchildren. *Acta paediatrica scandinavica,* **61**: 648–652 (1972).
14. **McLeod, B.E. & Robinson, M.F.** Dietary intake of manganese by New Zealand infants during the first six months of life. *British journal of nutrition,* **27**: 229–232 (1972).
15. **Mena, I. et al.** Chronic manganese poisoning: clinical picture and manganese turnover. *Neurology,* **17**: 128–136 (1967).
16. **Kostial, K. et al.** The effect of iron additive to milk on cadmium, mercury and manganese absorption in rats. *Environmental research,* **22**: 40–45 (1980).
17. **Thomson, A.B.R. et al.** Interrelation of intestinal transport system for manganese and iron. *Journal of laboratory and clinical medicine,* **78**: 642–655 (1971).
18. **Sumino, K. et al.** Heavy metals in normal Japanese tissues. Amounts of 15 heavy metals in 30 subjects. *Archives of environmental health,* **30**: 487–494 (1975).
19. **Kello, D. & Vrcic, H.** Manganese toxicity and homeostasis. *In: Proceedings of the Second International Symposium on Trace Elements, Human Health and Hair Analysis, Amsterdam, May 18–19, 1984.* Amsterdam, Elsevier Scientific Publishers (in press).
20. **Thompson, T.N. & Klaassen, C.D.** Presystemic elimination of manganese in rats. *Toxicology and applied pharmacology,* **64**: 236–243 (1982).
21. **Mena, I. et al.** Chronic manganese poisoning: clearance of tissue manganese concentrations with persistence of the neurological picture. *Neurology,* **18**: 376–382 (1968).

22. **Seth, P.K. & Chandra, S.V.** Neurotransmitters and neurotransmitter receptors in developing and adult rats during manganese poisoning. *Neurotoxicology,* **5**: 67–76 (1984).
23. **Lai, J.C.K. et al.** Differences in the neurotoxic effects of manganese during development and aging: some observations on brain regional neurotransmitter and non-neurotransmitter metabolism in a developmental rat model of chronic manganese encephalopathy. *Neurotoxicology,* **5**: 37–47 (1984).
24. **Suzuki, Y. et al.** Effects of the inhalation of manganese dioxide dust on monkey lungs. *Tokushima journal of experimental medicine,* **25**: 119–125 (1978).
25. **Coulston, F. & Griffin, T.** *Inhalation toxicology of airborne particulate manganese in rhesus monkeys.* Washington, DC, US Environmental Protection Agency, 1977 (Report No. EPA-600/1-77-026).
26. **Ulrich, C.E. et al.** Evaluation of the chronic inhalation toxicity of a manganese oxide aerosol. III — Pulmonary function, electromyograms, limb tremor, and tissue manganese data. *American Industrial Hygiene Association journal,* **40**: 349–353 (1979).
27. **Bergström, R.** Acute pulmonary toxicity of manganese dioxide. *Scandinavian journal of work, environment & health,* **3**(Suppl. 1): 1–41 (1977).
28. **Maigretter, R.Z. et al.** Potentiating effects of manganese dioxide on experimental respiratory infections. *Environmental research,* **11**: 386–391 (1976).
29. **Adkins, B., Jr. et al.** Increased pulmonary susceptibility to streptococcal infection following inhalation of manganese oxide. *Environmental research,* **23**: 110–120 (1980).
30. **Barbeau, A.** Manganese in extrapyramidal disorders. *Neurotoxicology,* **5**: 13–35 (1984).
31. **Sarić, M.** *Biological effects of manganese.* Research Triangle Park, NC, US Environmental Protection Agency, 1978 (Report No. EPA-600/1-78-001).
32. **Elstad, D.** Manganholdig fabrikkrök som medvirkende arsak ved pneumoni-epidemier i en industrybygd [Factory smoke containing manganese as a predisposing cause in pneumonia epidemics in an industrial district]. *Norsk magazin for lægevidenskaben,* **3**: 2527–2533 (1939).
33. **Nogawa, K. et al.** [Epidemiological studies on disturbance of the respiratory system caused by manganese air pollution. Report 1: Effects on respiratory system of junior high school students]. *Japanese journal of public health,* **20**: 315–326 (1973) (in Japanese).
34. **Sarić, M. et al.** Acute respiratory diseases in a manganese contaminated area. *In: Proceedings of the International Conference on Heavy Metals in the Environment.* Toronto, Institute for Environmental Studies, University of Toronto, 1975, pp. 389–398.
35. **Sarić, M. & Lučić-Palaić, S.** Possible synergism of exposure to airborne manganese and smoking habit in occurrence of respiratory symptoms. *In:* Walton, W.H., ed. *Inhaled particles IV.* Oxford & New York, Pergamon Press, 1977, pp. 773–779.

25

Mercury

General Description

Mercury exists in three oxidation states: Hg^{o} (metallic), Hg^{+} (mercurous) and Hg^{++} (mercuric) mercury. The last named forms a variety of inorganic as well as organometallic compounds. In the case of organometallic derivatives, the mercury atom is covalently bound to one or more carbon atoms.

Sources

The global cycle of mercury involves the emission of Hg^{o} from land and water surfaces to the atmosphere, transport of Hg^{o} in the atmosphere on a global scale, possible conversion to unidentified soluble species, and return to land and water by various depositional processes. The ultimate deposition of mercury, probably as cinnabar ore, is believed to be in ocean sediments.

Inorganic mercury undergoes methylation by microorganisms that are widespread in bodies of both fresh and ocean water and probably in soils. Methylmercury is avidly accumulated by fish and attains its highest concentrations in large predatory fish at the top of the aquatic food-chain. By this means, it enters the human diet. Certain microorganisms can demethylate CH_3Hg; others can reduce Hg^{++} to Hg^{o}. Thus, microorganisms are believed to play an important role in the fate of mercury in the environment and in affecting human exposure.

Anthropogenic sources of mercury are numerous and worldwide. Mercury is produced by the mining and smelting of cinnabar ore. It is used in chloralkali plants (producing chlorine and sodium hydroxide), in paints as preservatives or pigments, in electrical switching equipment and batteries, in measuring and control equipment (thermometers, medical equipment), in mercury vacuum apparatus, as a catalyst in chemical processes, in mercury quartz and luminescent lamps, in the production and use of high explosives using mercury fulminate, in copper and silver amalgams in tooth-filling materials and as fungicides in agriculture (especially as seed dressings).

In addition, many industrial activities not related to mercury production or use contribute significant amounts of this element to the environment; these include the burning of fossil fuels, the smelting of various metals, cement manufacture and waste disposal. Watson *(1)* has estimated that the total anthropogenic emission of mercury to air, water and land for the year 1985 amounted to approximately 9300 tonnes and that natural emissions accounted for 15 800 tonnes per year. In Europe, anthropogenic emissions in 1985 were expected to amount to 2630 tonnes, of which 765 tonnes would be emitted to the atmosphere. Total natural emissions would amount to 570 tonnes, nearly all (527 tonnes) being emitted to the atmosphere. Thus, Watson's figures indicate that the majority of emissions of mercury to the atmosphere in Europe are anthropogenic.

Occurrence in air

This topic has been reviewed by Lindqvist et al. *(2)*. Background levels in the troposphere of the northern hemisphere are now estimated at 2 ng mercury per m^3. Levels in the upper troposphere are only slightly lower, but few measurements have been reported. In areas of Europe remote from industrial activity, such as rural parts of southern Sweden and Italy, mean concentrations of total mercury in the atmosphere are reported to be usually in the range of 2–3 ng mercury per m^3 in summer and 3–4 ng mercury per m^3 in winter. Mean mercury concentrations in urban air are usually higher; for example, a value of 10 ng mercury per m^3 has been reported in Mainz, Federal Republic of Germany, and in an urban area of Italy. Individual determinations may cover a wide range of values. In the same urban area of Italy, the range was 2–31 ng mercury per m^3. A recent documentation of concentrations in Member States of the European Community gives the following ranges for atmospheric mercury: remote areas: 0.001–6 ng/m^3; urban areas: 0.1–5 ng/m^3; industrial areas: 0.5–20 ng/m^3 *(3)*.

"Hot spots" of mercury concentration have been reported in the atmosphere close to industrial emissions or above areas where mercury fungicides have been used extensively. Fujimura *(4)* reported air levels up to 10 000 ng/m^3 near rice fields where mercury fungicides had been used and values of up to 18 000 ng/m^3 near a busy motorway in Japan. Air values may rise to 600 and 1500 ng/m^3 near mercury mines and refineries *(5)*.

Few data are available on the speciation of mercury in the atmosphere. It is generally assumed that the vapour of metallic mercury is the predominant form *(2,5,6)*. Johnson & Braman *(7)* reported the presence of methylmercury compounds in the atmosphere above a highly polluted ocean bay in Florida, USA, and in a metropolitan area (Tampa, FA, USA). In the latter area, mercury vapour accounted for 50%, mercuric halide vapour and monomethylmercury for 25% and 21% respectively, and dimethylmercury only 1% of total gaseous mercury. The water-soluble fraction is, on the average, between 5% and 10% of the total gaseous mercury *(2)*. All published studies to date indicate that particulate mercury accounts for less than 5% of total mercury in the atmosphere.

Total mercury, assumed to be mainly vapour of metallic mercury, has a residence time of 0.4–3 years and is therefore globally distributed *(2)*. The "water-soluble" form may have a residence time of only a few weeks. This water-soluble form and particulate mercury, which also has a short residence time, although accounting for only a small fraction of total mercury, may nevertheless be important in the transport and depositional processes. Furthermore, the possibility that some conversion of water-insoluble to water-soluble forms of mercury takes place in the atmosphere has been raised.

No data are available on indoor air pollution due to mercury vapour. Fatalities and severe poisonings have resulted from heating metallic mercury and mercury-containing objects in the home. Incubators used to house premature infants have been found to contain mercury vapour at levels approaching occupational threshold limit values; the source was mercury droplets from broken mercury thermostats. Indoor air pollution caused by central-heating thermostats and by the use of vacuum cleaners after thermometer breakage, etc., also needs attention.

Routes of Exposure

Air

The atmosphere is the only source of human exposure to the vapour of metallic mercury. The daily amount absorbed as a result of respiratory exposure into the bloodstream in adults is about 32 ng mercury in rural areas and about 160 ng mercury in urban areas, assuming rural concentrations of 2 ng/m^3 and urban concentrations of 10 ng/m^3 (absorption rate 80%).

Drinking-water

Mercury in drinking-water is usually in the range of 5–100 ng mercury per litre, the average value being about 25 ng mercury per litre. The forms of mercury in drinking-water are not well studied, but Hg^{++} is probably the predominant species present as complexes and chelates with ligands.

Food

Concentrations of mercury in most foodstuffs *(6,8,9)* are often below the detection limit (usually 20 ng mercury per gram fresh weight). Fish and fish products are the dominant source, mainly in the form of methylmercury compounds (70–90% of the total). The normal concentrations in edible tissues of various species of fish cover a wide range, from 50 to 1400 ng mercury per gram fresh weight. Large predatory freshwater fish, such as pike and trout, have some of the highest average concentrations.

Relative significance of different routes of exposure

Human exposure to the three major forms of mercury present in the environment is summarized in Table 1. Although the choice of values given is somewhat arbitrary, the Table nevertheless provides a perspective on the relative magnitude of the contributions from various media. Humans may be exposed to additional quantities of mercury occupationally and in heavily polluted areas, and to additional forms of mercury, e.g. to aryl and alkoxyaryl compounds widely used as fungicides.

The intake from drinking-water is about 50 ng mercury per day, mainly as Hg^{++}; only a small fraction is absorbed. Intake of fish and fish products,

Table 1. Estimated average daily intake (retention) of mercury and its compounds

Media	Estimated average daily intake (retention) in ng of mercury per day		
	Mercury vapour	Inorganic mercury compounds	Methylmercury
Atmosphere	200[a] (160)[b]	0[c]	0[c]
Food:			
fish	0	80[d] (8)[b]	3200[d] (2880)[b]
other	0	?[e]	?
Drinking-water	0	50 (5)[b]	0

[a] Assumes an air concentration of 10 ng/m³ and a daily respiratory volume of 20 m³.

[b] Figures in parentheses are the amount retained that was estimated from the pharmacokinetic parameters discussed on pp. 276–278, i.e 80% of inhaled vapour, 90% of ingested methylmercury and 10% of inorganic mercury is retained.

[c] For the purposes of comparison, it is assumed that in the atmosphere concentrations of species of mercury other than mercury vapour are negligible.

[d] It is assumed that 80% of the total mercury in edible fish tissues is in the form of methylmercury and 20% in the form of inorganic mercury compounds. It is also assumed that 90% of the methylmercury and 10% of the inorganic mercury is absorbed. It should be noted that fish intake may vary considerably between individuals and across populations. Certain communities whose major source of protein is fish may exceed this estimated methylmercury intake by an order of magnitude or more.

[e] Levels of inorganic mercury in many food items are below the detection limit.

averaged over months or weeks, results in a daily absorption of methylmercury variously estimated to be between 2000 and 4700 ng mercury *(5)*. The absorption of inorganic mercury from foodstuffs is difficult to estimate because levels of total mercury are close to the limit of detection in many food items and the chemical species of mercury have not usually been identified.

The calculations in Table 1 assume that all the mercury is present in the atmosphere as the vapour of metallic mercury. If one takes the highest reported concentrations for the methylated form of mercury *(7)*, mercury vapour would account for 50%, methylmercury compounds approximately 25% and inorganic mercury 25%; the amount of mercury vapour absorbed daily would be reduced to 80 ng mercury per day and the amount of additional methylmercury absorbed by inhalation would be about 40 ng mercury per day, a negligible amount compared to intake from fish.

The intake of total dietary mercury in the USA has been measured *(10)* over a number of years for various age groups. The average daily intake over the period 1973–1982 was in the range of 2000–7000 ng for adults and up to 1000 ng for toddlers and infants. The most recent figures for the USA (fiscal year 1981–1982) were 3000 ng for adults, 1000 ng for toddlers, and less than 1000 ng for infants. These figures support the conclusion that fish and fish products are the major source of total mercury in the diet. In the Federal Republic of Germany, the total weekly intake from food has been estimated to be less than 100 μg *(11)*, or less than 14 μg/day.

In the ensuing sections information will be presented on each of the three main environmental forms and on phenylmercury compounds, the latter taken to be representative of the aryl and alkoxyaryl classes.

Kinetics and Metabolism

The pharmacokinetics and biotransformation of mercury depend upon its chemical and physical form.

Absorption

Approximately 80% of inhaled *mercury vapour* is absorbed via the lungs and retained in the body.

The deposition and absorption of inhaled aerosols of *inorganic mercury* will depend on particle size, solubility, etc. *(12)*. No data have been reported for humans. In dogs, Morrow et al. *(13)* have reported that 45% of deposited mercury(II) oxide aerosols (mean diameter 0.16 μm) was cleared in less than 24 hours and the remainder cleared with a half-time of 33 days. Approximately 15% of an oral, nontoxic dose is absorbed from the gastrointestinal tract in adults and retained in body tissues. In children, the gastrointestinal absorption is probably greater.

Information on the pulmonary retention of gaseous *dimethylmercury and aerosols of the monomethylmercury compounds* is lacking. To judge from cases of human poisoning via inhalation and from the ability of these mercurials rapidly to cross cell membranes, it seems likely that a large fraction of the inhaled alkylmercurial is absorbed into the bloodstream. The alkylmercurials are absorbed almost completely in the gastrointestinal tract and retained in the body. Certain methylmercury compounds are probably well absorbed across the skin.

Distribution

Mercury vapour is distributed via the bloodstream to all tissues in the body, the initial distribution process after single pulsed exposure being complete in about 3 days. Within the blood, concentrations in red cells tend to be higher than in plasma shortly after exposure to mercury vapour. Mercury in this form readily penetrates the blood–brain and placental barriers after exposure. However, the kidney becomes the major site of accumulation after oxidation to Hg^{++}.

Inhaled mercury vapour is oxidized to divalent ionic mercury in body tissues by the hydrogen-peroxide catalase pathway. This oxidation step is inhibited by low doses of ethanol and may have consequences for workers and pregnant women. Ethanol can lead to decreased retention of mercury vapour in workers and, according to animal experiments, result in increased transport to the fetus. Dissolved mercury vapour in plasma is believed to be the species transported to the brain and the fetus and may be important in transport and distribution to all tissues in the body. Inhalation of mercury vapour can lead to the induction and binding of mercury to metallothionein in the kidneys.

The kidney is the predominant site of *inorganic mercury* accumulation, but accumulation also occurs in the cells of the mucous membranes of the gastrointestinal tract. Inorganic mercury penetrates the blood–brain and placental barriers only to a small extent. Inorganic divalent mercury can induce and bind to metallothionein in the kidney *(14)* and probably forms complexes with selenium in tissues after chronic exposure.

Distribution of *methylmercury* takes place via the bloodstream to all tissues in the body, the initial phase being complete in about 3 days. The pattern of tissue distribution is much more uniform than after inorganic mercury exposure, except in red cells, where the concentration is 10–20 times greater than the plasma concentration. Methylmercury readily crosses the blood–brain and placental barriers, and, as with other forms, the kidneys retain the highest tissue concentration.

At the end of the initial phase of distribution, approximately 1% of the body's methylmercury is found in one litre of blood in the human adult weighing 70 kg *(5)*. The brain/blood concentration ratio is about 5 : 1. Methylmercury accumulates in hair in the process of formation of hair strands. The hair/blood concentration ratio is approximately 250 : 1 in humans at the time of incorporation into hair. Once incorporated into the hair, the concentration remains unchanged.

Elimination

The whole-body half-time of mercury after inhalation of *mercury vapour* is approximately 50 days, whereas the half-time in the brain is about 20 days *(10)*. After a single exposure, the decline of mercury levels in the blood is described by two exponential terms with equivalent half-times of about 3 and 30 days each. The lower half-time accounts for 90% of the mercury in the blood compartment. Blood levels can serve as indicators of recent mercury vapour exposure. Conversely, urine levels correlate with long-term exposure, although only on a group basis. Individual (including diurnal) variability of urine levels can be considerable.

Excretion of *inorganic mercury* is mainly via urine and faeces, the rates by each pathway being roughly equal. The whole-body half-time in adults is about 40 days.

Excretion of *methylmercury* is predominantly via the faeces. Methylmercury is slowly broken down to inorganic mercury and most, if not all, of the excreted mercury is in the inorganic form. Enterohepatic recirculation probably explains the absence of methylmercury in the faeces.

The whole-body half-time of methylmercury is usually between 70 and 80 days, but substantial individual differences occur *(9)*. The brain half-time is roughly the same as that of the whole body, whereas the half-time in the blood compartment is about 50 days. Blood concentration is a useful indicator of body-burden, and hair concentration, when measured along the length of a hair strand, is a useful indicator of past blood levels. In the analysis of blood, inorganic mercury and methylmercury should be analysed separately.

Health Effects

The toxic effects of mercury and its compounds depend upon the chemical form of mercury.

Mercury vapour

Damage is mainly to the nervous system, but effects are seen, depending on dose, in the oral mucosa and the kidneys *(10)*. Effects on the nervous system are manifested as tremor and a cluster of psychological difficulties labelled erethism. These include deficits in short-term memory and social withdrawal. The peripheral nervous system may also be involved, as evidenced by changes in nerve conduction velocity. The effects on the nervous system, especially those on motor functions, are usually reversible. The more insidiously developing cognitive decrements and emotional alterations may be the most harmful effects of current exposures in the workplace.

Mercury vapour can elicit the nephrotic syndrome, characterized by excessive loss of protein (mainly albumin) in the urine and oedema *(15,16)*. Differences in individual susceptibility appear to be great. Low levels of exposure to mercury vapour are associated with mild proteinuria and enzymuria *(15)*. In general, the renal effects of mercury are reversible after cessation of exposure.

Host-related factors may be specially important in the development of the nephrotic syndrome in workers exposed to mercury vapour or compounds of inorganic mercury *(17)*. Druet et al. *(18)* have demonstrated an immunological mechanism that appears to be genetically determined in various inbred strains of rats.

Limited information is available on the effects of mercury vapour on the early stages of the human life cycle. Effects on pregnancy and birth in women occupationally exposed to mercury vapour have been reported, but insufficient details were available to evaluate dose–response relationships.

Urine is the most frequently used indicator medium for assessing body-burden after chronic exposure to mercury vapour. Approximately 95% of all urine samples from people without known exposure to mercury are below 20 μg mercury per litre. Mild proteinuria may occur in the most sensitive adults at urine values between 50 and 100 μg/litre following chronic occupational exposures. Objective tremor and psychomotor disturbances appear in sensitive individuals usually at urine values above 300μg/litre, but no clear threshold has been established (Table 2). Most effects of mercury vapour usually disappear within a few months after cessation of exposure.

Inorganic divalent mercury compounds

These are corrosive poisons; acute single doses can cause death by kidney failure and systemic shock. Little information is available on the chronic effects of inorganic mercury compounds in humans or animals. Possible effects due to chronic exposure to compounds of inorganic mercury have been rarely reported.

Occupational exposure to mercuric oxide has been shown to damage the peripheral nervous system and to produce effects somewhat similar to amyotrophic lateral sclerosis; these effects are reversible. Exposure to inorganic divalent mercury compounds has been known to produce acrodynia (pink disease) in susceptible children. Dose–effect relationships have not been reported, but air concentrations associated with these effects are probably higher than those associated with the earliest effects of mercury vapour. In general, effects from chronic exposure are reversible.

Methylmercury compounds

Damage is almost exclusively limited to the nervous system. In human adults, the damage is selective to certain areas of the brain concerned with

Table 2. Concentrations of total mercury in air and urine at which effects are observed at a low frequency in workers subjected to long-term exposure to mercury vapour

Observed effects[a]	Mercury level		Reference
	Air[b] (μg/m³)	Urine (μg/litre)	
Objective tremor	100	300	*(4)*
Nonspecific symptoms	50	150	*(4)*
Changes in plasma lysomal enzymes	50[c]	150	*(15)*[d]
Proteinuria	15[c]	50	*(16)*

[a] These effects occur with low frequency in occupationally exposed groups. Many other effects have been reported, but air and urine levels are not available.

[b] The air concentrations measured by static air samplers are taken as a time-weighted average, assuming 40 hours per week for long-term exposure (at least five biological half-times, equivalent to 250 days).

[c] Calculated from the urine concentration, assuming that 100μg mercury per m³ measured by static samplers is equivalent to 300μg mercury per litre urine.

[d] Information on frequency is not available, but the statistical significance was high.

sensory and coordination functions, particularly neurons in the visual cortex and granule cells of the cerebellum. Effects such as constricted visual fields and ataxia appear to have a latent period of weeks to months. Such effects are usually irreversible. The first effect usually noted is a subjective complaint of paraesthesia in the extremities and in the circumoral region. Paraesthesias may or may not be irreversible. The peripheral nervous system may also be affected, especially at high doses, but usually after effects have already appeared in the CNS.

The prenatal stage is the stage of the human life cycle most susceptible to methylmercury exposure. Severe effects, such as cerebral palsy, may be seen in the offspring of mothers in whom effects have been minimal or absent. One report by Eyssen et al. *(19)* suggests that the mildest effects of prenatal exposure, such as psychomotor retardation, may be associated with maternal hair levels in pregnancy between 10–30 μg mercury per gram. These effects were seen only in prenatally exposed males but not in females. Data from the outbreak of acute mercury poisoning in Iraq are consistent with the report of Eyssen et al. The estimates of dose–response relationships at these low exposure levels are subject to considerable uncertainty *(20)*. All prenatal effects to date have been found to be irreversible.

Blood and scalp hair are the most commonly used media for indicating the body-burden of mercury after exposure to methylmercury compounds. Levels of methylmercury in these media are related to fish consumption *(5)*. Blood levels above 200 μg mercury per litre may be associated with health effects. Since the major route of exposure to methylmercury is oral, this aspect will not be discussed further *(5)*.

Evaluation of Human Health Risks

Exposure

In areas remote from industry, atmospheric levels of mercury are about 2–4 ng/m^3, and in urban areas about 10 ng/m^3.

Health risk evaluation

Sensitive population groups

With regard to exposure to mercury vapour, sensitive population groups have not been identified from epidemiological, clinical or experimental studies. Effects on the kidney are believed to occur first in a subgroup of individuals whose susceptibility may be genetically determined. The proportion of this subgroup in the general population is unknown. The same considerations apply to kidney effects induced by inorganic and phenylmercury compounds. Virtually nothing is known of the relative sensitivity of different stages of the life cycle to mercury vapour or inorganic compounds, except that the developing rat kidney is less sensitive than the mature tissue to inorganic mercury *(21)*.

The stage of the life cycle most sensitive to methylmercury is the prenatal stage, when sensitivity is believed to be at least three times greater than that in adults. A group of adults, accounting for about 19% of the sampled population in Iraq *(22)*, have unusually long biological half-times. These individuals may be at greater risk, as they should accumulate more methylmercury on chronic exposure than the rest of the population.

The role of the life cycle in determining susceptibility is well demonstrated in the case of methylmercury compounds, but little information is available for other forms of mercury.

Mercury vapour

Time-weighted air concentrations are the usual means of assessing human exposure. Reported air values depend on the type of sampling. Static sampling generally gives lower values than personal sampling. In order to convert the air concentrations quoted in Table 2 to equivalent concentrations in ambient air, two factors have to be taken into account. First, the air concentrations listed in Table 2 are those obtained in the working environment as determined by static samplers. The conversion factor is an approximation and may vary, depending on exposure conditions. These values should be increased by a factor of 3 to correspond to the true air concentrations inhaled by the workers as determined by personal samplers. Second, the total amount of air inhaled at the workplace per week is assumed to be 50 m^3 (10 m^3/day $\times$ 5 days) whereas the amount of ambient air inhaled per week would be 140 m^3 (20 m^3/day $\times$ 7 days). Thus the volume of ambient air inhaled per week is approximately three times the volume inhaled at the workplace. Thus, to convert the workplace air concentrations quoted in Table 2 to equivalent ambient air concentrations, they should first be multiplied by 3 to convert to actual concentrations in the workplace, and divided by 3 to correct for the greater amount of ambient air inhaled per week by the average adult. It follows that the mercury vapour concentrations quoted in Table 2 are approximately equivalent to ambient air concentrations.

Since these figures are based on observations in humans, a protection factor of 10 would seem appropriate. However, the lowest-observed-effect levels in Table 2 are rough estimates of air concentrations at which effects occur at a "low frequency." Because it seems unlikely that such effects would occur in occupationally exposed workers at air concentrations as low as one half of those given in Table 2, it seems appropriate to use a protection factor of 20. Thus, the estimated guidelines (using rounded figures) would be 1 μg mercury per m^3 for mild proteinuria, 2.5 μg mercury per m^3 for non-specific symptoms and 5 μg mercury per m^3 for objective tremors.

Inorganic compounds

Compounds of inorganic mercury are retained in the lungs about half as efficiently (40% versus 80% retained) as inhaled mercury vapour; thus, the estimated guideline would be 2 μg mercury per m^3, providing adequate protection against mild proteinuria.

Methylmercury compounds

It does not seem appropriate to set air quality guidelines for methylmercury compounds. Inhalation of this form of mercury, if it is present in the atmosphere, would make a negligible contribution to total human intake. However, it is possible that mercury in the atmosphere might ultimately be converted to methylmercury after deposition on soils or sediments in natural water-bodies, leading to an accumulation of that form of mercury in aquatic food-chains. In this eventuality, guidelines for food intake would be appropriate, such as those recommended in previous WHO/FAO documents *(23)*.

Conclusions

Since mercury and its compounds are neither mutagens nor carcinogens, it is assumed that a threshold atmospheric concentration exists below which the risks of health effects are negligible. As such, a threshold concentration cannot be identified with any precision; margins of protection have to be used to estimate an acceptable air concentration. Such protection factors must take into account the possibility of sensitive subgroups in the population.

In principle, it is necessary to take into account the different forms of mercury in the atmosphere and intake of these forms of mercury from other media. As stated above, the atmosphere is the sole source of exposure to mercury vapour, whereas the diet is the dominant source of methylmercury compounds. Therefore, it seems appropriate to consider a guideline for human inhalation of mercury vapour.

Present levels of mercury in outdoor air, except for "hot spots", are well below 1 $\mu g/m^3$ and should not have direct effects on human health. However, 1 $\mu g/m^3$ is probably far above an acceptable level if postdeposition effects (i.e. increase in methylmercury compounds in the diet) are taken into account. The possibility exists that an increase in ambient air levels of mercury will result in an increase in deposition in natural water-bodies, resulting, in turn, in elevated concentrations of methylmercury in freshwater fish. Such a contingency might have an important bearing on acceptable levels of mercury in the atmosphere. Unfortunately, the limited knowledge of the global cycle and of the methylation and bioaccumulation pathways in the aquatic food-chain does not allow any quantitative estimates of risks from postdeposition sequelae.

Monitoring of the atmosphere would be useful in situations of suspected contamination. In the case of postdeposition effects, monitoring of methylmercury levels in freshwater fish would be useful for those freshwater bodies where elevated methylmercury levels might be expected through mercury contamination for example, or for areas affected by acid rain, etc.

Guidelines

Because of the lack of quantitative information on the consequences of deposition, it is not possible to identify guidelines that would provide protection against possible postdeposition effects of air pollution by mercury, and therefore an ambient air quality guideline value is not proposed.

Adequate protection from indoor air pollution by mercury is provided by a guideline value of approximately 1 μg mercury per m^3 as an annual average, irrespective of the form of mercury in the air.

References

1. **Watson, W.D., Jr.** Economic considerations in controlling mercury pollution. *In:* Nriagu, J.O., ed. *The biogeochemistry of mercury in the environment.* Amsterdam, Elsevier, 1979, pp. 41–47.
2. **Lindqvist, O. et al.** *Mercury in the Swedish environment: global and local sources.* Solna, National Swedish Environmental Protection Board, 1979, pp. 1–105 (Sweden Report No. 1816).
3. **Lahmann, E. et al.** *Heavy metals: identification of air quality and environmental problems in the European Community.* Luxembourg, Commission of the European Communities, 1986, Vol. 1 & 2 (Report No. EUR 10678 EN/I and EUR 10678 EN/II).
4. **Fujimura, Y.** Studies on the toxicity of mercury (Hg Series No. 7). II. The present status of mercury contamination in the environment and foodstuffs. *Japanese journal of hygiene,* **18**: 402–411 (1964).
5. *Mercury.* Geneva, World Health Organization, 1976 (Environmental Health Criteria, No. 1).
6. **Matheson, D.H.** Mercury in the atmosphere and in precipitation. *In:* Nriagu, N.O., ed. *The biogeochemistry of mercury in the environment.* Amsterdam, Elsevier, 1979, pp. 113–129.
7. **Johnson, D.C. & Braman, R.S.** Distribution of atmospheric mercury species near ground. *Environmental science and technology,* **8**: 1003–1009 (1974).
8. **Piotrowski, J.K. & Inskip, M.J.** *Health effects of methylmercury.* London, University of London Monitoring and Assessment Research Centre, 1981 (MARC Report No. 24).
9. *Ambient water quality criteria for mercury.* Washington, DC, US Environmental Protection Agency, 1980 (Report No. EPA-440/5-80-058).
10. *Mercury health effects update. Health issue assessment.* Washington, DC, US Environmental Protection Agency, 1984 (Report No. EPA-600/8-84-019F).
11. **Umweltbundesamt** [Federal Office of the Environment]. *Umwelt- und Gesundheitskriterien für Quecksilber* [Environmental and health criteria for mercury]. Berlin (West), Erich Schmidt Verlag, 1980 (Berichte 5/80).
12. **Task Group on Lung Dynamics.** Deposition and retention models for internal dosimetry of the human respiratory tract. *Health physics,* **12**: 173–208 (1966).
13. **Morrow, P.E. et al.** Clearance of insoluble dust from the lower respiratory tract. *Health physics,* **10**: 543–555 (1964).
14. **Nordberg, M. et al.** Studies on metal-binding proteins of low molecular weight from renal tissue of rabbits exposed to cadmium or mercury. *Environmental physiology and biochemistry,* **4**: 149–158 (1974).
15. **Foa, V. et al.** Patterns of some lysosomal enzymes in the plasma and of proteins in urine of workers exposed to inorganic mercury. *International archives of occupational and environmental health,* **37**: 115–124 (1976).

16. **Buchet, J.P. et al.** Assessment of renal function of workers exposed to inorganic lead, cadmium or mercury vapor. *Journal of occupational medicine,* **22**: 741–750 (1980).
17. **Kazantzis, G. et al.** Albuminuria and the nephrotic syndrome following exposure to mercury and its compounds. *Quarterly journal of medicine,* **31**: 403–418 (1962).
18. **Druet, P. et al.** Nephrotoxin-induced changes in kidney immunobiology with special reference to mercury-induced glomerulonephritis. *In:* Bach, P.H. et al., ed. *Nephrotoxicity assessment and pathogenesis.* New York, Wiley, 1982, pp. 206–221.
19. **Eyssen, G.E.M. et al.** Methylmercury exposure in northern Quebec II. *Neurologic findings in children,* **118**: 470–478 (1983).
20. **Marsh, D.O. et al.** Fetal methylmercury poisoning; clinical and toxicological data on 29 cases. *Annals of neurology,* **7**: 348–355 (1980).
21. **Daston, G.P. et al.** Toxicity of mercuric chloride to the developing rat kidney. I. Post-natal ontogeny of renal sensitivity. *Toxicology and applied pharmacology,* **71**: 24–41 (1983).
22. **Al-Shahristani, H. & Shihab, K.M.** Variation of biological half-life of methylmercury in man. *Archives of environmental health,* **18**: 342–344 (1974).
23. WHO Technical Report Series, No. 505, 1972 (*Evaluation of certain food additives and the contaminants mercury, lead and cadmium:* sixteenth report of the Joint FAO/WHO Expert Committee on Food Additives).

26

Nickel

General Description

Nickel (Ni) is a silvery-white, hard metal. Although it forms compounds in several oxidation states, only the divalent ion is important for both organic and inorganic substances. Nickel compounds that are practically insoluble in water include carbonate, sulfides (the main forms being amorphous or crystalline monosulfide, NiS, and subsulfide Ni_3S_2) and oxides (NiO, Ni_2O_3). Soluble nickel salts include chloride, sulfate and nitrate. Nickel carbonyl ($Ni(CO)_4$) is a volatile, colourless liquid with a boiling-point of 43 °C; it decomposes at temperatures above 50 °C. In biological systems, nickel forms complexes with adenosine triphosphate, amino acids, peptides, proteins and deoxyribonucleic acid.

Sources

In the earth's crust, nickel is found in the proportion 75 mg nickel per kg (range: 58–94 mg/kg) *(1)*.

The natural background levels of nickel in water are relatively low; however, there are local or regional increases resulting from natural and anthropogenic sources. In 969 public water-supply systems in the USA, the average level of nickel in drinking-water was 4.8 μg/litre *(2)*: in only 11 systems did it exceed 25 μg/litre.

Average nickel concentrations in rocks range from 2 to 60 mg/kg (acidic rocks), 50 to 200 mg/kg (basic rocks) and 10 to 20 000 mg/kg (ultramafic rocks). Farm soils contain nickel concentrations in the range of 5–500 mg/kg, with a typical level of 50 mg/kg. In coal, nickel impurities (millerite, pyrite) up to 2 g/kg are observed *(3)*.

Nickel is enriched in petroleum. The nickel content of some western Canadian crude oils was found to be in the range of 0.29–76.6 mg/kg *(4)*.

There are two commercial classes of nickel ore: (*a*) the sulfide ores (pentlandite and pyrrhotite) and (*b*) the silicate garnierite. About 60% of all nickel produced is obtained from pentlandite *(5)*.

Forty-seven per cent of nickel is used in steel production and 21% in the production of other alloys; electroplating in the form of nickel sulfate accounts for about 12%. Nickel is also used in coinage and as a catalyst. Other commercial uses are in ceramics, in storage batteries (nickel hydroxide), in electronic components (nickel carbonate), in a dyeing process for polypropylene and for colouring glass *(6)*.

The burning of residual and fuel oils, nickel mining and refining, and municipal waste incineration form the majority of anthropogenic sources of nickel emissions to the atmosphere. The major nickel species in ambient air is nickel sulfate. This soluble ("leachable") form is estimated to comprise 60–100% *(7)* or 15–93% (average 54%) *(8)* of the nickel components emitted by fly ash from oil-fired utility boilers. The corresponding value for nickel sulfate emission arising from coal combustion is 20–80%. The insoluble fraction of fly ash emitted from both oil and coal combustion exists as nickel oxides and complex metal oxides (ferrites, aluminates, vanadates).

In the USA, coal and oil combustion are the chief sources of nickel emission. Of the total amount of nickel emitted from coal combustion and oil combustion, the latter has been estimated to account for 60–98% *(9)*.

Total annual nickel emissions from anthropogenic sources have been calculated to be 9.8×10^{10} g, whereas emissions from natural sources (continental dusts, volcanic dusts and gases) contribute 3×10^{10} g per year to global atmospheric emissions *(10)*; corresponding values of 4.3×10^{10} g and 0.85×10^{10} g per year were calculated by Schmidt & Andren *(11)*.

Occurrence in air

Considerable data on nickel concentrations are available from network or single-site studies. Average particulate concentrations have been found to range from 0 to 0.6 ng/m^3 in remote areas, from 9 to 50 ng/m^3 in semi-remote and from 60 to 300 ng/m^3 in urban locations. The daily dustfall rates corresponding to these levels range from 6 to 36 ng/m^3 *(12)*. Nickel concentrations of particulate matter as high as 3310 μg/m^3 have been found in the ambient air of nickel-processing centres. The following concentration ranges have been given for Member States of the European Community: 0.1–0.7 ng/m^3 in remote areas, 3–100 ng/m^3 in urban areas and 8–200 ng/m^3 in industrial areas *(13)*. Daily deposition rates indicated in the same survey were 0.2–10 μg/m^2 in remote areas, 2–10 μg/m^2 in urban areas and 7–70 μg/m^2 in industrial areas.

The particle size distribution of nickel in urban dustfall in the USA has been analysed *(2)*. In particles below 43 μm in diameter, 27.5% (by weight) of total nickel was found; particles smaller than 840 μm contained 75% nickel, whereas 25% of nickel was associated with particles of 840–2000 μm (actual size). Measurements obtained in both urban and remote areas indicate mass median aerodynamic diameters of 1.2 μm (Toronto, urban), 0.83–1.67 μm (USA, urban), 0.6 μm (England, rural) and 2 μm (Arizona, remote). At least 50% of airborne nickel is in particles of $<$2–3 μm diameter *(12)*. Nickel-containing particles released from oil combustion (California, urban area) are in the fine-size fraction below 1 μm *(14)*. In coal-fired power plants nickel enrichment occurs in the smaller particulate fraction ($<$1 μm) *(15)*.

Routes of Exposure

The major routes of nickel intake for man are inhalation, ingestion and percutaneous absorption.

Air

Assuming a daily respiratory rate of 20 m^3, the amount of airborne nickel entering the respiratory tract is in the range of 0.2–0.4 μg/day when concentrations are 10–20 ng nickel per m^3 ambient air. Cigarette-smoking increases these figures: the smoking of two packs of cigarettes per day will result in the inhalation of 3–15 μg nickel per day (1–5 mg nickel per year) *(2)*. The possibility cannot be ruled out that in the mainstream smoke nickel occurs perhaps in part as nickel carbonyl.

Drinking-water

Assuming a concentration of 5 μg/litre, a daily consumption of 2 litres of drinking-water would result in a daily nickel intake of 10 μg *(2)*.

Food

Nickel uptake by plants has been documented *(16)*; concentrations range from 0.05 to 5 mg/kg dry weight. Murthy et al. estimated a daily oral intake of 451 μg nickel by American children *(17)*, Myron et al. *(18)* calculated an average value of 165 μg/day (determination of institutional diets), and Nielsen & Flyvholm *(19)* indicated a comparable average value for Danish diet of 150 μg/day. Clemente et al. *(20)*, evaluating the available references, estimated a mean intake value of 200–300 μg nickel per day, and a value of 160 μg nickel per day was estimated by Bennett *(21)*. Vegetable products generally have a higher nickel content than animal products; thus, a vegetarian ingests 3–4 times the amounts of nickel quoted above.

Relative significance of different routes of exposure

Table 1 summarizes the levels of daily nickel intake by humans from different routes of exposure.

Percutaneous absorption of nickel has been shown, but no quantitative data are available. For individuals with hypersensitivity to nickel, this route of exposure is of significance.

Leaching of nickel from various implanted metals accounts for parenteral exposure to nickel as well as parenteral contamination by dialysis (100 μg nickel/dialysis treatment) *(22)*.

In the working environment, nickel concentrations in the air may be significantly increased compared with ambient air levels (1 : 10^3, maximum 1 : 10^6). Particulate nickel concentrations (24-hour average) in the ambient air of nickel-processing centres in Ontario, Canada, in 1976 were in the range of 0.07–0.77 $\mu g/m^3$ (geometric mean, minimum below detection limit of 0.004 $\mu g/m^3$, maximum: 2.22 $\mu g/m^3$) *(23)*. Data on nickel concentrations in alloy industries, welding operations and battery production have been compiled *(5,24)*. The range is fairly wide, from a few micrograms to several milligrams/m^3.

Table 1. Levels of daily nickel intake by humans from different types/routes of exposure

Type/route of exposure	Daily nickel intake	Absorption/retention
Foodstuffs	≦300 μg	30 μg (≦10%)
Drinking-water	≦ 10 μg	1 μg (≦10%)
Ambient air (urban dweller)	≦ 0.4 μg	0.4 μg? (100%?)
Ambient air (smoker)	≦ 15.4 μg	7.7 μg (50%)

Kinetics and Metabolism

There is considerable evidence that nickel is an essential trace element in several animal species, plants and prokaryotic organisms. Essentiality for humans is suggested, though no data are available concerning nickel deficiency in man.

Absorption

Pulmonary absorption of nickel carbonyl is rapid and extensive, the agent passing the alveolar wall intact *(25)*. Few data exist on the pulmonary absorption of nickel from particulate matter deposited in the lungs. Using the model of the International Commission on Radiological Protection, it can be calculated that the lung deposition level for nickel in the smallest particles of fly ash reaches 63%, with 25% reaching the pulmonary compartment, of which 5–6% undergo clearance *(15)*. Bennett *(21)* estimates a respiratory intake of 0.4μg/day (range 0.2–1.0μg/day) and 0.2μg/day (range 0.1–0.4μg/day) in urban and rural areas respectively, with an absorption rate of about 0.6.

In golden hamsters exposed to artificial nickel oxide aerosols, 20% of inhaled nickel oxide remained deposited after initial elimination, and 45% of this was still present after 45 days *(26)*; continuous inhalation (6 weeks) of 50μg nickel oxide per m^3 led to comparable figures *(27)*. Wehner et al. *(28)* exposed Syrian hamsters to nickel-enriched fly ash aerosols: nickel-leaching from the nickel-enriched fly ash does not seem to occur to any extent. Graham et al. *(29)* measured the clearance of nickel from mice lungs following a 2-hour exposure and found apparent first-order clearance kinetics, 72% of the deposited fraction disappearing by the fourth day after exposure.

Various studies presented by Mushak *(30)* indicate that 1–2% (up to 10%) of dietary nickel is absorbed; Sunderman *(22)* indicates a value of 5%. Percutaneous absorption of nickel is important in view of the dermopathological effects (contact dermatitis). Owing to lack of data it is not possible to calculate absorption values.

Distribution

Once nickel is absorbed into the organism, it is transported in the bloodstream. Nickel has been proved to cross the placental barrier in animals and man. The main carrier protein in sera is albumin *(31)*.

Zober et al. *(32)* report normal nickel levels in human body fluids and organs obtained from 45 autopsies in northern Bavaria, Federal Republic of Germany: blood 4.5μg/kg (range 0.2–115.4μg/kg), urine 2.7μg/kg (0.1–25.6μg/kg), lung 7.4μg/kg (0.2–70.0μg/kg), kidney 13.6μg/kg (0.2–165.0μg/kg). The authors also tabulate nickel values of normal urine and whole blood derived from studies which had been carried out during the period 1956–1982. Average urine levels are generally <3μg/litre. Average values in whole blood are estimated to be <5μg/litre. The metal concentrations in the different samples were not influenced by age or sex. Physiological stresses and various disease states influence the kinetics of nickel metabolism.

A two-compartment model describing mathematically the whole-body kinetics of nickel(II) has been formulated by Onkelinx & Sunderman *(33)*. Results based on single-injection as well as continuous-infusion and multiple-dosing experiments (rats: 82μg/kg body weight intravenously or 13μg/kg body weight intraperitoneally; rabbits: 240μg/kg body weight intravenously) revealed a typical distribution and elimination pattern, comprising a rapid and a slow clearance phase. A third compartment was made up in rats receiving single intramuscular injections of $^{63}Ni_3S_2$ (1.2 mg/animal), comprising quasi-permanently deposited (insoluble) nickel subsulfide (about 11% of dose).

Elimination

Unabsorbed nickel in the gastrointestinal tract is lost in the faeces (reflecting the daily dietary intake); a figure of 258μg nickel per day is reported.

Absorbed nickel is cleared by urinary excretion, the value being 2.7μg nickel per litre; excretion via sweat (important in hot climates), secretion via saliva and deposition in hair are reported. Urinary excretion, however, is the main clearance route. The biological half-time of nickel depends on the species tested. In humans (welders) a half-time of 17–53 hours has been described *(32,33)*.

Health Effects

Effects on laboratory animals and *in vitro* test systems

Toxicological effects

Inhalation of all types of nickel compound induces respiratory tract irritation and lesions and varied immunological responses, including an increase in the number of alveolar macrophages *(34)*, reduced ciliary activity and immunosuppression. All of these toxic effects correlate with an abnormal functioning of the respiratory defence system.

Nickel can cross the placental barrier, thus being able to affect prenatal development by direct action on the embryo. Fetal death and a wide variation of malformations have been reported to be included among the effects of various species of nickel compound in experimental animals *(5,24,35)*.

Mutagenic and carcinogenic effects
Negative mutagenicity data were obtained in most bacterial test systems, but many nickel compounds can induce *in vitro* mammalian cell transformation and are clastogenic to various degrees.

Cancer in experimental animals can be induced by the injection or implantation of many nickel compounds. The production of localized tumours at the site of injection, however, is not considered relevant to occupational exposure, although some important information has been obtained from studies. Great differences in carcinogenic potency are found, depending on the nickel compound and the animal species investigated. Some correlations exist between the carcinogenic potency of a given compound and its solubility in biological fluids, its surface oxidation reduction state, its ability to be phagocytized and its ability to stimulate erythropoiesis by intravenous injections.

Animal studies using different routes of administration suggest that at least some nickel compounds have a wide range of carcinogenic potency *(5,9,24)*. Nickel subsulfide and beta-nickel monosulfide seem to be the most potent carcinogens in these experiments. Inhalation and ingestion studies are most relevant for the assessment of potential human risk from environmental exposure to nickel. Ottolenghi et al. *(36)* described a significant increase in the number of lung tumours in rats following inhalation exposure to nickel subsulfide for about 2 years. Inhalation of nickel carbonyl produced a low incidence of lung tumours in rats *(31)*. On the other hand, nickel oxide inhalation studies in rats and hamsters have not shown an increase in respiratory tract tumours. There is no experimental evidence that nickel compounds are carcinogenic when administered orally or cutaneously.

The carcinogenicity of nickel compounds appears to be inversely correlated with their solubility in aqueous media. However, an exception to this is amorphous nickel monosulfide, which has a low solubility but is not carcinogenic. There is discussion as to whether particle surface properties of the various nickel compounds are responsible for a particle's ability to be phagocytized by a cell or not.

Effects on humans

Toxicological effects
Several monographs deal with the medical and biological effects of nickel and its compounds. In man, acute intoxication with nickel carbonyl, allergy dermatitis (most prevalent in women), asthma (in nickel workers) and mucosal irritations are reported *(5,37,38)*.

Allergic skin reactions to nickel (dermatitis and other dermatological effects) have been documented in both nickel workers and the general population. However, the significance of nickel as a cause of occupationally-induced skin reaction is decreasing. In contrast, there is evidence that nickel is increasingly a major allergen in the general population, especially in women. About 1–2% of males and 8–11% of females show a positive skin reaction to patch testing with nickel sulfate *(38)*.

The respiratory tract is also a target organ for allergic manifestations of nickel exposure. Allergic asthma has been reported among workers in the plating industry after exposure to nickel sulfate.

Carcinogenic effects

Studies linking nickel-uptake from the environment and cancer incidence in the general population are not available. There is agreement that nickel refinery workers in the past were at significantly higher risk for cancer of the lungs and the nasal cavity; laryngeal cancer (Norway) and gastric cancer and sarcoma (USSR) were reported *(39)*. Although details of the relationship between occupational exposure and cancer risks in the primary nickel industry are complex, the following undisputed facts emerge.

(*a*) At the Clydach refinery, Wales *(40)*, a very high relative risk of nasal and lung cancer has been associated with exposures in the calcining, roasting and leaching departments before 1920. For reasons which cannot be ascribed only to a decrease in nickel air concentrations in the workplace this risk considerably lessened after 1930.

(*b*) Very high relative risks of nasal and lung cancer have also existed in the calcining, roasting and leaching departments of other refineries (Canada, Norway, USSR). In Canada, those risks have certainly decreased in the last decade. In Norway, there is some evidence that the risk still persists, as evidenced by the high frequency of nasal dysplasia in current workers with 10 years' exposure.

(*c*) A high relative risk of nasal and lung cancer has been shown in electrolytic process workers in the Norwegian nickel refining industry, consistent with observations in the electrolytic departments of other refineries *(41)*, but no increase of cancer risk has been reported in a large study of electrolytic departments of the Canadian installations of INCO. This difference, as yet unexplained, is considered to be a crucial point calling for further investigation, as it is, of course, important to determine whether or not soluble salts of nickel are carcinogenic. Nickel subsulfide is regarded as the main suspected cause of increased cancer risk in other refinery departments.

(*d*) Studies in secondary nickel industries were generally not positive, with the exception of one study concerning electroplating processes.

(*e*) A Norwegian study by Magnus et al. *(42)* suggests that the effects of smoking and nickel-refining on lung cancer risk may be additive. The risk ratio for smokers/nonsmokers among employees of a nickel refinery during a follow-up period (1966–1977) was 2:0.

Evaluation of Human Health Risks

Exposure

Nickel is present throughout nature and is released into the air by both natural sources and human activity. Nickel levels in the ambient air are in the range of 1–10 ng/m^3 in rural areas and about 20 ng/m^3 in urban areas, although much higher levels (110–180 ng/m^3) have been recorded in heavily industrialized areas and large cities.

Health risk evaluation

Reliable quantitative data on exposure in the first half of this century are not available, although some attempts have been made to reconstruct historical levels. Several long-term effects have been seen in workers exposed to nickel (irritation of the nasal mucosa, pneumoconiosis, allergic dermatitis, asthma, cancer). Nickel has a very strong and prevalent allergenic potency, but there is no evidence that airborne nickel causes allergic reactions in the general population, although this reaction is well documented in the work environment. The key criterion for assessing the risk of nickel is its carcinogenic potential.

In general, nickel compounds give negative results in short-term mutagenicity tests. However, nickel compounds show a wide range of transformation potencies in mammalian cell assays, depending mainly on their bioavailability.

A lifetime inhalation study has demonstrated that nickel subsulfide is an effective lung carcinogen in rats. According to IARC's evaluation, evidence that nickel and certain nickel compounds are carcinogenic to humans is limited, while for nickel-refining processes the evidence is sufficient. Several epidemiological studies of workers exposed to nickel or nickel compounds clearly demonstrated excess incidences of cancer of the nasal cavity, the lung and possibly the larynx. However, it is still not possible to state with certainty which specific nickel compounds are carcinogenic in humans, although metallic nickel seems less likely to be so than nickel subsulfide or nickel oxides *(38,43)*.

Three data sets for nickel refinery workers can be used to assess the quantitative risk of lifetime exposure to nickel *(40,42,44)*. The average relative risk model was used to estimate the incremental unit risk.

The Kristiansand nickel refinery data *(42)* showed the average relative risk for all categories to be 3.7 for lung and larynx cancer. Assuming that earlier exposures must have been much higher than the values of 0.1–0.8 mg/m^3 measured in the early 1970s (most probably above 3 mg nickel per m^3 *(9)*), and assuming also that occupational exposure lasted for about one quarter of a lifetime, the average lifetime daily exposure can be determined as 164 μg/m^3 [$X = 3000 \mu g/m^3 \times 8/24 \times 240/365 \times 17.5/70$]. The estimated unit risk (UR) calculated from the Kristiansand refinery data is 5.9×10^{-4} [$UR = P_0(R - 1)/X$].

A risk assessment can also be made from the epidemiological data obtained at Copper Cliff, Ontario *(44)*. These data can be analysed in the same way as those of the Kristiansand refinery. The relative risk was estimated to be 8.7, assuming an average duration of exposure of 6 years and a daily (8-hour) time-weighted exposure of around 100 mg/m^3. The average lifetime daily exposure was calculated to be 1.9 mg/m^3 [X = 100 mg/m^3 × 8/24 × 240/365 × 6/70]. The estimated unit risk (UR) is 1.5×10^{-4}.

Finally, a unit risk estimate can also be calculated from the epidemiological data gathered at the Clydach refinery, Wales *(40)*. The average lifetime daily exposure, assuming that the average duration of exposure was 10.5 years, and that the average exposure was higher than 10 mg nickel per m^3, was calculated to be 329 μg/m^3 [X = 10 000 μg/m^3 × 8/24 × 240/365 × 10.5/70].

Because the relative risk estimated by Doll *(40)* was 6.2 for lung cancer, the incremental unit risk (UR) of death from lung cancer caused by nickel can be assumed to be 5.7×10^{-4}.

The results of the quantitative risk assessments from the various epidemiological data sets are presented in Table 2. The estimates, in the range of 1.5×10^{-4} to 5.9×10^{-4}, are fairly consistent.

Table 2. Incremental unit risks for lung cancer due to exposure to 1 μg nickel per m^3 (as nickel compounds occurring in nickel refineries) for a lifetime, based on extrapolations from epidemiological data sets and calculated using the average relative risk model

Location	Incremental unit risk
Copper Cliff, Ontario, Canada	1.5×10^{-4}
Clydach, Wales, United Kingdom	5.7×10^{-4}
Kristiansand, Norway	5.9×10^{-4}

Sufficient evidence exists of nickel subsulfide and nickel carbonyl being carcinogenic in animals. On the basis of the study by Ottolenghi et al. *(36)* in which rats were exposed to 970 μg nickel subsulfide per m^3 for 78 weeks, the US Environmental Protection Agency (EPA) has calculated an upper-bound unit risk of 4.8×10^{-3} for humans, using a multistage model *(9)*.

Taking the geometric mean of the three risk estimates calculated from human studies, the quantitative incremental unit risk estimate for exposure to 1 μg nickel refinery dust per m^3 is 4×10^{-4}. Using other studies and different risk assessment models, the US EPA has estimated the lifetime cancer risk from exposure to nickel dust to be 3.0×10^{-4}. This estimate placed nickel in the third quartile of the 55 substances evaluated by the US EPA Carcinogen Assessment Group with regard to their relative carcinogenic potency *(9)*.

Guidelines
Because of its carcinogenic properties, no safe level can be recommended for nickel. At an air nickel dust concentration of $1 \mu g/m^3$, a conservative estimate of the lifetime risk is 4×10^{-4}.

References

1. **Shaw, D.M. et al.** An estimate of the chemical composition of the Canadian precambrian shield. *Canadian journal of earth sciences,* **4**: 829–853 (1967).
2. *Nickel.* Washington, DC, National Academy of Sciences, 1975.
3. **Nicholls, G.D.** The geochemistry of coal-bearing strata. *In: Effects of nickel in the Canadian environment.* London, Oliver and Boyd, 1965, pp. 269–270.
4. **Hodgson, G.W.** Vanadium, nickel and iron trace metals in crude oils of Western Canada. *Bulletin of the American Association of Petroleum Geologists,* **38**: 2537–2554 (1954).
5. **Rigaut, J.-P.** *Rapport préparatoire sur les critères de santé pour le nickel* [Preparatory report on health criteria for nickel]. Luxembourg, Commission of the European Communities, 1983 (Document CCE/LUX/V/E/24/83).
6. **Stokinger, J.E.** Nickel. *In:* Clayton, G.D. & Clayton, F.E., ed. *Patty's industrial hygiene and toxicology,* 3rd ed. New York, Wiley, 1981, Vol. 2a, pp. 1820–1841.
7. **Henry, W.M. & Knapp, K.T.** Compound forms of fossil fuel fly ash emissions. *Environmental science and technology,* **14**: 450–456 (1980).
8. **Dietz, R.N. & Wieser, R.F.** *Sulfate formation in oil-fired power plant plumes.* Upton, NY, Brookhaven National Laboratory, 1983, Vol. 1 (Report N. EA-3231).
9. *Health assessment document for nickel.* Research Triangle Park, NC, US Environmental Protection Agency, 1985 (Report No. EPA-600/8-83-012F).
10. **Lantzy, R.J. & MacKenzie, F.T.** Atmospheric trace metals: global cycles and assessment of man's impact. *Geochimica et cosmochimica acta, 43*: 511–525 (1979).
11. **Schmidt, J.A. & Andren, A.W.** The atmospheric chemistry of nickel. *In:* Nriagu, J.O., ed. *Nickel in the environment.* New York, Wiley, 1980, pp. 93–135.
12. *Effects of nickel in the Canadian environment.* Ottawa, National Research Council Canada, 1981 (Report No. 18568).
13. **Lahmann, E. et al.** *Heavy metals: identification of air quality and environmental problems in the European Community.* Luxembourg, Commission of the European Communities, 1986, Vol. 1 & 2 (Report No. EUR 10678 EN/I and EUR 10678 EN/II).
14. **Cass, G.R. & McRae, G.J.** Source receptor reconciliation of routine air monitoring data for trace metals: an emission inventory-assisted approach. *Environmental science and technology,* **17**: 129–139 (1983).

15. **Natusch, D.F.S. et al.** Toxic trace elements: preferential concentration in respirable particles. *Science,* **183**: 202–204 (1974).
16. **Burton, K.W. & John, E.** A study of heavy metal contamination in the Rhondda Faur, South Wales. *Water, air and soil pollution,* **7**: 45–68 (1977).
17. **Murthy, G.K. et al.** Levels of copper, nickel, rubidium and strontium in institutional total diets. *Environmental science and technology,* **7**: 1042–1045 (1973).
18. **Myron, D.R. et al.** Intake of nickel and vanadium by humans: a survey of selected diets. *American journal of clinical nutrition,* **31**: 527–531 (1978).
19. **Nielsen, G.D. & Flyvholm, M.** Risks of high nickel intake with diet. *In: Nickel in the human environment. Proceedings of a joint symposium, 8–11 March 1983.* Lyon, International Agency for Research on Cancer, 1984, pp. 333–338 (IARC Scientific Publications, No. 53).
20. **Clemente, G.F. et al.** Nickel in foods and dietary intake of nickel. *In:* Nriagu, J.O., ed. *Nickel in the environment.* New York, Wiley, 1980, pp. 493–498.
21. **Bennett, B.G.** Environmental nickel pathways to man. *In: Nickel in the human environment. Proceedings of a joint symposium, 8–11 March 1983.* Lyon, International Agency for Research on Cancer, 1984, pp. 487–495 (IARC Scientific Publications, No. 53).
22. **Sunderman, F.W., Jr.** Potential toxicity from nickel contamination of intravenous fluids. *Annals of clinical and laboratory science,* **13**: 1–4 (1983).
23. *High volume particulate measurements.* Toronto, Ministry of the Environment, 1976.
24. *Occupational exposure to inorganic nickel.* Cincinnati, OH, US Department of Health, Education, and Welfare, 1977 (DHEW Publication (NIOSH) No. 77/164).
25. **Sunderman, F.W., Jr. & Selin, C.E.** The metabolism of nickel-63 carbonyl. *Toxicology and applied pharmacology,* **12**: 207–218 (1968).
26. **Wehner, A.P. & Craig, D.K.** Toxicology of inhaled NiO and CoO in Syrian golden hamsters. *American Industrial Hygiene Association journal,* **33**: 146–155 (1972).
27. **Oberdoerster, G. & Hochrainer, D.** Effect of continuous nickel oxide exposure on lung clearance. *In: Nickel toxicology.* London, Academic Press, 1980, pp. 125–128.
28. **Wehner, A.P. et al.** Acute and subchronic inhalation exposures of hamsters to nickel-enriched fly ash. *Environmental research,* **19**: 355–370 (1979).
29. **Graham, J.A. et al.** Influence of cadmium, nickel and chromium on primary immunity in mice. *Environmental research,* **16**: 77–87 (1978).
30. **Mushak, P.** Metabolism and systemic toxicity of nickel. *In:* Nriagu, J.O., ed. *Nickel in the environment.* New York, Wiley, 1980, pp. 499–524.
31. **Sunderman, F.W., Jr.** A review of the metabolism and toxicology of nickel. *Annals of clinical and laboratory science,* **7**: 377–398 (1977).

32. **Zober, A. et al.** Untersuchungen zum Nickel- und Chrom-Gehalt ausgewählter menschlicher Organe und Körperflüssigkeiten ["Normal values" of chromium and nickel in human lung, kidney, blood and urine samples]. *Zentralblatt für Bakteriologie, Mikrobiologie und Hygiene. I Abt. Orig. B.,* **179**: 80–95 (1984).
33. **Onkelinx, C. & Sunderman, F.W., Jr.** Modelling of nickel metabolism. *In:* Nriagu, J.O., ed. *Nickel in the environment.* New York, Wiley, 1980, pp. 525–545.
34. **Spiegelberg, T. et al.** Effects of NiO inhalation on alveolar macrophages and the humoral immune systems of rats. *Ecotoxicology and environmental safety,* **8**: 516–525 (1984).
35. **Léonard, A. & Jacquet, P.** Embryotoxicity and genotoxicity of nickel. *In: Nickel in the human environment. Proceedings of a joint symposium, 8–11 March 1983.* Lyon, International Agency for Research on Cancer, 1984, pp. 277–291 (IARC Scientific Publications, No. 53).
36. **Ottolenghi, A.D. et al.** Inhalation studies of nickel sulfide in pulmonary carcinogenesis of rats. *Journal of the National Cancer Institute,* **54**: 1165–1172 (1974).
37. **Izmerov, N.F., ed.** *Nickel and its compounds.* Moscow, Centre of International Projects, 1984 (Series "Scientific Reviews of Soviet Literature on Toxicity and Hazards of Chemicals", No. 58).
38. *Nickel in the human environment. Proceedings of a joint symposium, 8–11 March 1983.* Lyon, International Agency for Resarch on Cancer, 1983 (IARC Scientific Publications, No. 53).
39. *Ambient water quality criteria for nickel.* Washington DC, US Environmental Protection Agency, 1980 (Report No. EPA-440/5-80-060).
40. **Doll, R. et al.** Cancers of the lung and nasal sinuses in nickel workers: a reassessment of the period of risk. *British journal of industrial medicine,* **34**: 102–105 (1977).
41. **Pedersen, E. et al.** Second study of the incidence and mortality of cancer of respiratory organs among workers at a nickel refinery. *Annals of clinical and laboratory science,* **8**: 503–504 (1978).
42. **Magnus, K. et al.** Cancer of respiratory organs among workers at a nickel refinery in Norway. *International journal of cancer,* **30**: 681–685 (1982).
43. *Chemicals, industrial processes and industries associated with cancer in humans. IARC Monographs, Volumes 1 to 29.* Lyon, International Agency for Research on Cancer, 1982 (IARC Monographs on the Evaluation of the Carcinogenic Risk of Chemicals to Humans, Supplement 4).
44. **Chovil, A. et al.** Respiratory cancer in a cohort of nickel sinter plant workers. *British journal of industrial medicine,* **38**: 327–333 (1981).

27

Nitrogen dioxide

General Description

Many chemical species of nitrogen oxides exist, but the species apparently of most interest from the point of view of human health is nitrogen dioxide (NO_2). Nitrogen dioxide is soluble in water, reddish-brown in colour and a strong oxidant.

Sources

On a global scale, quantities of nitrogen oxides produced naturally by bacterial and volcanic action and by lightning far outweigh those generated by man's activities. However, as they are distributed over the entire surface of the earth, the resulting background atmospheric concentrations are very small. The major source of man-made emissions of nitrogen oxides into the atmosphere is the combustion of fossil fuels in stationary sources (heating, power generation) and in motor vehicles (internal combustion engines).

In most ambient situations, nitric oxide (NO) is emitted and transformed into nitrogen dioxide in the atmosphere. Oxidation of nitric oxide by atmospheric oxidants such as ozone occurs rapidly, even at the low levels of reactants present in the atmosphere. Altshuller *(1)* calculated that 50% conversion of nitric oxide would take less than one minute at a nitric oxide concentration of 0.1 ppm in the presence of an ozone concentration of 0.1 ppm. Consequently, this reaction is regarded as the most important route for nitrogen dioxide production in the atmosphere.

Other contributions to the atmosphere come from specific non-combustion industrial processes, such as the manufacture of nitric acid and the use of explosives and welding processes. Indoor sources include tobacco smoking and the use of gas-fired appliances and oil-stoves. Differences in the NO_x (nitric oxide and nitrogen dioxide) emissions of various countries are mainly due to differences in fossil-fuel consumption. Worldwide emissions of NO_x in 1970 were estimated at approximately 53 million tonnes *(2–4)*.

Occurrence in air

From outdoor monitoring data, maximum ½-hour values and maximum 24-hour values of nitrogen dioxide are found at concentrations up to

850 μg/m^3 (0.45 ppm) and 400 μg/m^3 (0.21 ppm), respectively. Annual mean nitrogen dioxide concentrations in urban areas throughout the world are generally in the range of 20–90 μg/m^3 (0.01–0.05 ppm). Urban outdoor levels of nitrogen dioxide vary according to the time of day, the season of the year, and meteorological factors. Typically, there is a low background level of nitrogen dioxide, on which are superimposed one or two peaks of higher levels that correspond to rush-hour traffic emissions of nitric oxide. Hourly averages near very busy roads often exceed 0.5 ppm *(5)*. Thus, the maximal hourly mean value may be 3–10 times the annual mean *(2,6)*. Long-term monitoring activities over the last two decades indicate an increase in concentrations of nitrogen oxides in urban areas throughout the world *(7,8)*. However, indoor sources such as cooking by gas or cigarette smoking may be the main contributors to individual exposure. For example, the tobacco smoke from one cigarette may contain nitric oxide concentrations of 150 000–226 000 μg/m^3 (80–120 ppm) and smaller quantities of nitrogen dioxide. Although some controversy exists about the relative proportions of nitric oxide and nitrogen dioxide, the latter is present initially, and other compounds are formed later *(2)*.

Owing to the widespread use of unvented combustion appliances, nitrogen dioxide concentrations in homes may considerably exceed those found outdoors. The average nitrogen dioxide concentration over a period of several days may exceed 200 μg/m^3 (0.1 ppm) when unvented gas-stoves are used for supplementary heating, clothes-drying, or when kerosene (paraffin) heaters are used *(9)*. Maximum 1-hour concentrations in kitchens are in the range of 470–1880 μg/m^3 (0.25–1.0 ppm) during cooking *(10)* and in the range of 1000–2000 μg/m^3 (0.53–1.06 ppm) when, in addition, unvented gas-fired water-heaters are used *(11)*. It is frequent commercial practice to burn propane or kerosene in order to enrich glasshouse atmospheres with carbon dioxide as well as to provide heat. Under such conditions, concentrations of nitrogen oxides (mainly nitric oxide) often reach 3.5 ppm and remain high because air change is minimized in well sealed glasshouses. In other words, certain crops are being exposed to concentrations of nitrogen oxides which far exceed the levels close to busy roads *(12)*.

Conversion factors

1 ppm = 1880 μg/m^3

1 μg/m^3 = 5.32×10^{-4} ppm

Routes of Exposure

In the environment, nitrogen dioxide exists as a gas. Thus, the only relevant route of exposure to humans is inhalation, whether the source is outdoor air, indoor air, or cigarette smoke. Occupational exposures are limited to a few industrial processes (see p. 297) and include a wide variety of nitrogen oxide exposures. Occupational exposures are relatively rare compared to outdoor and domestic exposures to nitrogen dioxide.

Kinetics and Metabolism

Owing, in part, to the high degree of scientific difficulties involved, very little research on the kinetics and metabolism of nitrogen dioxide has been conducted. The available information is very limited and only partially describes investigation of the deposition of nitrogen dioxide in the respiratory tract and its fate.

Upon inhalation, 80–90% of nitrogen dioxide can be absorbed. In monkeys, 50–60% of inhaled nitrogen dioxide was retained. A significant portion of the inhaled nitrogen dioxide is removed in the nasopharynx (about 40% in dogs and rabbits); thus, during exercise that shifts breathing more orally, an increased penetration of nitrogen dioxide to the lower respiratory tract can be expected *(3,4)*.

Mathematical modelling studies show that the deposition of nitrogen dioxide in the lower respiratory tract depends primarily on the physicochemistry of nitrogen dioxide, its biochemical interactions with mucus (and to a lesser degree surfactant), and the morphology of the lung. Nitrogen dioxide dose to the tissue is predicted to be maximal at about the junction of the conducting airways and the gas exchange region of the lungs in humans, rats, guinea pigs and rabbits. Although the actual tissue dose at a similar starting tracheal concentration differs across species, the shape of the deposition curve is roughly equivalent. The region predicted to receive maximal dose is that where the typical morphometric lesion is observed in several species of animals. Using this mathematical model, it can also be predicted that, as tidal volume increases in man (e.g. in exercise), the dose to the tissue of the gas exchange region increases substantially more than the dose to the conducting airways *(13)*.

Experimental studies have shown that nitrogen dioxide or its chemical products can remain within the lung for prolonged periods; radioactivity associated with labelled nitrogen, originally within nitrogen dioxide, was detectable in extrapulmonary sites. Nitric and nitrous acids or their salts have been observed in the blood and urine after exposure to nitrogen dioxide *(3,4)*.

Health Effects

Effects on experimental animals

Nitrogen dioxide may trigger biochemical changes at relatively low concentrations, beginning after a 30-minute exposure to about 380 $\mu g/m^3$ (0.2 ppm) *(3)*. The biological relevance of the change in prostaglandin metabolism observed at this level is not yet fully understood. The vast majority of biochemical studies show effects only after a week or more of exposure to levels of nitrogen dioxide exceeding 3160 $\mu g/m^3$ (2 ppm). Frequently observed features include the induction of lung oedema, an increase in antioxidant metabolism, an increase in lung enzymes associated with cell injury, and changes in lung lipids. These alterations, although still not understood fully, may be early signs of cell lesions, which become manifest only at higher concentrations or upon longer exposure *(3,4,14)*. Type I pneumocytes and the ciliated epithelium seem to be particularly sensitive to nitrogen dioxide. These cells are replaced by less sensitive cells (Type II cells, Clara cells). However, these substitute cells, too, show alterations of the cytoplasm and hypertrophy after short exposure (10 days) to nitrogen dioxide concentrations above 940 μg/m³ (0.5 ppm) *(3, 14, 15)*.

Long-term exposure leads to emphysema-like structural changes, characterized by thickening of the alveolar capillary membrane, loss of ciliated epithelium, and formation of collagen at atypical points of the lung. These irreversible changes have been observed even at low concentrations, such as continuous exposure to 190 μg/m³ (0.1 ppm) on which were superimposed peaks of 1880 μg/m³ (1 ppm) for 2 hours per day over a period of 6 months. Emphysema-like changes have been observed in mice, rats, dogs and monkeys. Dogs exposed for 5.5 years to a nitrogen dioxide–nitric oxide mixture of 1210 μg/m³ (0.64 ppm) and 310 μg/m³ (0.25 ppm) respectively exhibited several changes in pulmonary function, which continued to deteriorate during a 2.5-year post-exposure period in clean air. After this post-exposure period, lung morphometry studies showed changes which the investigators considered analogous to human centrilobular emphysema *(2–4)*.

The methods used for analysing pulmonary function in animal experiments are probably not sensitive enough to show effects due to lower nitrogen dioxide concentrations. There is only isolated evidence of disorders of the mechanics of breathing and of ventilatory function upon repeated exposure to concentrations between 1880 and 9400 μg/m³ (1–5 ppm). The only consistent observation is that nitrogen dioxide increases breathing frequency *(3, 4)*.

Several types of animal study have indicated that nitrogen dioxide increases susceptibility to bacterial lung infections and perhaps viral infections *(3)*. The most extensive set of data was collected using the infectivity model, which measures the total antibacterial defences of the lungs of mice. For long-term exposures, the lowest concentration tested which had an effect was 940 μg/m³ (0.5 ppm) for 6 months of exposure. After a 3-hour exposure, the lowest concentration tested with an effect was 3760 μg/m³ (2 ppm). Continuous exposure to concentrations ranging from 52 640 μg/m³ down to 940 μg/m³ (from 28 ppm down to 0.5 ppm) resulted in linear concentration-related increases in mortality due to pulmonary infection. Additional studies have shown that the increase in mortality is highly dependent on the exposure regimen. Concentration is far more important than length of exposure in increasing susceptibility to infection, although duration does play a role.

These, as well as other data, show that peak exposures are quite important in determining response *(3, 16)*. There are also several studies indicating that several months of exposure to around 940 μg nitrogen dioxide per m³ (0.5 ppm) can increase susceptibility to other bacteria and viruses and acute exposure to higher levels can decrease pulmonary bactericidal activity and alveolar macrophage function. In summary, the body of work shows that the effects of nitrogen dioxide are due more to concentration than to duration of exposure or to total dose (expressed as the product of concentration × time of exposure), that differences in species sensitivity exist, that the lowest effective concentration of nitrogen dioxide also depends on the microbe used in the test, and that low levels only cause effects after repeated exposures *(3,4,14)*. The extrapolation of these findings to man cannot be made directly, since most of the studies used pneumonia-induced mortality as an endpoint. However, the infectivity model used reflects alterations in defence mechanisms of mice that are shared by humans. This, together with other mechanism studies, implies that specific host defences of man (such as alveolar macrophages) have the potential to be altered by nitrogen dioxide. However, the quantitative relationship between effective nitrogen dioxide levels in animals and in man is unknown.

The development of tolerance to nitrogen dioxide, described by some authors, applies only to acute toxicity (oedema) and not to reduced resistance to infection nor to the development of structural changes in the lungs. To date, there are no reports that nitrogen dioxide causes malignant tumours, mutagenesis or teratogenesis. No evidence of formation of the potentially carcinogenic nitrosamines under normal physiological conditions has been obtained; however, there is a report that acute exposure to 380 $\mu g/m^3$ (0.2 ppm) in mice treated with morpholine caused an increase in N-nitrosomorpholine *(3,4)*.

In summary, acute exposures (hours) to low levels of nitrogen dioxide have rarely been observed to cause effects in animals. However, subchronic and chronic exposure (weeks to months) to low levels of nitrogen dioxide cause a variety of effects, including alterations of lung metabolism, structure and function; increased susceptibility to pulmonary infections; and emphysema-like lung effects. Extrapulmonary effects have also been observed. From the animal studies, the potential for human health effects is broad. However, at present it is difficult quantitatively to extrapolate effective pollutant concentrations from animals to man with confidence.

Effects on humans

Controlled clinical studies

There have been numerous investigations of the effect of nitrogen dioxide on the pulmonary function of normal, bronchitic and asthmatic humans, conducted under controlled conditions in the laboratory (Table 1). Previously, it had been clearly established that short (10–15 minutes) exposures to 3000–9400 μg nitrogen dioxide per m^3 (1.6–5 ppm) caused changes in the pulmonary function of normal and bronchitic subjects. There was an isolated report that exposure to 1300 $\mu g/m^3$ (0.7 ppm) for 10 minutes affected function, and that a 1-hour exposure to 190 $\mu g/m^3$ (0.1 ppm) might increase responsiveness to a pharmacological bronchoconstrictor in some asthmatics *(2–4)*.

The more recent controlled human exposure studies present mixed and conflicting results concerning respiratory effects in asthmatics and healthy individuals at nitrogen dioxide concentrations in the range of 190–7520 $\mu g/m^3$ (0.1–4.0 ppm) (see Table 1 for details). Some studies have reported an effect on airway resistance or other lung function from nitrogen dioxide alone or

Table 1. Controlled studies of the effects of human exposure to nitrogen dioxide[a]

Pollutant/concentration		Duration of exposure and activity	Number and type of subjects	Pulmonary effects	Symptoms	Reference
$\mu g/m^3$	ppm					
9400	5	14 hours	8, normal	Increase of *Raw* during the first 30 minutes of exposure, with decrease during the following 4 hours. Increase of *Raw* after 6, 8 and 14 hours of exposure. Reactivity to acetylcholine increased	Not described	*(17)*
9400	5	2 hours + intermittent light exercise	11, normal	Increase of *Raw* and decrease in $AaDO_2$; no further increase when combined with 200 μg O_3 per m^3 and 13.0 mg SO_2 per m^3	Not described	*(18)*
7520	4	75 minutes including light and heavy exercise	25, normal 23, asthmatic	No effect on *SRaw*, heart rate or skin conductance	Systolic blood pressure different. No symptoms	*(19)*
4700	2.5	2 hours	8, normal	Increase of *Raw*, no change in PaO_2 or $PaCO_2$	Not described	*(17)*
1880	1	2 hours	16, normal	Small changes in *FVC*	5 subjects complained of chest tightness	*(20)*

1880	1	2 hours	8, normal	No increase in *Raw*	Not described	*(17)*
940–9400	0.5–5	3–60 minutes	63, chronic bronchitic 25, chronic bronchitic	Increase of *Raw* at 3.0 mg/m³ Decrease of PaO_2 at 7.5 mg/m³; no change at 3.8 mg/m³	Not described Not described	*(21)* *(22)*
940	0.5	2 hours	10, normal 7, chronic bronchitic 13, asthmatic	None	7 out of 13 asthmatic subjects suffered from symptoms such as chest tightness	*(23)*
560	0.3	20 minutes at rest, followed by 10 minutes of moderate exercise (oral exposure, mouthpiece)	10, asthmatic	NO_2 plus exercise decrease in FEV_1 and partial expiratory flow rates at 60% *TLC*. After exposure at rest, no significant change in function	None	*(24)*
560	0.3	20 minutes at rest, followed by 10 minutes of moderate exercise (chamber exposure)	12, asthmatic	Decrease in FEV_1		*(25)*

Table 1 (contd)

Pollutant/concentration		Duration of exposure and activity	Number and type of subjects	Pulmonary effects	Symptoms	Reference
$\mu g/m^3$	ppm					
560 2000	0.3 1.06	1 hour	8, normal	Small increase in mean *SRaw* at 560 $\mu g/m^3$; no change at 2000 $\mu g/m^3$		*(26)*
380	0.2	2 hours' intermittent light exercise	31, asthmatic	No effect on forced expiratory function or total respiratory resistance observed with NO_2 alone. Small exacerbation by NO_2 of metacholine-induced bronchoconstriction in 17 of 21 subjects tested	Fewer symptoms during NO_2 exposure compared to air	*(27)*
280 290	0.15 NO_2 0.15 O_3 0.15 NO_2+O_3	2 hours' intermittent light exercise	6, normal	Decrease in *SGaw/Vtg* with O_3 for 5 of 6 subjects, and all 6 for combined O_3+NO_2; very small (<5%) decrease in *SGaw/Vtg* with NO_2 alone in 3 of 6 subjects	Cough with O_3 and O_3+NO_2, but not NO_2 alone	*(28)*
230 460 910	0.12 0.24 0.48	20 minutes at rest	8, normal 8, asthmatic	*Normal:* small increase in *SRaw* at 460 $\mu g/m^3$; decrease in *SRaw* at 910 $\mu g/m^3$; no change in reactivity to histamine *Asthmatic:* no effects in *SRaw*; increase in reactivity to histamine at 910 $\mu g/m^3$		*(29)*

190	0.1	1 hour at rest	20, asthmatic 20, normal	No effect on baseline *SGaw*, FEV_1 or V_{isov}; increased reactivity to carbachol in normal subjects and in asthmatics	None	*(30)*
190	0.1	1 hour at rest	9, asthmatic, hypersensitive to ragweed	No effect on baseline *SGaw*, FEV_1 and V_{isov} or reactivity to ragweed	None	*(31)*
190	0.1	1 hour at rest	15, normal 15, asthmatic (atopics)	No change in *SRaw* for either group; no change in sensitivity to methacholine	None	*(32)*
190	0.1	1 hour at rest	7, asthmatic	No change in response to grass pollen after exposure to NO_2		*(33)*
190	0.1	1 hour at rest	20, asthmatic	No effect on *SRaw*; increased sensitivity to carbachol in some subjects		*(34)*

[a] Indications of change only described if statistically significant. Abbreviations are as follows: *SRaw*, specific airway resistance; *Raw*, airway resistance; *SGaw*, specific airway conductance, the reciprocal of *SRaw*; FEV_1, forced expiratory volume at 1 second; *TLC*, total lung capacity; *Vtg*, total gas volume; V_{isov}, flow volume; PaO_2 and $PaCO_2$, arterial partial pressure of oxygen and carbon dioxide; $AaDO_2$, difference in partial pressure of oxygen in the alveoli as against the arterial blood; *FVC*, forced vital capacity.

Source: US Environmental Protection Agency *(3)*.

when challenged by a bronchoconstricting agent, at 190–850 μg/m³ (0.1–0.45 ppm) *(24–30,34)*, while other recent studies have reported no statistically significant effects from nitrogen dioxide alone or sometimes with a bronchoconstricting agent *(19,31–33)*. It is not possible, at present, to evaluate the reasons for these mixed results, although it has been suggested that nitrogen dioxide-induced increases in airway resistance at ambient concentrations may show a nonmonotonous concentration–effect relationship *(29)*. The same paradoxical phenomenon has been observed in an animal study on sensitivity to acetylcholine after short-term exposure to sulfur dioxide *(35)*. In addition, there are likely to be differences in methods, subject selection (e.g. diagnosis of asthma), and statistical power.

To date, the lowest observed level to affect pulmonary function without a bronchoconstrictor, reported in more than one laboratory, was a 30-minute exposure, with intermittent exercise, to a nitrogen dioxide concentration of 560 μg/m³ (0.3 ppm) *(24,25)*. Similar, but statistically nonsignificant, trends have been observed in other controlled human studies performed at 200–300 μg/m³ (0.1–0.15 ppm) *(29,32,34)*. The small size of the decrements reported after exposure to 190–280 μg/m³ (0.1–0.15 ppm), in conjunction with questions regarding the statistical significance of these results, suggests that caution should be exercised in accepting these findings as demonstrating acute nitrogen dioxide effects of concern on pulmonary function at these levels.

Evaluating the health impact of changes in human pulmonary function is difficult. At nitrogen dioxide levels above 3760 μg/m³ (2 ppm), it is clear that substantial changes occur in the pulmonary function of normal subjects. Below this level, there are both positive and negative results in normal subjects. It appears that bronchitics are not more responsive to nitrogen dioxide than normal subjects. However, asthmatics appear to be more responsive. Subjective complaints have been made by asthmatics exposed to 900 μg/m³ (0.48 ppm), whereas normal subjects have only made complaints at 1880 μg/m³ (1 ppm). At the lower levels of nitrogen dioxide (less than 940 μg/m³, 0.5 ppm), the changes in pulmonary function of asthmatics are small, although sometimes statistically significant. It should be understood that the asthmatic volunteers only had a mild level of disease. Also, human studies are designed to avoid causing significant harm to the volunteers during the study. Thus, small effects are to be expected in such research. The impact of nitrogen dioxide on subjects with more severe asthma is unknown.

It is clear that nitrogen dioxide increases reactivity to pharmacological bronchoconstrictor agents. As could be expected, asthmatics exposed to nitrogen dioxide typically have a greater response to these agents; the lowest level to cause such an effect remains unknown. Such studies were undertaken to evaluate the effects of nitrogen dioxide on mechanisms of bronchoconstriction, since these agents are naturally involved in regulating airway diameter. However, the doses and route of administration of these agents are not entirely natural, and therefore interpretation of such studies in relation to guidelines cannot be precise. Using more natural methods to elicit bronchoconstriction, Bauer et al. *(24)* found that cold-induced increased airway constriction in asthmatics was potentiated by a nitrogen dioxide concentration of 560 μg/m³ (0.3 ppm). No exacerbation was found when natural allergens were used in asthmatics, but this may be due to the low levels of nitrogen dioxide used (190 μg/m³, 0.1 ppm) *(31,33)*.

Findings on the effect of nitric oxide on human pulmonary function at concentrations higher than those described for nitrogen dioxide have been reported *(36)*. One recent study indicates an increase in airway resistance in healthy persons after 2 hours' exposure to nitric oxide concentrations of about 1200 μg/m³ with light exercise *(37)*. Such a concentration is not unusual in urban areas. As exposure to nitric oxide without simultaneous contamination with nitrogen dioxide is difficult to accomplish, this finding will need verification by other researchers.

Epidemiological studies

Since 1970 there have been a small number of epidemiological studies, principally undertaken in the Netherlands, the United Kingdom and the USA. Initial studies compared populations of children in communities that had contrasting levels of nitrogen oxide, sulfur oxide and particulates. Generally, epidemiological studies attempting to correlate outdoor pollutant levels and health effects are not sufficiently quantitative for use in determining guidelines. Most subsequent studies have taken the opportunity of comparing groups exposed to the much higher levels of nitrogen dioxide emitted from the combustion of gas inside buildings.

The major American study *(38)* involved schoolchildren aged 6–10 years in six cities. Children living in homes using gas for cooking were found to be significantly more likely to have had respiratory illness before the age of 2 years. In a similar study 4 years later in the same schools but with different children, the relationship was found to be weaker and not statistically significant *(39)*. A series of studies of schoolchildren in the United Kingdom have shown a very weak relation between respiratory illness reported by the mother and the use of gas for cooking at home. There was some evidence of a relation between the incidence of respiratory symptoms and nitrogen dioxide levels in bedrooms and living-rooms, but, after allowing for various interfering factors, the association was not statistically significant ($P = 0.11$) *(40)*. Studies of the frequency of respiratory illness during the first year of life have not shown significant effects of the use of gas-combustion appliances *(41,42)*.

Studies of adults have tended to show no relation between the use of gas for cooking and respiratory symptoms or lung functions. A few positive findings included a negative correlation between nitrogen dioxide levels indoors and spirometric measures of lung function and an increase in symptoms among men and women living in homes where gas was used for cooking *(43–45)*.

The overall impression given by the epidemiological studies is that the use of gas for cooking may have a very small effect on the human respiratory system, particularly in children, but that the effect (if it exists) disappears as the children grow older *(46)*. Although individual exposures have not been reported, the mean weekly and short-term peak values observed in homes lie in the same range as that covered by many laboratory experiments *(9–11)*, so it may be reasonable to extrapolate from the latter to common free-living conditions.

Sensory effects

Nitrogen dioxide has a stinging, suffocating odour. The odour threshold has been placed by various authors between 200 μg/m³ (0.11 ppm) and 410 μg/m³ (0.22 ppm). At a gradual increase of the concentration from 0 to 51 mg/m³ (27 ppm) within 15 minutes, no odour was perceived, due to an adaptation. There is an early report of changes in dark adaptation that were observed after 5 minutes' and 25 minutes' inhalation of nitrogen oxide levels as low as 140 μg/m³ (0.07 ppm). The health impact of these findings is not yet clear *(3,4)*.

Interaction of nitrogen oxides with other air pollutants

There are very few controlled studies of the interaction of nitrogen dioxide with other common air pollutants *(2–4)*. The results indicate that mixtures of nitrogen dioxide and ozone resulted in effects that were additive, synergistic, or due to ozone alone, depending upon the concentration and length of exposure and the parameter studied. From present knowledge of the toxicity of common air pollutants, it may be assumed that interactions occur. However, with so little information at present, evaluation relative to guidelines is not possible.

Evaluation of Human Health Risks

Exposure

Levels of nitrogen dioxide vary widely, since frequently a continuous baseline level is present, with peaks of higher levels superimposed. Natural background levels range from 0.4 to 9.4 μg/m³ (0.0002–0.005 ppm). Outdoor urban levels have an annual mean ranging from 20 to 90 μg/m³ (0.01–0.05 ppm) and an hourly mean ranging from 240 to 850 μg/m³ (0.13–0.45 ppm). Levels indoors where there are unvented gas combustion appliances may average more than 200 μg/m³ (0.1 ppm) over a period of several days. A maximum 1-hour peak may reach 2000 μg/m³ (1 ppm). A maximum-per-minute peak may reach 4000 μg/m³ (2.1 ppm).

Critical concentration–response data

Concentration–response data are only available from animal studies. Thus, this section will focus on lowest-observed-effect levels and their interpretation.

Short-term exposure effects

Available data from animal toxicology experiments rarely indicate effects of acute exposure to nitrogen dioxide concentrations of less than 1880 μg/m³ (1 ppm). Normal individuals exposed at rest or with light exercise for less than 2 hours to concentrations above 4700 μg/m³ (2.5 ppm) experience pronounced decrements in pulmonary function. The lung function of subjects with bronchitis is affected by a 5-minute exposure to 2820 μg/m^3 (1.5 ppm), and such people are usually as responsive to nitrogen dioxide as normal subjects. A wide range of findings has been reported; one study observed no effect due to a 75-minute exposure to 7520 μg/m^3 (4 ppm), while another showed an increase in airway resistance after a 20-minute exposure to 500 μg/m^3 (0.26 ppm).

Asthmatics are likely to be the most sensitive subjects, although uncertainties exist in the health data base. The lowest effects on pulmonary function were reported from two laboratories that exposed mild asthmatics for 30 minutes to a nitrogen dioxide concentration of 560 μg/m^3 (0.3 ppm) during intermittent exercise. One of these studies indicates that nitrogen dioxide can increase airway reactivity to cold air in asthmatic subjects. In most experiments involving a nitrogen dioxide concentration of 190 μg/m^3 (0.1 ppm) for 1 hour, the pulmonary function of asthmatics was not changed significantly.

Nitrogen dioxide increases bronchial reactivity as measured by pharmacological bronchoconstrictor agents in normal and asthmatic subjects, even at levels that do not affect pulmonary function directly in the absence of a bronchoconstrictor. For example, normal subjects exhibit an increased responsiveness to histamine after acute exposure to a nitrogen dioxide concentration of 910 μg/m^3 (0.48 ppm). Since the actual mechanisms are not fully defined and nitrogen dioxide studies with allergen challenges showed no effects at the lowest concentration tested (190 μg/m^3, 0.1 ppm), accurate evaluation of the health consequences of the increased responsiveness to bronchoconstrictors is not yet possible.

Long-term exposure effects

Studies with animals have clearly shown that several weeks to months of exposure to nitrogen dioxide concentrations of less than 1880 μg/m^3 (1 ppm) cause a plethora of effects, primarily in the lung, but also in other organs such as the spleen, liver and blood. Both reversible and irreversible lung effects have been observed. Structural changes range from a change in cell types in the tracheobronchial and pulmonary regions (lowest reported level 640 μg/m^3; 0.34 ppm) to emphysema-like effects (lowest reported level 190 μg/m^3 with peaks of 1880 μg/m^3; 0.1 ppm, peaks of 1 ppm). Biochemical changes often reflect the cellular alterations (lowest reported level 380–750 μg/m^3; 0.2–0.4 ppm). Nitrogen dioxide levels as low as 940 μg/m^3 (0.5 ppm) also increase susceptibility to bacterial infection of the lung.

There are no epidemiological studies related to outdoor exposures that can be used to evaluate quantitatively the risk of exposure to nitrogen dioxide. Homes with gas-cooking appliances have peak levels of nitrogen dioxide that are in the same range as levels causing effects in animal and human clinical studies. Many of the epidemiological studies of adults and children have shown no significant effect of the use of gas-cooking appliances on either respiratory symptoms or lung function. Nevertheless, a few have shown significant relationships in various subgroups, so that the precise relationship between health and gas-cooking (and by implication, nitrogen dioxide) remains unclear.

Health risk evaluation

Small, statistically significant, reversible effects have been confirmed in mild asthmatics exercising intermittently during a 30-minute exposure to a nitrogen dioxide concentration of 560 $\mu g/m^3$ (0.3 ppm). The sequelae of repetitive exposures of such individuals are not known. However, animals exposed for 1–6 months to nitrogen dioxide concentrations in the range of 190–940 $\mu g/m^3$ (0.1–0.5 ppm) show changes of lung structure, lung metabolism, and lung defences against bacterial infection. Further exposure of animals can lead to emphysematous changes. Thus, it is prudent to avoid repetitive exposures in man, since repetitive exposures in animals lead to adverse effects. Animal toxicology studies and epidemiological studies suggest that peak concentrations contribute more to the toxicity of nitrogen dioxide than does an integrated dose (the product of concentration and time of exposure).

The recommendation of a guideline is complicated by difficulties in establishing an appropriate margin of protection with regard to nitrogen dioxide. On the one hand, the replicated studies of asthmatics exercising during exposure to 560 $\mu g/m^3$ (0.3 ppm) indicate a change of about 10% in pulmonary function; although statistically significant, this change is within the range of physiological variation and therefore is not necessarily adverse. Furthermore, a number of epidemiological studies of relatively large populations exposed indoors (in kitchens) to 1-hour peak levels of nitrogen dioxide greater than 500 $\mu g/m^3$ (0.26 ppm) have not provided consistent evidence of adverse effects. Animal studies do not provide substantial evidence of biochemical, morphological or physiological effects in the lung following a single acute exposure to concentrations in the range of the lowest-observed-effect level in man. On the other hand, the mild asthmatics chosen for the controlled exposure studies do not represent all asthmatics and there are likely to be some individuals with greater sensitivity to nitrogen dioxide. Furthermore, subchronic and chronic animal studies do show significant morphological, biochemical and immunological changes. In addition, there is a published study of the effect of a 20-minute exposure to a nitrogen dioxide concentration of 460 $\mu g/m^3$ (0.24 ppm) on airway resistance of normal subjects, and one abstract describing increased sensitivity to a bronchoconstrictor in asthmatic and normal subjects after a 1-hour exposure to 190 $\mu g/m^3$ (0.1 ppm).

Guidelines

Nitrogen dioxide levels of 400 $\mu g/m^3$ (0.21 ppm) and 150 $\mu g/m^3$ (0.08 ppm) are recommended as 1-hour and 24-hour guidelines respectively. The 1-hour guideline is based on the judgement that the lowest-observed-effect level in asthmatics (560 $\mu g/m^3$, 0.3 ppm) is not necessarily adverse and a guideline somewhat lower provides a further margin of protection. The 24-hour guideline is based on the judgement that repeated exposures approaching the minimal repetitively observed effect level are to be avoided, so as to create a margin of protection against chronic effects.

References

1. **Altshuller, A.P.** Thermodynamic considerations in the interactions of NO_x (and its oxyacids) in the atmosphere. *Journal of the Air Pollution Control Association,* **6**: 97–100 (1956).
2. *Oxides of nitrogen.* Geneva, World Health Organization, 1977 (Environmental Health Criteria, No. 4).
3. *Air quality criteria for oxides of nitrogen.* Research Triangle Park, NC, US Environmental Protection Agency, 1982 (Report No. EPA-600/8-82-026F).
4. *Maximale Imissionskonzentrationen für Schwefeldioxid* [Maximum emission concentrations for nitrogen dioxide]. Düsseldorf, Verein Deutscher Ingenieure, 1985 (VDI 2310, Part 12).
5. **Hickman, A.J. et al.** *Atmospheric pollution from vehicle emissions: at four sites in Coventry.* Crowthorne, Department of the Environment, 1976 (Transport and Road Research Laboratory Report No. CR 695).
6. **Bevington, C.F.P. et al.** *Air quality standards for nitrogen dioxide: economic implications of implementing draft proposal for a council directive. Study phase II.* Luxembourg, Commission of the European Communities, 1983 (Report No. EUR 8680 EN).
7. **International Register of Potentially Toxic Chemicals.** *List of environmentally dangerous chemical substances and processes of global significance.* Geneva, United Nations Environment Programme, 1984 (UNEP Reports No. 1 & 2).
8. **Federal Ministry of the Interior.** *Bericht über Waldschäden und Luftverschmutzung* [Report on forest damage and air pollution]. Bonn, Council of Environmental Advisers, 1983.
9. **Goldstein, B.D. et al.** The relation between respiratory illness in primary schoolchildren and the use of gas for cooking. II. Factors affecting nitrogen dioxide levels in the homes. *International journal of epidemiology,* **8**: 339–345 (1979).
10. **Wade, W.A. et al.** A study of indoor air quality. *Journal of the Air Pollution Control Association,* **25**: 933–939 (1975).
11. **Lebret, E.** *Air pollution in Dutch homes: an exploratory study in environmental epidemiology.* Thesis, Wageningen Agricultural University, Netherlands, 1985.
12. **Law, R.M. & Mansfield, T.A.** Oxides of nitrogen and the greenhouse atmosphere. *In:* Unsworth, M.H. & Ormrod, D.P., ed. *Effects of gaseous air pollution in agriculture and horticulture.* London, Butterworth, 1982, pp. 93–112.
13. **Miller, F.J. et al.** Pulmonary dosimetry of nitrogen dioxide in animals and man. *In:* Schneider, T. & Grant, L., ed. *Air pollution by nitrogen oxides.* Amsterdam, Elsevier, 1982, pp. 377–386.
14. **Wagner, H.M.** *Update of a study for establishing criteria (dose/effect relationships) for nitrogen oxides.* Luxembourg, Commission of the European Communities, 1985 (Report No. EUR 9412 EN).

15. **Evans, M.J. & Freeman, G.** Morphological and pathological effects of NO_2 on the rat lung. *In:* Lee, S.D., ed. *Nitrogen oxides and their effects on health.* Ann Arbor, MI, Ann Arbor Science, 1980, pp. 243–265.
16. **Gardner, D.E. et al.** Influence of exposure mode on the toxicity of NO_2. *Environmental health perspectives,* **30**: 23–29 (1979).
17. **Beil, M. & Ulmer, W.T.** Wirkung von NO_2 im MAK-Bereich auf Atemmechanik und bronchiale Acetylcholinempfindlichkeit bei Normalpersonen [Effect of NO_2 in workroom concentrations on respiratory mechanics and bronchial susceptibility to acetylcholine in normal persons]. *International archives of occupational and environmental health,* **38**: 31–44 (1976).
18. **von Nieding, G. et al.** Zur akuten Wirkung von Ozon auf die Lungenfunktion des Menschen [Acute effects of ozone on human lung function]. *VDI-Berichte,* **270**: 123–129 (1977).
19. **Linn, W.S. et al.** Effects of exposure to 4 ppm nitrogen dioxide in healthy and asthmatic volunteers. *Archives of environmental health,* **40**: 234–239 (1985).
20. **Hackney, J.O. et al.** Experimental studies on human health effects of air pollutants. IV. Short-term physiological and clinical effects of nitrogen dioxide exposure. *Archives of environmental health,* **33**: 176–181 (1978).
21. **von Nieding, G. et al.** Grenzwertbestimmung der akuten NO_2-Wirkung auf den respiratorischen Gasaustausch und die Atemwegswiderstände des chronisch lungenkranken Menschen [Minimum concentrations of NO_2 causing acute effects on the respiratory gas exchange and airway resistance in patients with chronic bronchitis]. *Internationales Archiv für Arbeitsmedizin,* **27**: 338–348 (1971).
22. **von Nieding, G. et al.** Studies of the acute effects of NO_2 on lung function: influence on diffusion, perfusion and ventilation in the lungs. *Internationales Archiv für Arbeitsmedizin,* **31**: 61–72 (1973).
23. **Kerr, H.D. et al.** Effect of nitrogen dioxide on pulmonary function in human subjects: an environmental chamber study. *Environmental research,* **19**: 392–404 (1979).
24. **Bauer, M.A. et al.** 0.30 ppm nitrogen dioxide inhalation potentiates exercise-induced broncho-spasm in asthmatics (abstract). *American review of respiratory disease,* **129**: 151 (1984).
25. **Horstmann, D.H. et al.** Pulmonary effects in asthmatics exposed to 0.3 ppm NO_2 during repeated exercise. *American review of respiratory disease* (in press).
26. **Rehn, T. et al.** [Mucociliary transport in lung and nose, and airway resistance after exposure to nitrogen dioxide]. Stockholm, Coal–Health–Environment Project, 1982 (Technical report No. 40) (in Swedish with English summary).
27. **Kleinman, M.T. et al.** Effects of 0.2 ppm nitrogen dioxide on pulmonary function and response to bronchoprovocation in asthmatics. *Journal of toxicology and environmental health,* **12**: 815–826 (1983).

28. **Kagawa, J. & Tsuru, K.** Respiratory effects of 2-hour exposure to ozone and nitrogen dioxide alone and in combination in normal subjects performing intermittent exercise. *Japanese journal of thoracic disease,* **17**: 765–774 (1979).
29. **Bylin, G. et al.** Effects of short-term exposure to ambient nitrogen dioxide concentrations on human bronchial reactivity and lung function. *European journal of respiratory diseases,* **66**: 205–217 (1985).
30. **Ahmed, T. et al.** Effect of 0.1 ppm NO_2 on bronchial reactivity in normals and subjects with bronchial asthma (abstract). *American review of respiratory disease,* **125**: 152 (1982).
31. **Ahmed, T. et al.** Effect of NO_2 (0.1 ppm) on specific bronchial reactivity to ragweed antigen in subjects with allergic asthma (abstract). *American review of respiratory disease,* **127**: 160 (1983).
32. **Hazucha, M.J. et al.** Effects of 0.1 ppm nitrogen dioxide on airways of normal and asthmatic subjects. *Journal of applied physiology: respiratory, environmental and exercise physiology,* **54**: 730–739 (1983).
33. **Orehek, J. et al.** Réponse bronchique aux allergènes après exposition controllée au dioxyde d'azote [Bronchial reaction to allergens after controlled exposure to nitrogen dioxide]. *Bulletin européen de physiopathologie respiratoire,* **17**: 911–915 (1981).
34. **Orehek, J. et al.** Effect of short-term, low-level nitrogen dioxide exposure on bronchial sensitivity of asthmatic patients. *Journal of clinical investigation,* **57**: 301–307 (1976).
35. **Islam, M.S. et al.** Sulphur dioxide-induced bronchial hyperreactivity against acetylcholine. *Internationales Archiv für Arbeitsmedizin,* **29**: 221–232 (1972).
36. **von Nieding, G. & Wagner, H.M.** Vergleich der Wirkung von Stickstoffdioxid und Stickstoffmonoxid auf die Lungenfunktion des Menschen [Comparison of the effects of nitrogen dioxide and nitrogen monoxide on human lung function]. *Staub, Reinhaltung der Luft,* **35**: 175–177 (1975).
37. **Kagawa, J.** Respiratory effects of 2-hr exposure to 1.0 ppm nitric oxide in normal subjects. *Environmental research,* **27**: 485–490 (1982).
38. **Speizer, F.E. et al.** Respiratory disease rates and pulmonary function in children associated with NO_2 exposure. *American review of respiratory disease,* **121**: 3–10 (1980).
39. **Ware, J.H. et al.** Passive smoking, gas cooking and respiratory health of children living in six cities. *American review of respiratory disease,* **129**: 366–374 (1984).
40. **Florey, C. du V. et al.** The relation between respiratory illness in primary schoolchildren and the use of gas for cooking. III. Nitrogen dioxide, respiratory illness and lung function. *International journal of epidemiology,* **8**: 347–353 (1979).
41. **Melia, R.J.W. et al.** *The relation between respiratory illness in infants and gas cooking in the UK: a preliminary report. Proceedings of the VIth World Congress on Air Quality, Paris, 16–20 May 1983.* Paris, International Union of Air Pollution Prevention Associations, 1983, pp. 263–269.

42. **Ogsten, S.A., et al.** The Tayside infant morbidity and mortality study: effect on health of using gas for cooking. *British medical journal,* **290**: 957–960 (1985).
43. **Fischer, P. et al.** Indoor air pollution and its effect on pulmonary function of adult non-smoking women. II. Associations between nitrogen dioxide and pulmonary function. *International journal of epidemiology,* **14**: 221–226 (1985).
44. **Fischer, P. et al.** Associations between indoor exposure to NO_2 and tobacco smoke and pulmonary function in adult smoking and non-smoking women. *Environment international,* **12**: 11–16 (1986).
45. **Comstock, G. et al.** Respiratory effects of household exposures to tobacco smoke and gas cooking. *American review of respiratory disease,* **124**: 143–148 (1981).
46. **Melia, R.J.W. et al.** The relation between respiratory illness in primary schoolchildren and the use of gas for cooking. I. Results from a national study. *International journal of epidemiology,* **8**: 333–338 (1979).

28

Ozone and other photochemical oxidants

General Description

Sources

Ozone (O_3) is one of the strongest oxidizing agents. In the troposphere, it is formed indirectly by the action of sunlight on nitrogen dioxide. There are no significant anthropogenic emissions of ozone into the atmosphere. Existing ozone has been formed by chemical reactions that occur in the air. The ozone-producing and ozone-scavenging processes, which involve absorption of solar radiation by nitrogen dioxide, can be characterized by the following reactions:

$$NO_2 + h\nu \longrightarrow NO + O^{\bullet}$$

$$O^{\bullet} + O_2 \longrightarrow O_3$$

$$O_3 + NO \rightleftarrows NO_2 + O_2$$

In its steady state, the ozone concentration can be described thus:

$$O_3 \text{ concentration} = \text{constant} \times \frac{NO_2 \text{ concentration}}{NO \text{ concentration}}$$

The presence of hydroxyl radicals and volatile organic compounds in the atmosphere, either of natural or of anthropogenic origin, causes a shift in the atmospheric equilibrium towards much higher concentrations of ozone. Apart from ozone, other compounds are formed during the photochemical processes, such as the oxidants peroxyacyl nitrates, nitric acid and hydrogen peroxide, as well as secondary aldehydes, formic acid, fine particulates and an array of short-lived radicals.

The maximum ozone concentration that can be reached in polluted atmosphere appears to depend not only on the absolute concentrations of volatile organic compounds and nitrogen oxides but also on their ratio. At intermediate ratios (4 : 1–10 : 1), conditions are favourable for the formation of appreciable concentrations of ozone.

Since the ratio of volatile organic compounds to nitrogen oxides is not prone to large shifts in the densely populated and heavily industrialized European Region, the meteorological conditions appear to be the rate-limiting factor in photochemical processes in this area *(1–3)*.

Occurrence in air

Ozone

Natural background concentrations mainly caused by tropospheric infolds have been measured in places far from sources of air pollution. Maximal 24-hour mean values of 120 $\mu g/m^3$ (0.06 ppm) have been measured at sea-level on the Atlantic Ocean and at an altitude of 3000 m in the Federal Republic of Germany. The same concentration (4-hour mean) has been measured in rural areas of Canada *(3)*.

It is reasonable to assume that, under circumstances of increased probability of oxidant scavenging, the 24-hour background concentration of ozone is unlikely to exceed values of 120 $\mu g/m^3$. Both the 24-hour and the 1-hour 50-percentile background values of ozone range from 40 to 60 $\mu g/m^3$.

Elevated ozone concentrations have been measured in rural areas where local sources of ozone precursors are insignificant. The long-range transport of ozone and/or its precursors from upwind source areas or from a tropospheric origin is responsible for the high ozone concentration found. Maximum hourly ozone concentrations exceeding 200 $\mu g/m^3$ have been observed in rural areas of Norway and Sweden, and 300 $\mu g/m^3$ in rural areas of the USA *(2)*. Maximum hourly ozone values of 380 and 520 $\mu g/m^3$ were measured in the Netherlands and the United Kingdom, respectively.

In certain parts of Europe, urban 1-hour mean ozone concentrations exceed 350 $\mu g/m^3$, while in the USA, 1-hour mean concentrations often exceed 400 $\mu g/m^3$ *(2,3)*. Generally, ozone concentrations in city centres are lower than those in suburbs, mainly as a result of the scavenging of ozone by nitric oxide originating from traffic.

Diurnal patterns of ozone vary according to location, depending on the balance of the factors affecting ozone formation, transport and decomposition. In the early morning, some time is needed for the photochemical processes to develop. Ozone peak concentrations of 350 $\mu g/m^3$ can be reached in the afternoon. Mean hourly ozone concentrations may exceed values of 240 $\mu g/m^3$ for 10 hours or more. During the night, ozone is scavenged by nitric oxide. It should be noted that several of these typical days with increased photochemical activity can occur consecutively during an episode. Furthermore, it is conceivable that the length of the recovery period between two successive episodes and the number of episodes in a season may also be important factors in the nature and magnitude of the anticipated health effects.

Seasonal variations in ozone concentrations occur and are caused mainly by changes in meteorological processes. Quarterly mean ozone concentrations are highest during the second and third quarters of the year.

Ozone is a highly reactive pollutant, and for this reason indoor/outdoor ratios vary from 0.10 to 0.25 for houses with restricted ventilation. An indoor/outdoor ratio of 0.80 has been measured in a building where the air was changed 10 times per hour with 100% outside air *(4)*.

Peroxyacetyl nitrate

From data gathered in the Netherlands and the USA, the diurnal variation of peroxyacetyl nitrate concentrations appears to be qualitatively similar to the diurnal ozone cycle *(5)*. On the other hand, a time difference of about 5 hours between the peroxyacetyl nitrate and ozone peaks, peroxyacetyl nitrate being ahead of ozone, has been observed in Sweden *(2)*.

The peroxyacetyl nitrate/ozone ratio is not constant and is, moreover, season-dependent. The ratio varies from 2% to 20% for mean daily averages. The peroxyacetyl nitrate concentrations reach maximum values of 80 $\mu g/m^3$ in the USA and 90 $\mu g/m^3$ in the Netherlands. It is also clear that in the latter country an increase in peroxyacetyl nitrate concentrations has been observed during the last decade. The indoor/outdoor ratios for peroxyacetyl nitrate are higher than those for ozone. Measured values range from 0.90 to 1.50, depending mainly on the hour of the day.

Ozone–aerosol relationships

Ozone contributes to the formation of significant amounts of organic and inorganic aerosols. Correlations between levels of ozone and sulfuric acid, nitric acid, sulfates and nitrates have been observed *(2,3)*. Recent information obtained during an episode with increased photochemical activity revealed a steady increase in the daily maximum aerosol concentration with the lapse of the episode *(6)*.

Conversion factors

Ozone (O_3):	1 ppm	=	2 mg/m^3
Peroxyacetyl nitrate:	1 ppm	=	5 mg/m^3

Routes of Exposure

The chemical reactivity of the oxidants ozone and peroxyacetyl nitrate is so high that their half-time in solid and liquid media is virtually negligible; therefore the uptake of both gases, except by inhalation, is not expected.

Kinetics and Metabolism

Knowledge of the dosimetry of ozone (i.e. the mass of ozone that is absorbed per unit of tissue area at a particular site) is essential for quantitative extrapolation of experimental results from experimental animals to humans. It has been studied by means of animal experiments and deterministic mathematical modelling.

Nasopharyngeal removal of ozone in experimental animals *(7,8)* shows the following features: the fraction of ozone taken up is inversely related to flow rate and ozone concentration; uptake is greater for nose-breathing than for mouth-breathing; nasopharyngeal uptake and exposure chamber concentrations are positively correlated. Removal of ozone in the nasopharyngeal region of rabbits and guinea pigs appears to be of the order of 50% for ozone concentrations of from 0.2 to 4.0 mg/m^3 *(8)*.

Dosimetry models for the lower respiratory tract uptake of ozone must take into account convection, diffusion, wall losses, ventilatory patterns, morphometric data of airway dimensions and thickness of fluid lining layers *(9)*. In a recently refined model for the human lung, Miller et al. also took into account transport and chemical reactions in the mucous and surfactant layers as well as in the underlying tissue and capillaries *(10)*. The Miller model predicts the following characteristics for ozone uptake.

1. Ozone can penetrate to tissues everywhere in the pulmonary region, depending on the initial concentration, and decreasing radially.

2. The maximum tissue dose occurs at the first surfactant-lined transition zone between bronchioles and alveolar ducts in both man and laboratory animals.

3. The onset of ozone penetration of mucous-lined tissue of the tracheobronchial region, as well as dose in general, depends on the tracheal ozone concentration, the species of animal (including man) and the breathing pattern.

4. A very small fraction of inhaled ozone is taken up by the blood.

5. Increased minute volume has little effect on tracheobronchial uptake, but has a pronounced positive effect on pulmonary uptake.

Validation of the model, by measurements of actual ozone uptake in the lower respiratory tract, is extremely difficult. No uptake measurements in humans have been reported, although total uptake measurements can be easily performed.

Since peroxyacetyl nitrate is more soluble in water than ozone and chemically less reactive, it might be absorbed better in the nasopharyngeal and tracheobronchial regions.

Health Effects

Ozone is a powerful oxidant and, as such, it can react with virtually every class of biological substance. In general, ozone exerts its action mainly through two mechanisms:

(*a*) the oxidation of sulfhydryl groups and amino acids of enzymes, co-enzymes, proteins and peptides; and

(*b*) the oxidation of polyunsaturated fatty acids to fatty acid peroxides.

Membranes are composed of both proteins and lipids, and appear for that reason to be an obvious target for ozone. Cells or organelles with a large specific surface (surface/volume ratio) may be extremely vulnerable *(3)*.

Effects on experimental animals

Damage to all parts of the respiratory tract can occur, depending on the ozone concentration. At relatively low concentrations, damage is principally confined to the junction between the alveoli and the conducting airways. Type I cells, across which gas exchange occurs, are the first cells to be affected. Effects have been found at ozone concentrations of 400 μg/m^3 for exposure times as short as 7 days *(11,12)*. Higher concentrations cause similar effects after shorter exposure periods *(3)*.

In the nasopharyngeal and tracheobronchial regions, ciliated cells appear to be the most sensitive cell type. In several species studied, damage occurred after exposure to ozone concentrations of 400–1600 μg/m^3 for 8 or 24 hours a day, for 7 days *(12–16)*.

After cessation of exposure and even during exposure, pulmonary type I cells and ciliated cells are replaced by proliferating type II cells and nonciliated cells, respectively. Complete maturation of type II cells to type I cells, or of nonciliated cells to ciliated cells, does not occur during ozone exposure, regardless of the exposure period *(3,15)*.

Inflammation is a typical characteristic of ozone exposure and has been observed in all species examined *(1,3)*. The inflammatory response is seen after exposure to 1000 μg/m^3 for 4 hours. Longer exposure time will elicit the same response though at lower concentrations (400 μg/m^3). The contribution of the inflammatory cells to subsequent long-term effects of ozone has not been studied in detail, but should receive more attention, since apparently reversible alterations of this nature may ultimately result in irreversible effects.

Thickening of the septa of the centriacinar alveoli has been observed after exposure to ozone concentrations of 1000 μg/m^3 for several days. The thickening, which may be correlated with fibrosis, is probably caused by the inflammatory response *(13)*. Fibrosis has been observed by other authors *(3)*. The morphological alterations are preceded by or coincide with an array of biochemical changes. Increased activities of enzymes associated with the synthesis of reducing equivalents, glutathione, consumption of oxygen and collagen have been measured. These alterations were interpreted as the lung's response to loss of cell membrane integrity, cell damage, increased cell turnover, increased concentrations of oxygen radicals and increased rate of collagen synthesis. Animals fed a diet deficient in vitamin E were more vulnerable to ozone than animals consuming a standard diet *(17,18)*.

Acute exposure of a number of species for 2 hours to ozone concentrations of 440–1000 μg/m^3 resulted in rapid and shallow breathing, increased pulmonary resistance, decreased tidal volume, decreased pulmonary compliance, decreased diffusion capacity, and increased airway reactivity *(3,19)*. The effects disappeared 7–14 days after cessation of exposure *(1,3)*.

Long-term exposure of rats for 6 weeks or more to ozone concentrations of 400–1600 μg/m^3 resulted in increased lung distensibility, increased airway resistance and impaired stability *(20)*. Incomplete recovery of monkeys exposed to 1280 μg/m^3 for 1 year was found. During a 3-month recovery period, static lung compliance had decreased, suggesting ongoing injury and the development of central and peripheral airway constriction *(21)*.

Studies on the effect of ozone on the bactericidal pulmonary defence capacity have concentrated on mice. Short-term ozone exposure as low as 160 μg/m^3 impaired this capacity *(3)*. The response of the animals appeared to be concentration-dependent. The relevance of these findings for humans remains controversial, although man and animals share many antibacterial defence mechanisms. The low ozone concentrations that cause an effect, together with the potential seriousness of the effect, e.g. impairment of the immune system, call for further attention.

Animals treated with a sublethal ozone concentration are protected against subsequent exposure to an otherwise lethal ozone concentration *(1,3)*. This tolerance can last for weeks, but it is not developed for all kinds of effect. The mechanism by which ozone exerts this action has not been elucidated. Furthermore, there is no generally accepted opinion as to whether this phenomenon is to be judged adverse or desirable, inasmuch as the mechanism of tolerance for acute effects could be the basis of chronic problems.

A wide range of extrapulmonary effects has been identified after ozone exposure. Whether these effects are caused by ozone itself, reactive intermediates or biogenic compounds formed in the respiratory tract, or whether they are an expression of secondary reactions to pulmonary injury, is unknown. Ozone exposure causes alterations in circulating red blood cells and in various components of the serum. Biochemical and morphological changes in red blood cells have been observed in several animal species after exposure to 400 μg/m^3 for 4 hours *(3)*. Changes in enzyme activities, proteins and peptides have been reported at ozone concentrations of 400 μg/m^3 for 4 hours *(22)*. Ozone exposure decreases the activity of experimental animals at levels of 240 μg/m^3 for 6 hours *(23)*. It is unknown whether indirect effects, such as response to lung irritation, reaction to the odour of ozone or a direct effect on the CNS, cause this alteration.

Reproductive and teratogenic effects and effects on the parathyroid gland have been reported *(1,3)*. At present, these effects are difficult to judge. Uncertainty remains about the mutational and cytogenic effects of ozone and about its effects on the endocrine system *(3)*. An interesting feature of ozone exposure is the sex-specific increase in drug-induced sleeping time. Female, but not male, animals of several species responded after exposure to ozone concentrations of 2000 μg/m^3 for 5 hours *(24)*.

Peroxyacetyl nitrate is far less toxic than ozone with respect to its effect on lethality, behaviour, morphology and the pulmonary defence system *(25,26)*. The concentrations of peroxyacetyl nitrate used in these studies (25–750 mg/m^3) are considerably higher than maximum concentrations of peroxyacetyl nitrate in the ambient air in regions with high oxidant levels (0.09 mg/m^3).

Effects on humans

In a large number of controlled human studies significant impairment of pulmonary function has been reported, usually accompanied by respiratory and other symptoms *(3,27)*. Exposure to ozone generally lasts 1–3 hours. In most studies humans are exposed once to concentrations ranging from 200 to 2000 μg/m³ *(3)*. Not only healthy adults but also asthmatics and subjects with chronic obstructive lung disease have been exposed. In many studies, a pattern of 15 minutes of intermittent exercise alternating with 15 minutes of rest was employed for the duration of the exposure. Minute ventilation has a profound influence on the onset and magnitude of response to ozone exposure. An increased level of exercise results in an increase in the volume of inhaled ozone and in deeper penetration of ozone into the periphery of the lung *(3)*. Pulmonary function measurements have been made before, during and directly after exposure *(3,27–35)*.

Changes in pulmonary function associated with 1–3 hours of ozone exposure in normal subjects during exercise have been reported for the following parameters[a] *(3,28–31)*:

— forced expiratory volume for 1 second (235 μg/m³)

— airway resistance (470 μg/m³)

— forced vital capacity (235 μg/m³)

— respiratory frequency (increased) (470 μg/m³).

The severity of respiratory and other symptoms parallels the impairment of pulmonary function both in magnitude and time-scale *(3)*. Symptoms that have been reported are cough, throat dryness, thoracic pain, increased mucous production, râles, chest tightness, substernal pain, lassitude, malaise and nausea. In addition to causing functional changes and symptoms, ozone is capable of inducing increased nonspecific airway sensitivity to acetylcholine, metacholine and histamine *(3,31,32)*.

Functional recovery from a single exposure takes place in two phases: a 50% improvement within 1–3 hours and a return to pre-exposure values within 24–48 hours. Recovery of other regulatory systems does not parallel the functional recovery, and effects may persist for days (e.g. airway hyperreactivity) *(3,32)*.

Repeated daily short-term exposures show that the decrements in pulmonary function are maximal after the second exposure day; thereafter responsiveness to ozone is attenuated. Very small or no changes are observed after the fourth and fifth exposure days. Attenuation of respiratory symptoms has also been observed, lasting up to 3 weeks. Such attenuation apparently only occurs when the initial exposure is sufficient to cause functional changes, and it is not long-lasting *(3,34)*.

A number of studies of patients with chronic obstructive lung disease and asthma have tended to show that these subjects are similar to normal

---

[a] Lowest-observed-effect levels under conditions of very strenuous exercise *(28)*.

subjects in their response to ozone *(3,33,35)*. Cigarette smokers apparently are not substantially more responsive either *(3)*. Although these subpopulations are more or less equally responsive compared with normal subjects, they are thought to be at increased risk since an equivalent decrement in lung function would have more serious health consequences in lungs that were already compromised. There are insufficient data to determine whether males or females are more responsive *(3)*. Within an apparently normal population, there is a range of responsiveness to ozone that is reproducible, i.e. an individual who is highly responsive remains highly responsive at least over the period of study (months) *(28)*.

Minimal, or no, pulmonary effects of peroxyacetyl nitrate were observed after 4 hours of exposure with exercise to a concentration (1200 $\mu g/m^3$) about 15 times higher than the daily maximum concentrations occurring in regions with high photochemical air pollution *(3)*.

Field and epidemiological studies have indicated a number of acute effects of ozone and other photochemical oxidants. Eye, nose and throat irritation, chest discomfort, cough and headache have been associated with hourly average oxidant levels beginning at about 200 $\mu g/m^3$ *(3,36)*. Pulmonary function decrements in children and young adults have been reported at hourly average ozone concentrations in the range of 160–300 $\mu g/m^3$ *(3,37–40)*. Such symptoms may be responsible for the impairment of athletic performance in the range of 240–740 $\mu g/m^3$ *(41)*. In addition, an increased incidence of asthmatic attacks and respiratory symptoms has been observed in asthmatics exposed to similar levels of ozone *(3)*. Furthermore, epidemiological studies suggest that both the ozone concentration and the sulfate concentration are correlated with an increase in pulmonary effects. Hence, it is not clear what the precise contribution of ozone alone was to the observed effects. In general, the very limited number of human clinical studies have not shown interactions with other pollutants *(3)*.

Photochemical air pollution is a causative factor in eye irritation. This is due to non-ozone components of the photochemical mixture and occurs when ozone levels are at about 200 $\mu g/m^3$ (0.1 ppm).

Evaluation of Human Health Risks

Exposure

Ozone and other photochemical oxidants are formed by the action of sunlight on nitrogen dioxide. Natural background levels of ozone are usually less than 30 $\mu g/m^3$ (0.015 ppm) but can be as high as 120 $\mu g/m^3$ (0.06 ppm) for 1 hour. In Europe, maximum hourly ozone concentrations may exceed 300 $\mu g/m^3$ (0.15 ppm) in rural areas and 350 $\mu g/m^3$ (0.18 ppm) in urban areas. The highest 30-minute value in Europe, recorded in the Federal Republic of Germany, is 664 $\mu g/m^3$ (0.33 ppm). Submaximal levels (85%) can occur for 8–12 hours a day, for many consecutive days.

Health risk evaluation

Ozone toxicity occurs in a continuum in which higher concentrations cause greater effects. Several studies suggest no threshold for an observed ozone effect. Short-term acute effects are notable, beginning with eye irritation due to non-ozone oxidants at ozone levels of 200μg/m^3 (0.1 ppm) or perhaps lower, and symptomatic chest and upper respiratory tract effects at higher levels, particularly in susceptible populations. There is no question that substantial acute adverse effects will occur at concentrations of 1000μg/m^3 (0.5 ppm) for 1 hour.

Epidemiological field studies in children have indicated that pulmonary function decrements can occur at ozone concentrations of 220μg/m^3 (0.11 ppm) or somewhat lower. Other studies have associated changes in pulmonary function in children or asthmatics with ozone concentrations of 160–340μg/m^3 (0.08–0.17 ppm), but there were also associations with changes in temperature or with other pollutants. Various symptoms, including cough and headache, have been associated with ozone concentrations as low as 160–300μg/m^3 (0.08–0.15 ppm).

Exposure of heavily exercising adults or children to an ozone concentration of 240μg/m^3 (0.12 ppm) for 2.5 hours resulted in decrements in pulmonary function. Higher concentrations are required to elicit pulmonary effects when exposure time is shortened or when the level of exercise is decreased. Inhalation of ozone, with or without other ambient oxidants, demonstrates that the functional and symptomatic responses at low ozone levels are due to ozone alone.

A number of studies evaluating animals exposed to ozone for a few hours or days have shown alterations in lung biochemistry or other endpoints in which the lowest-observed-effect levels were in the range of 160–400μg/m^3 (0.08–0.2 ppm). These included the potentiation of bacterial lung infections, increased mitochondrial oxygen consumption in vitamin E deficient and normal rats, morphological alterations in the lung, increases in the function of certain lung enzymes active in oxidant defences, and increases in collagen content. Suggestions of changes in pulmonary function in animals following long-term exposure to ozone in the range of 240–400μg/m^3 (0.12–0.2 ppm) have also been made.

Guidelines

Existing data on the health effects of ozone, considered in conjunction with its high natural background level, lead to the recommendation of a 1-hour guideline in the range of 150–200μg/m^3 (0.076–0.1 ppm).

To lessen the potential for adverse acute and chronic effects and to provide an additional margin of protection, an 8-hour guideline for exposure to ozone of 100–120μg/m^3 (0.05–0.06 ppm) is recommended.

A guideline for peroxyacetyl nitrate is not warranted at present since it does not seem to pose a significant health problem.

References

1. *Photochemical oxidants*. Geneva, World Health Organization, 1979 (Environmental Health Criteria, No. 7).

2. **Grennfelt, P., ed.** *Ozone. Proceedings of an international workshop on the evaluation and assessment of the effects of photochemical oxidants on human health, agricultural crops, forestry, materials and visibility, Gothenburg, 29 February – 2 March 1984.* Gothenburg, Swedish Environmental Research Institute, 1984 (Document No. IVL-EM 1570).
3. *Air quality criteria for ozone and other photochemical oxidants.* Washington, DC, US Environmental Protection Agency, 1986, 4 volumes (Report No. EPA-600/8-84-020F).
4. **Yocom, J.E.** Indoor–outdoor air quality relationships: a critical review. *Journal of the Air Pollution Control Association,* **32**: 500–520 (1982).
5. **Altshuller, A.P.** Measurements of the products of atmospheric photochemical reactions in laboratory studies and in ambient air — relationships between ozone and other products. *Atmospheric environment,* **17**: 2383–2427 (1983).
6. **de Leeuw, F.A.A.M. et al.** *Meetplan aerosolen: interpretatie episode metingen* [Measurement plan for aerosol: interpreting episode measurements]. Bilthoven, National Institute of Public Health and Environmental Hygiene, 1986.
7. **Yokoyama, E. & Frank, R.** Respiratory uptake of ozone in dogs. *Archives of environmental health,* **25**: 132–138 (1972).
8. **Miller, F.J. et al.** Nasopharyngeal removal of ozone in rabbits and guinea pigs. *Toxicology,* **14**: 273–281 (1979).
9. **Miller, F.J. et al.** Similarity between man and laboratory animals in regional pulmonary deposition of ozone. *Environmental research,* **17**: 84–101 (1978).
10. **Miller, F.J. et al.** A model of the regional uptake of gaseous pollutants in the lung. 1. The sensitivity of the uptake of ozone in the human lung to lower respiratory tract secretions and exercise. *Toxicology and applied pharmacology,* **79**: 11–27 (1985).
11. **Plopper, C.G. et al.** Pulmonary alterations in rats exposed to 0.2 and 0.1 ppm ozone: a correlated morphological and biochemical study. *Archives of environmental health,* **34**: 390–395 (1979).
12. **Dungworth, D.L. et al.** Effects of ambient levels of ozone on monkeys. *Federal proceedings,* **34**: 1670–1674 (1975).
13. **Boorman, G.A. et al.** Pulmonary effects of prolonged ozone insult in rats. Morphometric evaluation of the central acinus. *Laboratory investigation,* **43**: 108–115 (1980).
14. **Moore, P.F. & Schwartz, L.W.** Morphological effects of prolonged exposure to ozone and sulfuric acid aerosol. *Experimental and molecular pathology,* **35**: 108–123 (1981).
15. **Evans, M.J. et al.** Cell renewal in the lungs of rats exposed to low levels of ozone. *Experimental and molecular pathology,* **24**: 70–83 (1976).
16. **Castleman, W.L. et al.** Acute respiratory bronchiolitis: an ultrastructural and autoradiographic study of epithelial cell injury and renewal in rhesus monkeys exposed to ozone. *American journal of pathology,* **98**: 811–840 (1980).

17. **Menzel, D.B.** Ozone: an overview of its toxicity in man and animals. *In:* Miller, F.J. & Menzel, D.B., ed. *Fundamentals of extrapolation modelling of inhaled toxicants: ozone and nitrogen dioxide.* Washington, DC, Hemisphere, 1984, pp. 3–24.
18. **Mustafa, M.G. et al.** Effects of ozone exposure on lung metabolism: influence on animal age, species and exposure conditions. *In:* Lee, S.D. et al., ed. *International symposium on the biomedical effects of ozone and related photochemical oxidants.* Princeton, NJ, Princeton Scientific Publishers, 1983, pp. 57–73.
19. **Amdur, M.O. et al.** Respiratory response of guinea pigs to ozone alone and with sulfur dioxide. *American Industrial Hygiene Association journal,* **39**: 958–961 (1978).
20. **Costa, D.L. et al.** A subchronic multi-dose ozone study in rats. *In:* Lee, S.D. et al., ed. *International symposium on the biomedical effects of ozone and related photochemical oxidants.* Princeton, NJ, Princeton Scientific Publishers, 1983, pp. 369–393.
21. **Wegner, C.D. et al.** The effects of long-term, low level ozone exposure on static, dynamic and oscillatory pulmonary mechanics in monkeys. *American review of respiratory disease,* **125**: 147 (1982).
22. **Veninga, T.S. et al.** Distinct enzymatic responses in mice exposed to a range of low doses of ozone. *Environmental health perspectives,* **39**: 153–157 (1981).
23. **Tepper, J.L. et al.** Microanalysis of ozone depression of motoractivity. *Toxicology and applied pharmacology,* **64**: 317–326 (1982).
24. **Graham, J.A. et al.** Influence of ozone on pentobarbital-induced sleeping time in mice, rats, and hamsters. *Toxicology and applied pharmacology,* **61**: 64–73 (1981).
25. **Dungworth, D.L. et al.** Pulmonary lesions produced in A-strain mice by exposure to peroxyacetylnitrates. *American review of respiratory disease,* **99**: 565–574 (1969).
26. **Thomas, G.B. et al.** Effects of exposure of peroxyacetylnitrate on susceptibility to acute and chronic bacterial infection. *Journal of toxicology and environmental health,* **8**: 559–574 (1981).
27. **von Nieding, G. et al.** Controlled studies of human exposure to single and combined action of NO_2, O_3 and SO_2. *International archives of occupational and environmental health,* **43**: 195–210 (1979).
28. **McDonnell, W.F. et al.** Reproducibility of individual responses to ozone exposure. *American review of respiratory disease,* **131**: 36–40 (1985).
29. **Folinsbee, L.J. et al.** Pulmonary function changes after 1-hour continuous heavy exercise in 0.21 ppm ozone. *Journal of applied physiology: respiratory, environmental and exercise physiology,* **57**: 984–988 (1984).
30. **Horvath, S.M. et al.** Pulmonary function and maximum exercise responses following acute ozone exposure. *Aviation, space and environmental medicine,* **40**: 901–905 (1979).
31. **König, G. et al.** Änderung der bronchomotorischen Reagibilität des Menschen durch Einwirkung von Ozon [Changes in the bronchial activity of humans caused by the influence of ozone]. *Arbeitsmedizin, Sozialmedizin, Präventivmedizin,* **151**: 261–263 (1980).

32. **Holtzman, M.I. et al.** Effect of ozone on bronchial reactivity in atopic and nonatopic subjects. *American review of respiratory disease,* **120**: 1059–1067 (1979).
33. **Linn, W.S. et al.** Response to ozone in volunteers with chronic obstructive pulmonary disease. *Archives of environmental health,* **38**: 278–283 (1983).
34. **Horvath, S.M. et al.** Adaption to ozone: duration of effect. *American review of respiratory disease,* **123**: 496–499 (1981).
35. **Linn, W.S. et al.** Health effects of ozone exposure in asthmatics. *American review of respiratory disease,* **117**: 835–843 (1978).
36. **Linn, W.S. et al.** Short-term human health effects of ambient air in a pollutant source area. *Lung,* **160**: 219–227 (1982).
37. **Kagawa, J. & Toyama, T.** Photochemical air pollution: its effects on respiratory function of elementary school children. *Archives of environmental health,* **30**: 117–122 (1975).
38. **Lippmann, M. et al.** Effects of ozone on the pulmonary function of children. *In:* Lee, S.D. et al., ed. *International symposium on the biomedical effects of ozone and related photochemical oxidants.* Princeton, NJ, Princeton Scientific Publishers, 1983, pp. 423–446.
39. **Zagraniski, R.T. et al.** Ambient sulfates, photochemical oxidants, and acute health effects: an epidemiological study. *Environmental research,* **19**: 306–320 (1979).
40. **Lioy, P.J. et al.** Persistence of peak flow decrement in children following ozone exposures exceeding the national ambient air quality standard. *Journal of the Air Pollution Control Association,* **35**: 1068–1071 (1985).
41. **Wayne, W.S. et al.** Oxidant air pollution and athletic performance. *Journal of the American Medical Association,* **199**: 901–904 (1967).

29

Radon

General Description

Radon (Rn) is a radioactive noble gas in several isotopic forms. Only two of these isotopes occur in significant concentrations in the general environment: radon-222, a member of the radioactive decay chain of uranium-238, and radon-220 (often referred to as thoron), from the decay chain of thorium-232. The contribution made by thoron to human exposures in indoor environments is usually small compared with that due to radon, and it will only occasionally be referred to here.

Sources

Uranium (U) occurs widely in the earth's crust. The average radium (Ra) concentration in soil is of the order of 25 Bq/kg *(1)*. (The Becquerel (Bq) is the SI unit of activity of a radionuclide and is equal to 1 disintegration per second.) The average level of radon gas concentration in the atmosphere at ground level is given as 3 Bq/m^3, with a range of 0.1 (over oceans) to 10 Bq/m^3. Higher radon concentrations in the outdoor air have been reported near substantial radon sources, such as mine tailings *(2)*. The radon daughter concentration in effective equilibrium with 3 Bq/m^3 of radon gas is of the order of 1–1.5 Bq/m^3 equilibrium equivalent radon (EER) concentration.

The observed radon daughter concentration is usually not in equilibrium, owing to various removal mechanisms acting on radon daughters. To calculate the radon daughter concentration, an equilibrium factor F has to be applied. Values reported for the equilibrium factor normally are in the range of 0.3–0.5 for residential buildings in different areas.

The radon concentration in the indoor air in most buildings is likely to be a multiple of the outdoor concentration. The main sources of naturally occurring radiation indoors are the ground of the site and building materials. Tap-water and the domestic gas supply are usually of minor importance, with a few exceptions. In houses with very high concentrations, the main source of radon is the soil gas of the subjacent ground. The outdoor air also contributes to the radon concentration indoors, via the ventilation air.

Soil

For those who live close to the ground, e.g. in detached houses or on the ground floor of apartment buildings without cellars, the most important radon source is the ground *(3)*. The potential for radon entry from the ground depends mainly on the activity level of radium-226 in the subsoil and its permeability with regard to air flow. Examples of terrains with a high radon potential are alum shales and some granites, due to high concentrations of radium-226 and the presence of eskers (gravel, sand and rounded stone deposited from subglacial streams during the ice ages), all these being characterized by high permeability. The ground could also be contaminated with waste tailings from uranium mining operations (4–18 Bq/kg) with enhanced activity levels *(4)*. The ingress of radon from the soil is predominantly one of pressure-driven flow, with diffusion playing a minor role. The magnitude of the inflow varies with several parameters, the most important being the air pressure difference between soil air and indoor air, the tightness of the surfaces in contact with the soil on the site, and the radon exhalation rate of the underlying soil *(4)*.

If there is no airtight layer between the basement and the ground, the underpressure indoors causes radon to be sucked from the ground under the building. Underpressure occurs in most houses if either the adjustment of inlet and outlet of air in forced ventilation systems or the outdoor air supply for vented combustion appliances is inappropriate. The underpressure may be considerable for all types of ventilation systems when the inlet air is restricted too much.

The tightness of the structures in contact with the ground varies with different building regulations and techniques, and is very dependent on cracks, openings and joints. Structures are hardly ever so airtight that radon inflow is completely prevented. For example, to get a radon daughter EER concentration of less than 100 Bq/m^3 in a house with a volume of 500 m^3 and an air change rate of 0.5 air changes per hour, not more than 1 m^3 per hour must be allowed to leak into the house if the radon gas concentration in soil air is about 50 000 Bq/m^3. These values are quite typical *(3)*.

Building materials

Radon exhalation from building materials depends not only on the radium concentration, but also on factors such as the fraction of radon produced which is released from the material, the porosity of the material and the surface preparation and finish of the walls. In general, no action need be taken concerning traditional building materials *(5)*. Building materials containing by-product gypsum and concrete containing alum shale may have elevated radium concentrations. The activity concentrations in brick and concrete may also be high if the raw materials were taken from locations with high levels of natural radioactivity.

Tap-water

In wells drilled in rock the radon concentration of the water may be high. When such water is used in the household, radon will be released into the indoor air, causing an increase in the average radon concentration. In a few regions, such as Finland *(6)* and Maine, USA *(7)*, the tap-water from wells drilled in rock has been shown to contribute significantly to radon concentrations indoors. Radon concentrations in tap-water from deep wells can range from 1 to 10 000 Bq/litre *(7)*. The concentration in these regions may already be high due to high rates of radon entry from the ground.

Domestic gas

In some regions, natural gas used for cooking and heating contains elevated concentrations of radon, which is released on combustion *(8)*. Normally this source is not significant, and it can be monitored at transmission and distribution points. Levels of the order of 1800 Bq/m^3 natural gas have been reported *(8)*. Natural gas as supplied usually contains gas from a number of wells and fields and thus can vary over time, depending on the proportions supplied by different sources.

Observed levels of radon and radon daughters in indoor air

The radon daughter concentration in indoor air is a function of the source strength, the effective ventilation rate removing radon from the building, and the rate at which radon daughters settle or plate onto surfaces. The effective ventilation rate in a building can be as low as 0.1 air changes per hour or as high as 3 air changes per hour, or even higher. The rate of radon gas influx can vary by at least three orders of magnitude, depending on the radon content of the soil gas, the permeability of the soil and the foundation, and the pressure difference between the building atmosphere and the soil gas.

As a result, the radon concentrations even of adjacent buildings can be quite different, and reliable estimates are not possible without making an actual measurement. In addition, in any given building, diurnal and seasonal differences occur in the radon daughter concentrations and measurements should take this into account.

A sizeable number of buildings have been monitored in a number of regions and countries, and the reports show consistently that the concentrations have a nearly log-normal distribution. Table 1 shows the range of mean concentrations, and the relative frequencies of radon daughters EER concentrations above 100, 200 and 400 Bq/m^3 in a number of countries. Most of the values are based on measurements of radon gas concentrations. The equilibrium factors reported in the different studies have been applied in Table 1.

Clearly, the mean radon daughter concentrations vary between the different countries by as much as a factor of 5, according to the distribution of soil composition in settled areas, building styles, etc. It is also clear that a small fraction of dwellings in each country has concentrations 10 times higher than the national average of that country.

Countries that have maps showing terrestrial gamma radiation levels of their territory can use such maps for initial guidance in locating areas of concern, but a high gamma flux level at a location does not guarantee high radon daughter levels, nor does a low terrestrial gamma flux assure low radon daughter levels. In general, the average radon daughter level in buildings

Table 1. Distribution of radon daughter EER concentrations in different countries

Country	Arithmetic mean (Bq/m^3)	Percentage of dwellings exceeding:		
		100 Bq/m^3	200 Bq/m^3	400 Bq/m^3
Sweden, Finland, Norway	50	10	3	1
United States	27	5	1	0.1
Canada	25	3	1	0.2
United Kingdom	10	<0.1	—	—
Germany, Federal Republic of	25	0.7	0.1	—

can be expected to be higher in a region with high terrestrial gamma flux as has been found for Finland *(6)*, and also for Cornwall as compared with the whole of the United Kingdom *(9)*. The values given in Table 1 for Canada and the United States were obtained almost exclusively in detached one-family dwellings, which tend to have somewhat higher concentrations than multi-family buildings.

Conversion factors

The concentration of radon gas in air is given in Bq/m^3. Radon daughter concentrations are given in Bq/m^3 EER after multiplying the radon gas concentration by the equilibrium factor F, which in most studies is found to be between 0.3 and 0.5. Concentrations are also often given in pCi/litre which converts to Bq/m^3 as: 1 pCi/litre = 37 Bq/m^3. Occupational exposures are described in working levels (WL). One WL is any combination of short-lived radon daughters in one litre of air that will result in the emission of 1.3×10^5 MeV of potential alpha energy. The working-level month (WLM) is the unit of cumulative exposure and is defined as the exposure to 1 WL for 170 hours (one occupational month). One WL can be equated to 3700 Bq/m^3. The dose delivered to tissue by radon daughters is expressed in mSv and is discussed in more detail on p. 335.

Routes of Exposure

Radon, an inert gas, is not significantly retained in the respiratory tract. It will exist wherever radium is found. Some part of the total effect of ingested and absorbed radium will be due to radon and its daughters produced *in situ* by bound radium.

Drinking-water

Water from wells drilled deep in rock can contain significant amounts of radium and radon. Such waters have been described in Finland *(6)*, and in Maine *(7)* and Iowa *(10)* in the USA, where levels exceed 3.7×10^6 Bq/m^3. Most of the concern about radon in drinking-water stems from the radon it releases into indoor air *(6)*. The efficiency of radon gas release depends on water temperature and agitation.

Food

Radon has not been reported to be present in food in any significant quantities.

Air

The major concern regarding radon daughter intake is the air route. Approximately 90% of radon daughters may attach initially to airborne particles which may be inhaled. Depending on the size of the particles, these daughters may collide with airway walls before reaching the lung. The unattached fraction (10%) has a higher rate of deposition and is more efficient in delivering a dose to the critical cells in the lung.

The amount of radon daughter intake depends on the breathing rate, which affects both the total quantity inhaled per unit time and the pattern of deposition in the respiratory tract. The volume of air inhaled at rest is of the order of 0.5 m^3/hour; during physical activity this volume can reach 1.5 m^3/hour or more. Of the attached radon daughters inhaled, the fraction that is deposited depends on the size of the particles and on breathing patterns, but has been estimated as 30%.

Other exposure routes

Outdoor environments

A minimum and irreducible exposure to radon daughters occurs in the outdoor environment, of the order of 1.5 Bq/m^3 EER.

Residential and non-radiation-connected occupational environments

Most people spend as much as 80–90% of the 24-hour day in these environments, which constitute their major source of exposure. The concentrations of radon daughters in these environments range from 10 to several hundred Bq/m^3 EER, and sometimes even higher.

Occupational exposures

Underground mine workers, especially those in uranium mines, are exposed to substantial concentrations of radon daughters. Historically, these exposures have been much higher, but today regulations and guidelines have produced upper-bound exposure levels of 4.8 WLM per year, which are equivalent to a 5 days-a-week, 8 hours-a-day exposure to 1500 Bq/m^3 EER. The annual upper limit of exposure thus becomes 1500 Bq/m^3 for 2000 hours, or 3×10^6 Bq/m^3 hours per year. For residential and nonoccupational indoor exposures, this would correspond to 428 Bq/m^3 for an exposure factor of 80% indoor occupancy. Other occupational exposures can occur whenever radium-rich materials are stored or processed in indoor environments.

Tobacco smoking

Tobacco smoking has an interaction with exposure to radon daughters that is usually considered to be more than additive. Thus, people exposed to radon daughters can, in particular, reduce their risk of lung cancer by abstaining from tobacco smoking.

Relative significance of different routes of exposure

One organ, the lung, is clearly very critical as far as exposure to radon daughters is concerned; the uptake of radon gas by all other routes is minor. In practice, radon gas would enter the body through inhalation, or by ingestion of radon-bearing water. Deposition of radon daughters in the bronchial region is by far the most important source of risk: a very small amount of tissue receives a concentrated exposure owing to the deposition pattern of the unattached daughters or the particles to which radon daughters are attached, combined with high density from the alpha particles emitted *(11)*.

Although risks to other tissues may occur, they are very small compared with that from bronchial exposure. All other pathways distribute a smaller amount of radon daughters over a very much larger tissue mass, with correspondingly lower absolute and relative risks.

Kinetics and Dose Considerations

Biological and radiological half-life

After deposition of radon daughters in the bronchial region, the resulting dose of ionizing radiation is delivered to the bronchial epithelium with a radiological half-life of the order of 30 minutes (see Table 2).

Once deposited, the alpha emissions will impact on the immediate region *(11)* and deliver a very densely concentrated ionization in that tissue. If radon gas and some of the daughters are considered soluble, they will dissolve in the blood flowing through the lung with a half-life of the order of 10–15 minutes, and some of the dose will be distributed over a very much larger tissue mass. If radon daughters are considered insoluble and if up to 30% of inhaled radon daughters are deposited, substantial bronchial doses will be delivered. The deposition patterns and particle size distribution of attached radon daughters have received considerable study *(11)*.

These dose conversion coefficients refer to a residence probability of 100% indoors (occupancy factor = 1). In practice, an occupancy factor should be used which is appropriate for the conditions being considered. For the indoor exposure in homes, an average occupancy factor of 0.65 seems to be typical. Taking into account the exposure from time spent in other buildings, e.g. at the workplace, a total occupancy factor of 0.8 indoors is a reasonable estimate.

Estimation of dose from indoor exposure to radon daughters

After reviewing different dosimetric models for the inhalation of short-lived radon daughters, an expert group of the Nuclear Energy Agency of

Table 2. The uranium series

Element	Symbol	Half-life	Energy of radiation (MeV)		
			alpha	beta	gamma
Uranium	$_{92}U^{238}$	4.3 × 10^9 years	4.21	—	—
Thorium	$_{90}Th^{234}$	24.1 days	—	0.21	0.09
Protactinium	$_{91}Pa^{234}$	1.14 minutes	—	2.32	0.80
Uranium	$_{92}U^{234}$	2.35 × 10^5 years	4.76	—	—
Thorium	$_{90}Th^{230}$	8.0 × 10^4 years	4.66	—	—
Radium	$_{88}Ra^{226}$	1612 years	4.79	—	0.19
Radon	$_{86}Rn^{222}$	3.825 days	5.49	—	—
Polonium	$_{84}Po^{218}$	3.05 minutes	5.99	—	—
Lead	$_{82}Pb^{214}$	26.8 minutes	—	0.65	0.29, 0.35
Bismuth	$_{83}Bi^{214}$	19.7 minutes	5.50	3.15	1.8, 0.61
Polonium	$_{84}Po^{214}$	1.5 × 10^{-4} seconds	7.68	—	—
Thallium	$_{81}Tl^{210}$	1.32 minutes	—	1.80	—
Lead	$_{82}Pb^{210}$	22.2 years	—	0.026	0.047
Bismuth	$_{83}Bi^{210}$	4.97 days	—	1.17	—
Polonium	$_{84}Po^{210}$	138 days	5.30	—	—
Lead	$_{82}Pb^{206}$	stable	—	—	—

the Organisation for Economic Co-operation and Development has proposed reference values for the expected radiation dose to target tissues in the lung and for the resulting effective dose-equivalent from indoor exposure to radon daughters *(12)*. Table 3 lists the corresponding dose conversion coefficients that resulted from this study.

In the 1982 report of the United Nations Scientific Committee on the Effects of Atomic Radiation (UNSCEAR) *(1)*, a mean indoor concentration of 15 Bq/m^3 EER was estimated as an average for the total population of the temperate regions of the world. Applying the occupancy factors mentioned above, and the dose conversion factors in Table 3, a mean dose-equivalent of about 15 mSv/year to the bronchial and 2 mSv/year to the pulmonary region from indoor exposure to radon daughters should be expected, corresponding to an effective dose-equivalent of 1 mSv/year. The indoor exposure at home accounts for 70–80% of this dose. The mean dose to other body tissues from all sources of natural exposure is of the same order of magnitude, about 1 mSv/year. This means that, owing to indoor exposure to radon daughters, the mean natural radiation exposure of the bronchial epithelium is about 15 times higher than that of other tissues. One arrives, then, at the total of about 2 mSv/year for the effective dose-equivalent, of which about 50% is due to inhaled radon daughters *(1)*.

Table 3. Reference dose conversion coefficients giving the annual dose-equivalent (in mSv per year) per unit of indoor concentration of radon daughters (in Bq/m^3 EER) for an occupancy factor of 1 (100%)

Target tissue	Dose-equivalent (mSv per year) per Bq/m^3 EER
Bronchial dose-equivalent[a]	1.2
Pulmonary dose-equivalent	0.15
Effective dose-equivalent	0.081

[a] Averaged over the basal cell layer in the ciliated tracheobronchial airways.

In population groups living in houses with strongly enhanced radon levels, considerably higher levels of dose to the target tissues in the lung will occur. These dose values can be derived from the dose conversion coefficients listed in Table 3.

Health Effects

Effects on animals

Rats inhaling radon daughters show increased rates of lung cancer, often of a type resembling the one that develops in humans *(13)*. Exposures as low as 20 WLM over the lifetime of the rat constitute a dose that doubles the background incidence of lung cancer. The controls were exposed to an irreducible minimum of 2 WLM of radon daughters over their lifetime. The life span of the rats was not affected. The controls had a lung cancer incidence of 0.83%; at 40 WLM of exposure the incidence rose to 3.82%.

Effects on humans

At the highest levels of radon daughters in indoor air (several thousand Bq/m^3 EER), effects in the lung that result from cell killing cannot be excluded (nonstochastic effects). Such effects are believed to be of far less importance than the lung cancer risk.

The increased risk of lung cancer among miners, especially uranium miners, has been well documented in a number of large epidemiological studies. More than 25 000 radon-exposed miners have been followed in these studies, the results of which have been summarized and reviewed in several reports *(1,5,14,15)*.

These results are consistent with the conclusion that, at least for the lower dose range in the miners, the relationship between cumulative radon daughter exposure and excess frequency of lung cancer is linear. This conclusion is also supported by the results of animal exposure studies and human studies on the effects of densely ionizing radiation.

The risk of lung cancer, as defined by the slope of the exposure–risk relationship that results from the three largest studies on radon-exposed uranium miners, agrees within a factor of three for the relative risk and within a factor of four for the absolute risk. For the induction of lung cancer by inhaled radon daughters, a relative risk model seems more appropriate than an absolute risk model, a suggestion further confirmed by the findings on lung cancer among atomic bomb survivors in Hiroshima and Nagasaki.

Evaluation of Human Health Risks

Dose–response and dose–effect relationships

At present, the epidemiological findings in radon-exposed miners can be regarded as the most suitable basis for estimating the attributable lung cancer risk from indoor exposures to radon daughters. In transferring the data from miners to exposure conditions in the residential environment, several correction factors have to be taken into account. These concern the different state of radon daughter atoms in mines and in indoor air, differences in breathing rates in the two environments which affect the deposition patterns in the bronchial tree, and the differences in occupancy factor in each of the environments.

In recent years, several evaluations and estimates have been made of the dose–response relationship and the attributable lung cancer risk from inhaled radon daughters *(15,16)*. The resulting best estimates are in reasonable agreement. For a typical population with a life expectancy of 70–75 years, averaged over males and females, the absolute annual lung cancer incidence will be between $C \times F_0 \times 10^{-6}$ and $3 \times C \times F_0 \times 10^{-6}$, where C = average indoor radon daughter concentration in Bq/m^3 EER; and F_0 = occupancy factor (between 0 and 1).

The dosimetric approach, which uses the reference risk coefficients recommended by the International Commission on Radiological Protection *(17,18)*, yields a value of the lung cancer risk at the lower end of the above range.

Health risk evaluation

In the 1982 UNSCEAR report *(1)*, a mean indoor radon daughter concentration of 15 Bq/m^3 EER was estimated, averaged over the whole population in the temperate region. Assuming an occupancy factor of 0.8 for indoor environments, an annual incidence of 10–40 cases of lung cancer per million persons should be expected that is attributable to radon daughter exposure.

The relative risk model leads to the conclusion that under these conditions about 5–15% of the observed lung cancer frequency or of the lifetime risk may be attributable to indoor radon daughters. This relative risk is nearly equal for males and females, and for smokers and nonsmokers. The combined effect of radon daughter exposure and of smoking can thus approximately be described as multiplicative.

Guidelines

In general, simple remedial measures should be considered for buildings with radon daughter concentrations of more than 100 Bq/m^3 EER as an annual average, with a view to reducing such concentrations wherever possible.

Remedial action in buildings with radon daughter concentrations higher than 400 Bq/m^3 EER as an annual average should be considered without delay (total exposure before corrective measures are taken should not be allowed to exceed 2000 Bq $\times$ years/m^3 EER).

Building codes should include sections designed to ensure that radon daughter levels do not exceed 100 Bq/m^3 EER in new buildings, and appropriate practices should be prescribed.

The relevant authorities should consider conducting representative surveys of radon daughter concentrations in their building stock to identify regions that may have excessive levels.

References

1. **United Nations Scientific Committee on the Effects of Atomic Radiation (UNSCEAR).** *Ionizing radiation: sources and biological effects.* New York, United Nations, 1982.
2. **Schmitz, J. & Urban, M.** Mine dumps as a source of radon impact on buildings. *Radiation protection dosimetry,* **7**: 63–67 (1984).
3. **Akerblom, G. et al.** Soil gas radon – a source for indoor radon daughters. *Radiation protection dosimetry,* **7**: 49–54 (1984).
4. **Wilkening, M.H. et al.** Radon-222 flux measurements in widely separated areas. *In:* Adams, J.A.S. et al., ed. *Natural radiation environment II.* Washington, DC, US Department of Energy, 1972, pp. 717–730 (Report CONF 26-720805).
5. **United Nations Scientific Committee on the Effects of Atomic Radiation (UNSCEAR).** *Sources and effects of ionizing radiation.* New York, United Nations, 1977.
6. **Castrén, O. et al.** Studies of high indoor radon areas in Finland. *Science of the total environment,* **45**: 311–318 (1985).
7. **Hess, C.T. et al.** Environmental radon and cancer correlations in Maine. *Health physics,* **45**: 339–348 (1983).
8. **Gesell, T.F.** Some radiological aspects of radon-222 in liquified petroleum gas. *In:* Stanley, R.E. & Moghissi, A.A., ed. *Noble gases.* Washington, DC, US Department of Energy, 1973, pp. 612–629 (Report CONF-730915).
9. **Wrixon, A.D. et al.** Indoor radiation surveys in the UK. *Radiation protection dosimetry,* **7**: 321–325 (1984).
10. **Bean, J.A. et al.** Drinking water and cancer incidence in Iowa. II. Radioactivity in drinking water. *American journal of epidemiology,* **116**: 924–932 (1982).
11. **Hofmann, W. et al.** Dose calculations for the respiratory tract from inhaled natural radioactive nuclides as a function of age. I. Compartmental deposition, retention and resulting dose. *Health physics,* **37**: 517–532 (1979).

12. *Dosimetry aspects of exposure to radon and thoron daughter products:* report by a NEA group of experts. Paris, OECD Nuclear Energy Agency, 1983.
13. **Chameaud, J. et al.** Influence of radon daughter exposure at low doses on occurrence of lung cancer in rats. *Radiation protection dosimetry,* **7**: 385–388 (1984).
14. **International Commission on Radiological Protection.** *Limits for inhalation of radon daughters by workers.* Oxford, Pergamon Press, 1981 (ICRP Publication No. 32).
15. *Evaluation of occupational and environmental exposures to radon and radon daughters in the United States.* Bethesda, MD, National Council on Radiation Protection, 1984 (NCRP Report No. 78).
16. **Jacobi, W.** Possible lung cancer risk from indoor exposure to radon daughters. *Radiation protection dosimetry,* **7**: 395–402 (1984).
17. **International Commission on Radiological Protection.** *Recommendations of the International Commission on Radiological Protection.* Oxford. Pergamon Press, 1977 (ICRP Publication No. 26).
18. **International Commission on Radiological Protection.** *Principles for limiting exposure of the public to natural sources of radiation.* Oxford, Pergamon Press, 1984 (ICRP Publication No. 39).

30

Sulfur dioxide and particulate matter

General Description

Sulfur dioxide (SO_2) and particles derived from the combustion of fossil fuels are major air pollutants in urban areas of the world. Sulfur oxides (SO_x) and particulate matter are parts of a complex pollutant mixture. For guideline purposes, a division into three categories is appropriate:

(*a*) sulfur dioxide,

(*b*) the acid aerosols that may result from the oxidation of sulfur dioxide in the atmosphere, and

(*c*) sulfur dioxide plus particles.

Sulfur dioxide. Sulfur dioxide is a colourless gas that reacts on the surface of a variety of airborne solid particles. It is readily soluble in water and can be oxidized within airborne water droplets.

Sulfur dioxide results from the combustion of sulfur-containing fossil fuels, the smelting of sulfur-containing ores, and other industrial processes. Domestic fires can also produce emissions containing sulfur dioxide.

Acid aerosol. Sulfuric acid (H_2SO_4) is a strong acid that is formed from the reaction of sulfur trioxide gas (SO_3) with water. Sulfuric acid is strongly hygroscopic. As a pure material, it is a clear colourless liquid with a boiling-point of 330 °C. Ammonium bisulfate (NH_4HSO_4), which is less acidic than sulfuric acid as a pure material, is a crystalline solid, with a melting-point of 147 °C.

Particulate matter. Airborne particulate matter represents a complex mixture of organic and inorganic substances. Mass and composition tend to divide into two principal groups: coarse particles larger than 2.5 μm in aerodynamic diameter, and fine particles smaller than 2.5 μm in aerodynamic diameter. The smaller particles contain the secondarily formed aerosols (gas to particle conversion), combustion particles and recondensed organic and metal vapours. The larger particles usually contain earth crustal materials and fugitive dust from roads and industries. The acid component of particulate matter, and most of its mutagenic activity, is generally contained in the fine fraction, although in fog some coarse acid droplets are also present.

Because of the complexity of particulate matter and the importance of particle size in determining exposure, multiple terms are used to describe particulate matter. Some terms are derived from and defined by sampling methods, e.g. suspended particulate matter, total suspended particulates, black smoke. Other terms refer more to the site of deposition in the respiratory tract, e.g. inhalable, thoracic particles that deposit primarily in the lower respiratory tract below the larynx. Other terms, such as PM_{10} (particulate matter with an aerodynamic diameter of $10 \mu m$), have both physiological and sampling components.

Methods for sampling and analysing suspended particulate matter were discussed by WHO *(1)* and the US Environmental Protection Agency (EPA) *(2)*. These methods included "smoke" measurements, which may represent the darkness of stain obtained on a white filter-paper through which air has been passed (according to the British smoke method, sometimes referred to as the black smoke method), and also total suspended particulate measurements (gravimetric measurement of particulates of all sizes collected on a glass fibre filter by a high volume sampler according to the method of the US Department of Health, Education, and Welfare *(3)*, as well as by several other methods).

Respirable particles *(1)*, typically with a $4.5 \mu m$ aerodynamic diameter (50% cut-off point), are collected by the black smoke method and its variations; some particles up to $7–9 \mu m$ are also collected.

Methods to measure total suspended particulates (by high volume sampler) have been used extensively in the USA. There are problems with this method, however, in that the size range of particles sampled extends well beyond those particles that are able to penetrate the upper respiratory tract, and in arid regions the method is liable to sample wind-entrained dust of noncombustive origin. This problem has been recognized by US EPA who recommended that particulate matter of less than $10 \mu m$ aerodynamic diameter (PM_{10}) be measured, as a better indicator of health-related particles.

Recommendations have been made by the International Organization for Standardization (ISO) regarding the aerodynamic particle size range corresponding with thoracic penetration *(4)*, and samplers that have acceptance characteristics that approximate that curve are being increasingly used. Such thoracic particle measurements according to the ISO standard (ISO-TP) are roughly equivalent to the sampling characteristics for particulate matter with a 50% cut-off point at $10 \mu m$ diameter.

Sources

Sulfur dioxide

While there are some natural sources of sulfur dioxide (such as volcanoes) that contribute to environmental levels in the European Region, man-made contributions from the combustion of fossil fuels are of prime concern in relation to human exposures. Over the past 10–20 years there has been a tendency towards declining emissions in much of the Region, due to changes in the types or amounts of fuel used. More importantly, however, the types of sources have changed even more, away from small multiple sources (domestic, commercial or industrial) towards large single sources such as power stations, which disperse pollutants at higher altitudes. The net result has been a marked reduction in concentrations of sulfur dioxide in many large cities that were at one time highly polluted. A more widespread distribution, by long-distance transport within the Region, is now the dominant pattern.

Acid aerosol

The major proportion of sulfur emissions from combustion sources is emitted as sulfur dioxide, which is further oxidized to sulfur trioxide in the atmosphere at a rate of 0.5–10% per hour. As a result of the presence of moisture, sulfuric acid is formed; this is present as an aerosol, often associated with other pollutants in droplets or solid particles extending over a wide range of sizes. Most of the sulfuric acid in ambient air results from sulfur dioxide emitted by combustion. Other direct or primary point sources of sulfuric acid include acid manufacturing plants and consuming industries, such as fertilizer and pigment factories.

Sulfuric acid and its partial atmospheric neutralization product, ammonium bisulfate, represent almost all of the strong acid content in the ambient aerosol. The ultimate neutralization product, ammonium sulfate ($(NH_4)_2SO_4$), is only weakly acidic. Other strong acids in the ambient air, e.g. nitric acid (HNO_3) and hydrochloric acid (HCl), will be present as vapours, except when incorporated into fog droplets.

Because of its hygroscopic property, sulfuric acid in ambient air will always be present as a solution droplet whose H^+ concentration varies with ambient humidity. Pure ammonium bisulfate can be present as a salt crystal at humidities up to 80%. However, once it is dissolved into droplet form it will not become a crystal again until the humidity falls below 69%. Once inhaled into the moist respiratory tract, it will take up water vapour and deposit as dilute droplets.

Particulate matter

Suspended particulate matter is a term used to cover a range of finely divided solids or liquids that originate from a number of natural or man-made sources.

Particulate matter of respirable size may be emitted from a number of sources, some of them natural (e.g. volcanoes and dust storms) and many others that are more widespread and more important (e.g. power plants and industrial processes, vehicular traffic, domestic coal burning, industrial incinerators). The majority of these non-natural sources are concentrated in limited portions of the territory, i.e. the urbanized areas, where populations are also concentrated *(1,5)*.

Occurrence in air

Sulfur dioxide

As a result of the changes in sources, annual mean levels of sulfur dioxide in major cities of Europe, stated earlier by WHO *(1)* to be within the range 100–200 $\mu g/m^3$, are now largely below 100 $\mu g/m^3$. Similarly, there has been a decline in maximum daily mean values, which are now mainly in the range 250–500 $\mu g/m^3$. Peaks over shorter averaging periods, such as 1 hour, extend to 1000–2000 $\mu g/m^3$ and in some situations higher transient peaks may also occur. Indoor concentrations of sulfur dioxide are generally lower than outdoor concentrations, since absorption of sulfur dioxide occurs on walls, furniture, clothes and in ventilation systems. An exception is occupational exposure, where concentrations of several thousand micrograms may occur regularly *(1)*.

Data on European concentrations of sulfur dioxide and deposition of other sulfur compounds are based either on national monitoring networks, which are largely concentrated in urban areas, or on cooperative programmes for the study of the long-range transport of pollutants *(6,7)*. Natural concentrations of sulfur dioxide are normally below 5 $\mu g/m^3$. The annual mean sulfur dioxide concentrations in most rural areas of Europe are between 5 $\mu g/m^3$ and 25 $\mu g/m^3$. However, as a result of the common practice of using high chimneys to disperse emissions, there are also large rural areas in Europe where average concentrations now exceed 25 $\mu g/m^3$. Sulfur dioxide is often accompanied by elevated levels of nitrogen oxides (NO_x) *(8)*.

Acid aerosol

Current average acid aerosol levels in Europe and North America are not known. The highest current levels reported in recent years have been summarized by Lioy & Lippmann *(9)*. They are in the range of 20–30 μg sulfuric acid per m^3 (6–12 hours average) in various parts of North America, and 28 μg sulfuric acid per m^3 in Europe (Berlin (West)). The highest reported level in the United Kingdom was 680 μg sulfuric acid per m^3 (1-hour average) in London in 1962. Higher levels were almost certainly present in London in earlier years. Maximum ambient concentrations are likely to occur in urban fogs or downwind of coal- and oil-fired power plants and industrial sources. The distribution of secondary acidic aerosol is much more general, ambient levels depending on the rates of sulfur dioxide oxidation and the subsequent neutralization of sulfuric acid in the ambient air by ammonia (NH_3). Rates of sulfur dioxide oxidation depend on ambient temperature, humidity, and concentrations of oxidants and catalytic components of particles in the atmosphere and cloud droplets. Rates of ammonia neutralization depend on the strength of ammonia sources and atmospheric mixing. Ammonia emissions are lowest over water and afforested regions, and higher over urban and agricultural regions. Indoor sources of sulfuric acid are generally not significant except in some occupational environments.

Particulate matter
In rural areas within Europe, black smoke values range from near zero to about 10 $\mu g/m^3$. In the larger cities, annual mean concentrations of smoke range from 10 to 40 $\mu g/m^3$. Where gravimetric measurements of particulates are made, the annual values lie between about 50 and 150 $\mu g/m^3$. Corresponding maxima are 100–250 $\mu g/m^3$ (black smoke) and 200–400 $\mu g/m^3$ (suspended particulate matter gravimetric).

Conversion factors

Sulfur dioxide

1 ppm = 2860 $\mu g/m^3$
1 mg/m^3 = 0.35 ppm

Acid aerosol
Acidic aerosol concentrations can be expressed as μmols of H^+/m^3 or as sulfuric acid equivalent in $\mu g/m^3$. There are 98 μg per μmol.

Particulate matter
As indicated, no generally applicable conversion factors can be set between black smoke values and various gravimetric particulate matter values (e.g. total suspended particulates or ISO-TP).

Routes of Exposure

Inhalation is the only route of exposure that is of interest in relation to the effects of sulfur dioxide, acidic aerosol and suspended particulate matter on human health. For some special substances, which are constituents such as lead and some highly toxic organic compounds, other routes of uptake such as the alimentary tract may also be of interest. In this context, however, only health effects on the respiratory tract will be considered.

Kinetics and Metabolism

Sulfur dioxide
Absorption of sulfur dioxide in the mucous membranes of the nose and upper respiratory tract occurs as a result of its solubility in aqueous media. The absorption is concentration-dependent, with 85% absorption in the nose at 4–6 mg/m^3 and about 99% at 46 mg/m^3. Only minimal amounts reach the lower respiratory tract *(2, 10, 11)*. From the respiratory tract, sulfur dioxide enters the blood. Elimination occurs (after biotransformation to sulfate in the liver), mainly by the urinary route.

Acid aerosol
The deposition pattern within the respiratory tract is dependent on the size distribution of the ambient droplets and humidity. Acidic ambient aerosol typically has a mass median aerodynamic diameter of 0.3–0.6 μm. Thus,

even with hygroscopic growth in diameter in the respiratory airways by a factor of between 2 and 4, particles remain within the fine-particle range and deposit preferentially in the distal lung airways and airspaces. Some neutralization of the droplets can occur before deposition, due to the normal excretion of endogenous ammonia into the airways. Deposited free H^+ reacts with components of the mucus of the respiratory tract, changing its viscosity *(12)*. The unreacted part of H^+ diffuses into surrounding tissues. The capacity of the mucus to react with H^+ is dependent on the H^+ absorption capacity, which is reduced in acidic saturated mucus as found, for example, in asthmatics.

Under fog conditions the ambient acid is incorporated into droplets, with average droplet sizes in the range of 10–15 μm. Such droplets can also contain dissolved nitric acid and other acidic vapours. Inhaled fog droplets will deposit primarily in the upper respiratory tract; very little will penetrate to the deeper lung airways, where most of the fine acidic aerosol will deposit.

Particulate matter

As discussed elsewhere *(1,11,13,14)*, a portion of the inhaled aerosol is deposited by contact with airway surfaces and the remainder is exhaled. In inhalation toxicology, the term "deposition" refers to removal from inspired air of inhaled particles. "Clearance" refers to the subsequent removal of deposited material from the respiratory tract. Within a species, deposition of inhaled particles in the respiratory tract depends mainly on breathing pattern and particle size (aerodynamic diameter). Larger particles (10 μm and above) are mainly deposited in the extrathoracic part of the respiratory tract (above the epiglottis) and the main proportion of particles 5–10 μm in size are deposited in proximity to the fine airways (respiratory bronchioles) with normal nasal breathing. With mouth breathing, the regional deposition pattern changes markedly, extrathoracic deposition being reduced and tracheobronchial and pulmonary deposition enhanced. The proportion of mouth breathing to nose breathing increases with exercise and conversation *(15)*.

During mouth breathing, fine particles (<2.5 μm aerodynamic equivalent diameter (D_{ae})) deposit primarily in the pulmonary region; between about 3 and 5 μm D_{ae} significant deposition in both the pulmonary and the tracheobronchial regions occurs; at larger sizes (about 7–15 μm D_{ae}), deposition is predominantly in the tracheobronchial region as opposed to the pulmonary region *(16)*.

Health Effects

Sulfur dioxide

Acute effects

High concentrations of sulfur dioxide can give rise to severe effects in the form of bronchoconstriction and chemical bronchitis and tracheitis, as seen in animal experiments *(1)* and in occupational exposures to more than 10 000 μg/m³. Concentrations of sulfur dioxide in the range 2600–2700 μg/m³ give rise to frank effects with bronchospasm in asthmatics *(17)*.

The effects of concern in relation to short-term exposures are those on the respiratory tract. There is an extremely large variability of sensitivity to sulfur dioxide exposure among individuals. This is true for normal persons, but especially so if asthmatics are included *(12)*. Asthmatics have very labile airways and resistance is likely to change in response to many other stimuli, including pollens *(1,2,11)*. Effects observed in asthmatics at relatively low concentrations of sulfur dioxide under laboratory exposure situations are listed in Table 1.

Effects of repeated and/or long-term exposures

Repeated short-term occupational exposure to high concentrations of sulfur dioxide combined with long-term exposure to lower concentrations can give rise to an increased prevalence of chronic bronchitis, especially in cigarette smokers. A possible contribution of simultaneously occurring sulfuric acid aerosol has, however, not been examined in these studies *(24)*. Several epidemiological studies have associated the occurrence of pulmonary effects in communities with combined exposure to sulfur dioxide and particulates.

A continuum of response to sulfur dioxide exposures at relatively low concentrations has been observed in laboratory investigations of human volunteers. The magnitude of the effects was much enhanced when subjects increased their breathing rates through exercise. The findings in a wide range of studies among asthmatics (Table 1) are consistent with a linear relationship *(25)* between magnitude of effect (in terms of proportionate increase in airway resistance) and dose of sulfur dioxide delivered to the airways (after allowing for removal of a substantial proportion in the nose or mouth). Thus, in a strict sense it would be difficult to define a lowest-adverse-effect level since the effect appears to be a function of the sensitivity of the subject, concentration, duration of exposure (10 minutes being the most usual duration of test exposure), level of activity and mucus rheological properties. It was, nevertheless, considered that effects of concern to the health of exercising asthmatic subjects were demonstrable down to sulfur dioxide levels of about 1000 μg/m³, with discernible effects of less certain consequence below that level.

Another aspect, of greater importance to public health, is the proportion of the population liable to be affected. Detailed information regarding the proportion of asthmatic or otherwise sensitive people in the community is not available, although estimates of around 5% have been suggested.

Sensory effects

At concentrations of 10 000 μg/m³, sulfur dioxide has a pungent, irritating odour. Since the odour threshold of sulfur dioxide is several thousand μg/m³, this criterion is not critical in relation to public health.

Acid aerosol: effects on experimental animals

Acute exposures

Respiratory mechanical function. Alterations of pulmonary function, particularly increases in pulmonary flow resistance, occur after acute exposure. Reports of the irritant potency of various sulfate species are variable *(2, 11)*, owing in part to differences in animal species and strains, and also to differences in particle size, pH, composition and solubility. Sulfuric acid is more potent than any of the sulfate salts in terms of increased airway irritancy. For short-term (1-hour) exposures, the lowest concentration of sulfuric acid reported to increase airway resistance was $100 \mu g/m^3$ (in guinea pigs). The irritant potency of sulfuric acid depends in part on particle size, with smaller particles having more effect.

Particle clearance function. Donkeys exposed by inhalation for 1 hour to $0.3–0.6 \mu m$ sulfuric acid at concentrations ranging from 100 to $1000 \mu g/m^3$ exhibited slowed bronchial mucociliary clearance function at concentrations of $\geqq 200 \mu g/m^3$, while rabbits undergoing similar exposures showed an acceleration of clearance at concentrations between 100 and $300 \mu g/m^3$, and a progressive slowing of clearance at concentrations of $\geqq 500 \mu g/m^3$ *(26)*.

Subchronic exposures

Particle clearance function. Donkeys exposed for 1 hour per day, 5 days per week, for 6 months to an aerosol ($0.3–0.6 \mu m$) of sulfuric acid at a concentration of $100 \mu g/m^3$ developed highly variable clearance rates, and a persistent shift from baseline rate of bronchial mucociliary clearance during the exposures and for 3 months after the final exposure. During the 3 months of follow-up, 2 animals had much slower clearance than the baseline rate, while 2 had rates faster than the baseline *(26)*. Rabbits exposed for 1 hour per day, 5 days per week for 20 days to $0.3 \mu m$ sulfuric acid at $250 \mu g/m^3$ developed variable mucociliary clearance rates during the exposure period, and their clearance during a 2-week period following the exposure was substantially faster than their baseline rates *(26)*.

Histology. In the study cited above, in which rabbits were exposed to $250 \mu g/m^3$ for 4 weeks and sacrificed 2 weeks later, histological examinations of the airways showed increased numbers of secretory cells in distal airways, and thickened epithelial cell layers in airways extending from medium-sized airways to terminal bronchioles. There were no corresponding changes in the trachea or other large airways *(26)*. In a study in which dogs were exposed daily for 5 years to $1100 \mu g$ sulfur dioxide per m^3 plus $90 \mu g$ sulfuric acid per m^3 and were then allowed to remain in unpolluted air for 2 years, there were small changes in pulmonary functions during the exposure, which continued following the termination of exposure. Morphometric lung measurements made at the end of the two-year post-exposure period showed changes analogous to an incipient stage of human centrilobular emphysema *(14)*.

Table 1. Effects observed in asthmatic subjects during laboratory conditions of exposure to sulfur dioxide

Sulfur dioxide concentration[a] (ppm)	Duration of exposure (min)	Number and type of subject	Type of exposure	Type of activity	Effects[b]	Reference
1, 3, 5	10	7, normal 7, atopic 7, asthmatic	Mouthpiece	Rest	*SRaw* increased significantly at all concentrations for asthmatic subjects, only at 5 ppm for normal and atopic subjects. Some asthmatics exhibited marked dyspnoea requiring bronchodilation therapy	*(18)*
1.0 0.1, 0.25, 0.5	5 10	6, asthmatic 7, asthmatic	Mouthpiece	Exercise	*SRaw* significantly increased in the asthmatic group at 0.5 and 0.25 ppm of sulfur dioxide and at 0.1 ppm in the two most responsive subjects. At 0.5 ppm three asthmatic subjects developed wheezing and shortness of breath	*(19,20)*
0.50	180	40, asthmatic	Oral chamber Nose clips	Rest	*MMFR* significantly decreased 2.7%; recovery within 30 minutes	*(21)*
0.5	10	5, asthmatic	Mouthpiece	Exercise	*SRaw* increases were observed over exercise baseline rates for 80% of the subjects	*(22)*
0.25, 0.5	60	24, asthmatic	Chamber	Exercise	No statistically significant changes in *FVC* or *SRaw*	*(22)*

0.30	120	19, asthmatic	Chamber	Exercise	No pulmonary effects seen with 0.3 ppm of sulfur dioxide and 0.5 ppm of nitrogen dioxide exposure compared to exercise baseline	(23)

[a] 0.1 ppm of sulfur dioxide ≅ 262 μg/m³; 0.5 ppm ≅ 1310 μg/m³; 1.0 ppm ≅ 2620 μg/m³; 5.0 ppm ≅ 13 100 μg/m³; 10 ppm ≅ 26 200 μg/m³; 50 ppm ≅ 131 000 μg/m³.

[b] Significant increase or decrease noted here refers to "statistically significant" effects, independent of whether the observed effects are "medically significant" or not. Abbreviations are as follows: *SRaw*, specific airway resistance; *MMFR*, maximum mid-expiratory flow rate; *FVC*, forced vital capacity.

Acid aerosol: effects on humans

Acute effects

Respiratory mechanical function. Sulfuric acid and other sulfates have been found to affect both the sensory and the respiratory function in humans.

Respiratory effects from exposure to sulfuric acid (350–500 $\mu g/m^3$) have been reported to include increased respiratory rate and decreased maximal inspiratory and expiratory flow rates and tidal volume *(2,11)*. However, other studies of pulmonary function in nonsensitive healthy adult subjects indicated that pulmonary mechanical function was little affected when subjects were exposed to 100–1000 μg sulfuric acid per m^3 for 10–120 minutes. In one study, the bronchoconstrictive action of carbachol was potentiated by sulfuric acid and other sulfate aerosols, more or less in relation to their acidity. Asthmatics are substantially more sensitive in terms of changes in pulmonary mechanics than healthy people, and vigorous exercise potentiates the effects at a given concentration. The lowest-demonstrated-effect level for sulfuric acid was 100 $\mu g/m^3$ via mouthpiece inhalation in exercising adolescent asthmatics. The effects were relatively small and disappeared within about 15 minutes. In adult asthmatics undergoing similar protocols, the lowest-observed-effect level was 350 $\mu g/m^3$ *(11,27)*.

Particle clearance function. In healthy nonsmoking adult volunteers exposed to 0.5 μm sulfuric acid at rest at 100 $\mu g/m^3$ for 1 hour, there was an acceleration of bronchial mucociliary clearance of particles which deposited primarily in large thoracic airways, and a slowing of clearance when the exposure was raised to 1000 $\mu g/m^3$. For particles that deposited primarily in medium-sized and small airways, there was a small but significant slowing of clearance at 100 $\mu g/m^3$ and a greater slowing at 1000 $\mu g/m^3$. These changes are consistent with the greater deposition of acid in medium-sized to smaller airways. Exposures to 100 $\mu g/m^3$ for 2 hours produced slower clearance than the same exposure for 1 hour, indicating a cumulative relationship to dose *(26)*.

Effects of longer-term exposure

Kitagawa *(28)* identified sulfuric acid as the probable causal agent for approximately 600 cases of respiratory disease in the Yokkaichi area of central Japan between 1960 and 1969. The patients' homes were concentrated within 5 km of a titanium dioxide plant with a 14 m stack that emitted from 100 000 to 300 000 kg sulfuric acid per month in the period 1961–1967. The average concentration of sulfur trioxide in February 1965 in Isozu, a village 1–2 km from the plant, was 130 $\mu g/m^3$, equivalent to a sulfuric acid concentration of 159 $\mu g/m^3$. Kitagawa estimated that peak concentrations might be up to 100 times as high when a north wind was blowing. Electrostatic precipitators were installed to control aerosol emissions in 1967, and after 1968 the number of newly found patients with "allergic asthmatic bronchitis" or "Yokkaichi asthma" gradually decreased. Kitagawa's quantitative estimates of exposure to sulfuric acid and the criteria used to describe cases of respiratory disease may differ from current methods. The unique aspect of this report is the identification of sulfuric acid as the likely causal agent for excess morbidity.

Other evidence of links between high concentrations of ambient sulfuric acid and effects on human health is more circumstantial. Sulfuric acid concentrations in the ambient air were certainly much higher than current levels during the classic episodes in London, the Meuse valley, and Donora, but so were concentrations of many other pollutants. Similarly, the decline in the prevalence of chronic bronchitis in the United Kingdom over the past three decades could have been due to the decline in emissions of any of several pollutants. However, on mechanistic grounds and in view of known exposure–response relationships, sulfuric acid is a more plausible candidate than sulfur dioxide, carbonaceous particles and other known constituents *(29)*.

In an analysis of 1980 cross-sectional mortality for the USA *(30)*, predictors of mortality due to air pollution were expressed in terms of four aerosol pollutant surrogates, i.e. total suspended particulates, inhalable particles $< 15\mu m$, fine particles $< 2.5\mu m$, and sulfate (SO_4^{2-}). Among these, only fine particles and sulfate had statistical significance as predictors of response, but these two surrogates' P values were typically < 0.01.

The measured sulfate includes strong acids (sulfuric acid), the less acidic salt (ammonium bisulfate) and the fully neutralized salt (ammonium sulfate). Since the H^+/SO_4^{2-} ratio is highly variable in time and location and is often close to zero, sulfate is a relatively poor surrogate for acid aerosol concentration. The conclusion that sulfate is a better surrogate for the active component of fine particles than the other three surrogates does not necessarily make it a good one *(29)*. It does, however, lend support to the hypothesis that H^+ is the active agent *(12)*. Unfortunately, epidemiological studies are not available by which mortality and/or morbidity can be related to the acidity (i.e. H^+ ion concentration) of respirable particles *(29)*. This would be expected to constitute a more appropriate measurement *(12)*.

Sensory effects

The odour threshold for sulfuric acid has been estimated to be $750\mu g/m^3$ on the basis of one study and $3000\mu g/m^3$ on the basis of another *(2)*.

Sulfur dioxide and particulate matter

Short-term health effects related to 24-hour average values of sulfur dioxide and particulate matter

Variations in 24-hour average concentrations of sulfur dioxide, black smoke and total suspended particulates have been associated with increased mortality, morbidity and deficits in pulmonary function tests *(1,2,11)*. Regression analysis of daily pollution variables in relation to urban death rates results in significant coefficients, even after accounting for temperature and other associations. These relationships cannot clearly establish a threshold effect. However, on the basis of the London studies *(31)* in which 24-hour concentrations of sulfur dioxide and black smoke were above $500\mu g/m^3$, the daily mortality increased significantly above baseline rates. This does not preclude the possibility that mortality effects occur below these concentrations. In fact, recent time-series analyses of New York City mortality data over 15 years *(32)* suggest that variations in fine particle measures can explain approximately 5% of the fluctuation in mortality, regardless of weather effects. Concentrations in a range below 500 μg black smoke per m^3 were reported in the London analysis, but a different measurement method was used in the report from the USA. Short-term effects of air pollution have been investigated in several studies involving responses in "sensitive" populations. Panel studies of asthmatic individuals have been the most frequently used design *(11)*. Some of the earlier studies, using the responses of asthmatics to varying daily pollution levels, have not been relied upon, primarily because of their small sample size and inadequate exposure measurements. In addition, incidences of illness within a population of bronchitic patients have been studied with respect to daily air pollution concentrations. Significant changes in patients' conditions were observed when black smoke exceeded 250 μg/m^3 and sulfur dioxide exceeded 500 μg/m^3 *(33)*. Taking into account indications from some other studies, as in the earlier WHO report *(1)*, the minimum level of smoke and sulfur dioxide needed to produce effects was taken as 250 μg/m^3.

In some studies, deviations in pulmonary function measures have been observed in children and adults that are associated with short-term fluctuations in particulate concentrations *(1,2,11,34,35)*. In another study of approximately 200 children living in an industrialized community, a statistically significant negative mean slope of forced vital capacity (FVC) and forced expiratory volume (FEV) was found for total suspended particulates (11–272 μg/m^3) and sulfur dioxide (0–281 μg/m^3), with a correlation coefficient $r = 0.75$ *(36)*. In this study total suspended particulate measurements were complemented by parallel inhalable particle measurements *(37,38)*. Since inhalable particle values are generally similar to thoracic particle values, it was possible to estimate total suspended particulates/ISO-TP ratios. From the data collected by Dockery et al. *(36)* it can be calculated that in those 25% of children who were most sensitive, there was at least a four times greater deficit in pulmonary function compared with those of average sensitivity (for this subgroup a decrease in FEV of 0.39 ml/μg per m^3 was observed). Those effects are associated with concentrations of total suspended particulates in the range of 150–200 μg/m^3 (in the presence of sulfur dioxide), although total suspended particulate concentrations have frequently exceeded 260 μg/m^3. Minimum levels for effects were judged to be 180 μg/m^3 in the presence of sulfur dioxide. Relating total suspended particulates to ISO-TP would result in the same deficit in pulmonary function at concentrations of thoracic particles above 110 μg/m^3 in the presence of sulfur dioxide. These values are estimated using specific total suspended particulates/ISO-TP ratios *(37)*.

Although these changes are of health concern, the physiological significance of such apparently reversible effects on the immediate or long-term health of the individual is unknown.

In Table 2 the evidence on short-term health effects is summarized in terms of the lowest-observed-effect levels of air pollutants on health.

Table 2. Summary of effects on human health of lowest-observed-effect levels of sulfur dioxide and particulate matter (short-term exposure)

Effect	24-hour mean exposure to:			
	SO_2 (μg/m³)	smoke (μg/m³)	total suspended particulates (μg/m³)	thoracic particles (μg/m³)
Excess mortality	500	500		
Increased acute respiratory morbidity (adults)	250	250		
Decrements in lung function (children)			180	110

Long-term health effects related to annual means of sulfur dioxide and particulate matter

Mortality. Variations in mortality (all causes) and, more specifically, in mortality from cardiorespiratory diseases have been found during comparison of the findings from different cities in several countries *(1)*. Multiple-regression analyses, using various indices of pollution (as long-term means), together with socioeconomic factors, indicate associations with pollutants (particulates and sulfate being the ones generally incorporated in analyses in the USA) that account for a small proportion (about 4%) of the variation in death rates between cities *(30,39–41)*. Thus, it could be said that there are discernible effects of long-term exposure to the pollution complex of the particulate matter/sulfur dioxide type at relatively low annual mean levels, but it is considered that no firm guidance on lowest-observed-effect levels can be given on the basis of relationships of this type.

Morbidity. Further epidemiological studies on differences in the prevalence of respiratory symptoms (adults and children) and the frequency of respiratory illness (children) between communities with differing levels of pollution have provided results that are consistent with the conclusions reached earlier by WHO *(1)*, indicating detectable increases where annual mean concentrations of both black smoke and sulfur dioxide exceed 100μg/m³ *(42,43)*. Other pollutants, such as sulfates (or acid sulfates) may be relevant, but no measurements were available in the studies in question. The more recent studies have mainly been analysed using multiple-regression models, taking confounding variables into account as far as possible *(44,45)*. In this way, the relative importance of different factors has been shown more clearly and relationships are taken to be continuous, indicating that effects may well extend below the pollution levels quoted.

Community-based health studies are useful in attributing excess illness rates or differences in pulmonary performance to air pollution. Communities differ for a variety of cultural, social, economic and other factors that can result in different frequencies of illness. While air pollution may contribute to elevated illness rates, it is difficult to describe with certainty a level, an averaging time or even a specific contaminant that is unequivocally associated with a threshold effect level. Increased age-adjusted illness rates are associated with indices for sulfur dioxide, black smoke, total suspended particulates, and fine particles in several studies *(1,2,11)*. Community differences in illness rates can be discerned in several more contemporary studies conducted in the late 1970s and early 1980s. It is of interest to note that the annual sulfur dioxide and total suspended particulate concentrations are lower than the concentrations associated with effects in earlier studies. For instance, in the USA *(46)* differences in community illness rates have been associated with annualized total suspended particulate concentrations ranging from 30 to 100$\mu g/m^3$ (20–55$\mu g/m^3$ when the particles measure less than 10μm in diameter). The two communities with the highest illness rates had particle concentrations (for particles less than 10μm in diameter) of 35 and 55$\mu g/m^3$ (annual means).

In the Netherlands, a decreasing difference in respiratory symptom rates between a polluted and a cleaner area was observed *(47)*. Initially, annual average sulfur dioxide concentrations above 200$\mu g/m^3$ were observed in the polluted area, but after the mid-1970s sulfur dioxide levels were between 45 and 80$\mu g/m^3$, while black smoke decreased from 34–45 to 25–35$\mu g/m^3$. In the cleaner area sulfur dioxide values, measured after 1975, were 10–25$\mu g/m^3$, and black smoke levels, measured after 1982, were 10–15$\mu g/m^3$. In France *(48)* differences in symptom rates are associated with annual averages of sulfur dioxide over a range of 13–127$\mu g/m^3$, measured by acidimetry, or a range of 22–85$\mu g/m^3$, as measured by a specific technique.

Decrements in lung function. Measurements of respiratory physiology were included in several of the studies referred to above. Several of these observations have been reviewed by WHO *(1)*, EPA *(2)* and Ericsson & Camner *(11)*. Studies that have been conducted in the same communities over a period of years show associations between the magnitude of lung function changes and the levels of pollution. One series of such studies, carried out in the USA *(49–51)*, indicated effects associated with particulates (measured as total suspended particulates) at an annual mean of 180$\mu g/m^3$, though documentation of pollution levels in the series as a whole was incomplete and other pollutants could have been involved. From a more extensive series carried out in the Netherlands *(47)* it has been concluded that consistently lower lung function values in an urban, as compared with a rural, area might point to long-term effects of pollution. While much current information on a wide range of pollutants was available, it was considered that the effect could have related to earlier higher levels, and no firm guidance can be given at this stage in relation to lowest-observed-adverse-effect levels.

Sensory effects
Community exposure to urban air pollutants, including sulfur oxides, nitrogen oxides and particulate matter, may give rise to feelings of discomfort, which can only be assessed subjectively by those persons who are affected *(1,52)*. Annoyance reactions to urban air pollutants are common phenomena. In a Swedish study *(52)* of population groups in central Stockholm, 60% of the population reported annoyance of this kind. One quarter of those were classed as being very annoyed. Comparative studies in suburban areas and smaller Swedish towns disclosed lower prevalence figures for annoyance. Surveys of annoyance are fraught with many problems *(1)*. Since annoyance reactions have a large sociocultural component, prevalence figures in relation to air pollution levels may vary from place to place and should be determined for each locality.

Evaluation of Human Health Risks

Sulfur dioxide

When using the evidence from human experimental studies of sulfur dioxide to draw up recommendations for guideline values aimed at protecting people from the risk of adverse effects, the need to avoid brief exposures to peak values is implied. Some protection (safety) factor may have to be incorporated when using information on the lowest-observed-effect level in order to protect especially sensitive asthmatic patients (who have not been subject to testing), though they would be less likely to be involved in exercise at the levels used in the experimental exposures. In relation to a lowest-observed-effect level of concern to health of $1000\mu g/m^3$ (10 minutes), it appears reasonable to apply a protection (safety) factor of 2 for the protection of public health; this would give a concentration of $500\mu g/m^3$ (10 minutes). The occurrence of such concentrations can often be predicted from the frequency distribution of locally measured concentrations, by using some existing models for averaging values over different time periods in the case of diffuse or multiple sources *(53)*.

Predictions for point sources can also be made if the characteristics of the source and the local diffusion conditions are known *(54)*. Frequency distribution characteristics can also help in guiding authorities towards solutions. These frequency distributions are known for a large number of towns in Europe *(55)* and the USA *(56)*. As an example, if the aim were to ensure that the 10-minute mean value of $500\mu g/m^3$ was not exceeded, then on the basis of calculations of multiple-source situations in the Netherlands *(55)*, the corresponding 1-hour value that should not be exceeded would be $350\mu g/m^3$.

Acid aerosol

While the data currently available are insufficient to establish a numerical guideline, they do raise serious concern that acidic aerosol could account for past associations between particulate air pollution and the exacerbation and development of chronic bronchitis.

Recent 1-hour acute experimental inhalation exposure data on humans and two animal species (donkeys and rabbits) show similar exposure–response relationships in terms of transient and reversible changes in the rate of tracheobronchial mucociliary clearance. Comparable exposures, when repeated on a daily basis in the two animal species, produced persistent changes in clearance rates, and in the one species in which histological examinations were made, changes in the airways after only 20 days of exposure were of a similar character to those seen in young human smokers examined at autopsy. The analogy with cigarette smoke, which is a known causal factor in chronic bronchitis, has been pointed out by Lippmann *(29)*.

The association shown in Japan (Yokkaichi) between sulfuric acid aerosols and respiratory morbidity gives support to the hypothesis that acid aerosol is an important component of urban air pollution. This hypothesis is also consistent with the results of cross-sectional studies of daily mortality in major cities in the USA, which indicate that sulfate is a better predictor for mortality than any of the nonspecific gravimetric indices that have been used.

More data on human exposures are clearly needed to test the hypothesis of causality. Situations that would be of concern for monitoring purposes would be those where humans were exposed repeatedly to concentrations at or above 10 $\mu g/m^3$ (sulfuric acid or equivalent acidity of aerosol).

Sulfur dioxide and particulate matter

The lowest-observed-effect levels for short-term and long-term (annual mean) average air pollution measurements are summarized in Tables 2 and 3. Evaluation of the measured components of air pollution in relation to public health is, however, difficult for a number of reasons noted in the

Table 3. Summary of effects on human health of lowest-observed-effect levels of sulfur dioxide and particulate matter (long-term exposure)

Effect	Annual mean exposure to: SO_2 ($\mu g/m^3$)	smoke ($\mu g/m^3$)	total suspended particulates ($\mu g/m^3$)
Increased respiratory symptoms or illness	100	100	
Decrements in lung function			180

WHO publication *(1)*. A number of these points still remain largely unresolved. For example, it is not clear whether long-term effects can be related simply to annual mean values or to repeated exposures to peak values. Similarly, it remains uncertain which components of the sulfur dioxide/particulates complex are involved in the adverse effects, though increasingly attention is being given to the role of secondary products such as acid sulfates. Arbitrary protection (safety) factors of 2 in relation to the morbidity and mortality data, and 1.5 for decrements in lung functions (considered to represent a less severe effect), seem to be appropriate according to the present state of knowledge.

Measurements of black smoke can no longer be interpreted in terms of $\mu g/m^3$ in many localities, and decisions have already been made (by ISO) to abandon any attempt at mass equivalence. The method is still of value as an index of soiling capacity and of the type of pollution (coal smoke) that has been associated in the past with adverse health effects, and to provide continuity with any further epidemiological studies. Therefore, observations should be continued.

Various direct gravimetric measurements have been used in recent decades, notably the total suspended particulate measurements (by high volume sampler) in the USA. There are problems, however, with the wide size range of particles sampled and the influence of wind-entrained dust. Although a large body of data on such measurements exists, it is now considered misleading to attempt to specify guidelines in terms of total suspended particulates.

Total suspended particulate measurements may, nevertheless, be used for comparison with newer indices of pollution, and they may be of value as a supplement to gravimetric ISO-TP measurements, especially in areas where there is special concern about larger particles.

Efforts should now be made to establish a method of gravimetric measurement representing more realistically the size range of particles that can be inhaled into the thoracic region, even though uncertainties must remain about the component or components most relevant to health. Recommendations have already been made by ISO regarding the (aerodynamic) particle size range corresponding with thoracic penetration, and it is proposed that samplers should have acceptance characteristics that approximate to that curve.

The inclusion of the somewhat wider size range of particles than those sampled by the black smoke method would mean that, even in areas where coal smoke still forms a dominant part of the suspended particulates, results from these gravimetric instruments would be somewhat higher than might be obtained from co-located smoke samplers. Thus, in those circumstances the corresponding guidelines would be a little higher in true gravimetric terms (possibly by about 10%). Now that the characteristics of present-day pollution differ from those of coal smoke pollution, the old data cannot be used with any confidence as a basis for guidelines.

In view of the considerable uncertainties involved in formulating guidelines for particulate matter, there is a need for further epidemiological studies, particularly in those areas where high concentrations will occur, using well defined methods for particulate measurement and epidemiological assessment, including the control of possible confounding factors such as smoking.

Guidelines

Sulfur dioxide

It appears reasonable to apply a protection factor of 2 for the protection of public health; a guideline value of 500$\mu g/m^3$ (10 minutes, not to be exceeded) is recommended. A 1-hour maximum value that conforms with this guideline can be calculated as approximately 350$\mu g/m^3$.

Acid aerosol

Recommendations for air quality guideline values for the strong acid content of ambient aerosol cannot now be made owing to the sparsity of current data on effects and ambient exposure levels. However, monitoring is warranted when levels (sulfuric acid or equivalent acidity of aerosol) exceed 10$\mu g/m^3$. Therefore, ambient air should be regularly monitored for the H^+ ion concentration of the aerosol (which should be sampled in a size-fractionating particulate sampler) when levels of this magnitude are likely to occur.

Combined effects

In proposing guidelines based on the present knowledge of exposure to both sulfur dioxide and particulate matter, an arbitrary protection (safety) factor of 2 has been used in relation to morbidity and mortality, and a factor of 1.5 has been used for the decrement in lung function, which is considered to be a less severe effect. The recommended guideline values are shown in Table 4.

References

1. *Sulfur oxides and suspended particulate matter.* Geneva, World Health Organization, 1979 (Environmental Health Criteria, No. 8).
2. *Air quality criteria for particulate matter and sulfur oxides,* Vol. I, II & III. Research Triangle Park, NC, US Environmental Protection Agency, 1982 (EPA-600/8-82-029a, b & c).
3. *Air pollution measurements of the National Air Sampling Network, 1957–1961.* Cincinnati, OH, US Department of Health, Education, and Welfare, 1962, pp. 3–4.
4. *Air quality — particle size fraction definitions for health-related sampling.* Geneva, International Organization for Standardization, 1983 (Technical Report ISO/TR 7708-1983 (E)).
5. Proceedings of the Symposium on Biological Tests in the Evaluation of Mutagenicity and Carcinogenicity of Air Pollutants with Special Reference to Motor Exhausts and Coal Combustion Products. *Environmental health perspectives,* **47**: 1–324 (1983).
6. *Manual for sampling and chemical analysis.* Lillestrøm, Norwegian Institute for Air Research, 1977 (EMEP/CHEM 3/77).

Table 4. Guideline values for combined exposure to sulfur dioxide and particulate matter[a]

	Averaging time	Sulfur dioxide ($\mu g/m^3$)	Reflectance assessment: black smoke[b] ($\mu g/m^3$)	Gravimetic assessment	
				Total suspended particulates (TSP)[c] ($\mu g/m^3$)	Thoracic particles (TP)[d] ($\mu g/m^3$)
Short term	24 hours	125	125	120[e]	70[e]
Long term	1 year	50	50	—	—

[a] No direct comparisons can be made between values for particulate matter in the right- and left-hand sections of this table, since both the health indicators and the measurement methods differ. While numerically TSP/TP values are generally greater than those of black smoke, there is no consistent relationship between them, the ratio of one to the other varying widely from time to time and place to place, depending on the nature of the sources.

[b] Nominal $\mu g/m^3$ units, assessed by reflectance. Application of the black smoke value is recommended only in areas where coal smoke from domestic fires is the dominant component of the particulates. It does not necessarily apply where diesel smoke is an important contributor.

[c] TSP: measurement by high volume sampler, without any size selection.

[d] TP: equivalent values as for a sampler with ISO-TP characteristics (having 50% cut-off point at 10 μm): estimated from TSP values using site-specific TSP/ISO-TP ratios.

[e] Values to be regarded as tentative at this stage, being based on a single study (involving sulfur dioxide exposure also).

7. *International operations handbook for measurement of background atmospheric pollution.* Geneva, World Meteorological Organization, 1978.
8. **Fowler, D. & Cape, J.N.** Air pollutants in agriculture and horticulture. *In:* Unsworth, M.H. & Ormrod, D.P., ed. *Effects of gaseous air pollution in agriculture and horticulture.* London, Butterworth, 1982, pp. 3–26.
9. **Lioy, P.J. & Lippmann, M.** Measurement of exposure to acidic sulfur aerosols. *In:* Lee, S.D. et al., ed. *Aerosols.* Chelsea, MI, Lewis, 1986.
10. *Air quality in selected urban areas, 1973–1974.* Geneva, World Health Organization, 1976 (WHO Offset Publication No. 30).
11. **Ericsson, G. & Camner, P.** Health effects of sulfur oxides and particulate matter in ambient air. *Scandinavian journal of work, environment & health,* **9**(Suppl. 3): 1–52 (1983).
12. **Holma, B.** Influence of buffer capacity and pH-dependent rheological properties of respiratory mucus on health effects due to acidic pollution. *Science of the total environment,* **41**: 101–123 (1985).
13. **Lippmann, M. et al.** Deposition, retention, and clearance of inhaled particles. *British journal of industrial medicine,* **37**: 337–362 (1980).
14. **Stara, J.F. et al.** *Long-term effects of air pollutants in canine species.* Research Triangle Park, NC, US Environmental Protection Agency, 1980 (Report No. EPA-600/8-80-014).
15. **Camner, P. & Bakke, B.** Nose or mouth breathing? *Environmental research,* **21**: 394–398 (1980).
16. *Second addendum to air quality criteria for particulate matter and sulfur oxides.* Washington, DC, US Environmental Protection Agency, 1986 (EPA-600/8-86-020A).
17. **Islam, M.S. & Ulmer, W.T.** Untersuchungen zur Schwellenkonzentration von Schwefeldioxyd bei besonders Gefährdeten [Borderline concentrations of SO_2 for patients with oversensitivity of the bronchial system]. *Wissenschaft und Umwelt,* **1**: 41–47 (1979).
18. **Sheppard, D. et al.** Exercise increases sulfur dioxide-induced bronchoconstriction in asthmatic subjects. *American review of respiratory disease,* **123**: 486–491 (1981).
19. **Sheppard, D. et al.** Lower threshold and greater bronchomotor responsiveness of asthmatic subjects to sulfur dioxide. *American review of respiratory disease,* **122**: 873–878 (1980).
20. **Sheppard, D. et al.** Inhibition of sulfur dioxide-induced bronchoconstriction by disodium cromoglycate in asthmatic subjects. *American review of respiratory disease,* **124**: 257–259 (1981).
21. **Jaeger, M.J. et al.** Effect of 0.5 ppm sulfur dioxide on the respiratory function of normal and asthmatic subjects. *Lung,* **156**: 119–127 (1979).
22. **Linn, W.S. et al.** Respiratory responses of young adult asthmatics to sulfur dioxide exposure under simulated ambient conditions. *Environmental research,* **29**: 220–232 (1982).
23. **Linn, W.S. et al.** Respiratory effects of mixed nitrogen dioxide and sulfur dioxide in human volunteers under simulated ambient exposure conditions. *Environmental research,* **22**: 431–438 (1980).

24. **Stjernberg, N. et al.** Prevalence of bronchial asthma and chronic bronchitis in a community in Northern Sweden; relation to environmental and occupational exposure to sulphur dioxide. *European journal of respiratory diseases,* **67**: 41–49 (1985).
25. **Kleinman, M.T.** Sulfur dioxide and exercise: relationships between response and absorption in upper airways. *Journal of the Air Pollution Control Association,* **34**: 32–37 (1984).
26. **Schlesinger, R.B.** The effects of inhaled acid aerosols on lung defenses. *In:* Lee S.D. et al., ed. *Aerosols.* Chelsea, MI, Lewis, 1986.
27. **Utell, M.J. & Morrow, P.E.** Effects of inhaled acid aerosols on human lung function: studies in normal and asthmatic subjects. *In:* Lee, S.D. et al., ed. *Aerosols.* Chelsea, MI, Lewis, 1986.
28. **Kitagawa, T.** Cause analysis of the Yokkaichi asthma episode in Japan. *Journal of the Air Pollution Control Association,* **34**: 743–746 (1984).
29. **Lippmann, M.** Airborne acidity: estimates of exposure and human health effects. *Environmental health perspectives,* **63**: 63–70 (1985).
30. **Evans, J.S. et al.** Cross-sectional mortality studies and air pollution risk assessment. *Environment international,* **10**: 55–83 (1984).
31. **Martin, A.E. & Bradley, W.H.** Mortality, fog and atmospheric pollution. An investigation during the winter of 1958–59. *Monthly bulletin of the Ministry of Health Public Health Laboratory Service,* **19**: 56–72 (1960).
32. **Schimmel, H.** Evidence for possible acute health effects of ambient air pollution from time series analysis: methodology questions and some new results based on New York City daily mortality, 1963–1976. *Bulletin of the New York Academy of Medicine,* **54**: 1052–1108 (1978).
33. **Lawther, P.J. et al.** Air pollution and exacerbations of bronchitis. *Thorax,* **25**: 525–539 (1970).
34. **Cohen, A.A. et al.** Symptom reporting during recent publicized and unpublicized air pollution episodes. *American journal of public health,* **64**: 442–449 (1974).
35. **van der Lende, R. et al.** A temporary decrease in the ventilatory function of an urban population during an acute increase in air pollution. *Bulletin européen de physiopathologie respiratoire,* **11**: 31–43 (1975).
36. **Dockery, D.W. et al.** Change in pulmonary function in children associated with air pollution episodes. *Journal of the Air Pollution Control Association,* **32**: 937–942 (1982).
37. **Spengler, J.D. et al.** *Comparison of hi-vol, dichotomous, and cyclone samplers in four US cities.* 73rd Annual Meeting of the Air Pollution Control Association, Montreal, Quebec, 22–27 June, 1980 (unpublished paper).
38. **Spengler, J.D. & Thurston, G.D.** Mass and elemental composition of fine and coarse particles in six US cities. *Journal of the Air Pollution Control Association,* **33**: 1162–1171 (1983).
39. **Lave, L.B. & Seskin, E.P.** Air pollution, climate, and home heating: their effects on US mortality rates. *American journal of public health,* **62**: 909–916 (1972).
40. **Lave, L.B. & Seskin, E.P.** *Air pollution and human health.* Baltimore, MD, Johns Hopkins University Press, 1977.

41. **Chappie, M. & Lave, L.** The health effects of air pollution: a reanalysis. *Journal of urban economics,* **12**: 346–376 (1982).
42. **Lunn, J.E. et al.** Patterns of respiratory illness in Sheffield infant schoolchildren. *British journal of preventive and social medicine,* **21**: 7–16 (1967).
43. **Lunn, J.E. et al.** Patterns of respiratory illness in Sheffield junior schoolchildren — a follow-up study. *British journal of preventive and social medicine,* **24**: 223–228 (1970).
44. **Ostro, B.** The effects of air pollution and community health. *Journal of the Royal College of Physicians (London),* **5**: 362–368 (1983).
45. **Wojtyniak, B. et al.** Importance of urban air pollution in chronic respiratory problems. *Zeitschrift für Erkrankungen der Atmungsorgane,* **163**: 274–284 (1984).
46. **Ware, J.H. et al.** Effects of ambient sulfur oxides and suspected particles on respiratory health of preadolescent children. *American review of respiratory disease,* **133**: 834–842 (1986).
47. **van der Lende, R. et al.** Longitudinal epidemiological studies on effects of air pollution in the Netherlands. *In:* Lee, S.D. et al., ed. *Aerosols.* Chelsea, MI, Lewis, 1986.
48. **PAARC Cooperative Group.** Atmospheric pollution and chronic or recurrent respiratory diseases. I. Methods and material. II. Results and discussion. *Bulletin européen de physiopathologie respiratoire,* **18**: 87–99, 101–116 (1982).
49. **Ferris, B.G., Jr & Anderson, D.O.** The prevalence of chronic respiratory disease in a New Hampshire town. *American review of respiratory disease,* **86**: 165–185 (1962).
50. **Ferris B.G. et al.** Chronic non-specific respiratory disease, Berlin, New Hampshire, 1961–67: a cross-section study. *American review of respiratory disease,* **104**: 232–244 (1971).
51. **Ferris, B.G. et al.** Chronic non-specific respiratory disease in Berlin, New Hampshire 1967–1973. A further follow-up study. *American review of respiratory disease,* **113**: 475–485 (1976).
52. **Sörensen, S. et al.** Annoyance reactions. *In:* Ewetz, L. & Camner, P., ed. *Health risks resulting from exposure to motor vehicle exhausts:* report to the Swedish Government Committee on Automotive Air Pollution. Stockholm, National Institute of Environmental Medicine, 1983.
53. **Larsen, R.I.** *A mathematical model for relating air quality measurements to air quality standards.* Research Triangle Park, NC, US Environmental Protection Agency, 1971 (Report No. AP-89).
54. **Pasquill, F.** Atmospheric dispersion of pollution. *Quarterly journal of the Royal Medical Society,* **97**: 369–395 (1971).
55. **National Air Pollution Monitoring Network 1984.** Bilthoven, National Institute of Public Health and Environmental Hygiene, 1984 (Report Nml-RIUm No. 28).
56. *Health consequences of sulfur oxides: a report from CHESS, 1970–71.* Research Triangle Park, NC, US Environmental Protection Agency, 1974 (Report No. EPA-650/1-74-004).

31

Vanadium

General Description

Vanadium (V) is a bright white ductile metal belonging to group V of the periodic system of elements. It forms compounds mainly in valence states + 3, + 4 and + 5. In the presence of oxygen, air or oxygenated blood, or oxidizing agents, vanadium is always in the + 5 oxidation state. In the presence of reducing agents, vanadium compounds are in the + 4 oxidation state *(1)*. Vanadium forms both cationic and anionic salts, and can form covalent bonds to yield organometallic compounds which are mostly unstable.

Sources

Vanadium is an ubiquitous metal. The average concentration of vanadium in the earth's crust is 150 μg/g *(2)*; concentrations in soil vary in the range 3–310 μg/g *(3)* and may reach high values (up to 400 μg/g) in areas polluted by fly ash *(4)*. The concentration of vanadium in water is largely dependent on geographical location and ranges from 0.2 to more than 100 μg/litre in freshwater *(2)*, and from 0.2 to 29 μg/litre in seawater *(3)*. The ocean floor is the main long-term sink of vanadium in the global circulation *(4)*. The concentrations of vanadium in coal and crude petroleum oils vary widely (1–1500 mg/kg) *(2)*.

The world production of vanadium was about 35 000 tonnes in 1981 *(5)*, the major producing countries being Chile, Finland, Namibia, Norway, South Africa, USSR and the United States.

Most of the vanadium produced is used in ferrovanadium and, of this, the majority is used in high-speed and other alloy steel (usually combined with chromium, nickel, manganese, boron and tungsten).

It has been estimated that around 65 000 tonnes of vanadium annually enter the environment from natural sources (crustal weathering and volcanic emissions) and around 200 000 tonnes as a result of man's activities *(6)*. The major anthropogenic point sources of atmospheric emission are metallurgical works (30 kg per tonne of vanadium produced), followed by the burning of crude or residual oil and coal (0.2–2 kg per 1000 tonnes and 30–300 kg per 10^6 litres) *(7)*.

Global vanadium emissions into the atmosphere from coal combustion in 1968 were estimated to range from 1730 to 3760 tonnes. The contribution of vanadium to the atmosphere from residual-fuel combustion was estimated at 12 400–19 000 tonnes in 1969 and 14 000–22 000 tonnes in 1970. In the production of ferrovanadium for alloy additions in steel-making, vanadium emission to the atmosphere was estimated at 144 tonnes in 1968 *(2)*. The burning of wood, other vegetable matter and solid wastes probably does not result in significant vanadium emission. In 1972, about 94% of all anthropogenic emissions of vanadium to the atmosphere in Canada (2065 tonnes) resulted from the combustion of fuel oil and only 1.2% from metallurgical industries *(8)*.

Occurrence in air

Currently the ambient air levels of vanadium are quite low, but it is suspected that air concentrations will increase in the future as a function of accelerated fossil-fuel combustion in rural areas.

Concentrations vary from tenths of nanograms to a few nanograms. In the air of big cities the annual average was between 50 and 100 ng/m^3 *(8,9)* with significant seasonal variations (winter averages are more than three times the summer averages). More recent data from a five-year study in four Belgian cities show yearly averages of 41–179 ng/m^3 *(10)*. A recent survey of measurements in Member States of the European Community indicates the following concentration ranges: remote areas, 0.001–3 ng/m^3; urban areas, 7–200 ng/m^3; industrial areas, 10–70 ng/m^3 *(11)*. Concentrations up to 2 μg/m^3 have been reported in several cities in the north-eastern USA. A marked decline of vanadium concentration in ambient air was reported after the introduction of low-sulfur fuels *(4)*. Air pollution by industrial plants may be less than that by power stations and heating equipment. The average concentration in the area surrounding a steel-plant site in Pennsylvania, USA, was 72 ng/m^3 *(2)*. Vanadium pentoxide (V_2O_5) was detected near a large metallurgical works in concentrations ranging from 0.98 to 1.49 μg/m^3 in 87% of samples and exceeding 2 μg/m^3 in 11% of samples *(12)*.

With combustion of the carbonaceous matrix, vanadium is released to the atmosphere as fine fly ash with a long atmospheric residence time. Vanadium concentrates on the surface of particulate matter from coal combustion. The concentration of vanadium in fly ash from residual fuel-oil burning averages around 8% (range 2–18%), representing 56% (39–74%) of the vanadium concentration in fuel. The soluble fraction of fly ash vanadium averages 70% (42–96%). About 80% of the vanadium-containing particles have a mass median aerodynamic diameter smaller than 0.5 μm at the chimney, representing 10–20% of the total aerosol mass of vanadium *(13)*.

A large number of occupations involve exposure to vanadium. Concentrations in the working environment range from 0.01 to 30 mg/m^3 *(14)*, with varying particle distribution patterns and different degrees of solubility for different vanadium compounds. Very high levels of vanadium in the air were reported in boiler cleaning (17–60 mg/m^3) *(15)*.

Routes of Exposure

Air

Concentrations of airborne vanadium have increased in recent years, probably because of the increasing direct combustion of crude oil residues in power plants and community-heating systems. The average annual values for urban areas are reported to be in the range of 0.05–0.18 μg/m^3. Maximum concentrations of vanadium as high as 2 μg/m^3 occur in areas of greatest population density, during the coldest part of the year and during the late evening hours.

Drinking-water

Concentrations of vanadium in drinking-water may range from about 0.2 to more than 100 μg/litre *(16)*; typical values appear to be between 1 and 6 μg/litre *(8)*. The concentration of vanadium in drinking-water depends significantly on geographical location.

Food

The few data on vanadium concentrations in food vary considerably; this may be explained by differences in the food itself and in the analytical methods used. The mean vanadium concentration in the diet was reported to be 32 μg/kg (range 19–50 μg) *(17)* and the mean daily intake was estimated to be 20 μg/day *(18)*. Byrne & Kosta *(19)* estimated that the daily dietary intake of vanadium amounts to several tens of micrograms.

The concentration of vanadium in human milk was found to be 0.1–0.2 ng/g *(18)*. An infant drinking one litre of human milk per day would thus have a daily vanadium intake of 0.1–0.2 μg *(8)*.

Relative significance of different routes of exposure

According to a pathway analysis *(8)*, an individual living in a rural environment (assumed air concentration 8 ng vanadium per m^3) will have a body-burden of around 100 μg, over 80% of which is derived from the diet; an urban dweller (assumed air concentration 70 ng/m^3) will have a total body-burden of around 200 μg (in the latter case, the inhalation route contributes approximately one half of the total burden).

It was estimated that the daily intake by ingestion is about 20 μg and by inhalation 1.5 μg in an urban area and 0.2 μg in a rural area *(8)*.

Kinetics and Metabolism

Absorption

The lungs are a significant site of entry of vanadium in the case of community exposure. The distribution pattern of particles and the solubility of vanadium compounds, as well as alveolar and mucociliar clearance, are important factors that determine the rate of absorption in the respiratory tract. Moreover, the irritative effect of vanadium compounds can significantly modify the absorption of vanadium by the lungs. Vanadium accumulates in the human lung with age, reaching approximately 6.5 μg/g in the over-65 years age group *(3)*. The lung clearance of vanadium pentoxide is relatively rapid in animals after acute exposure, but substantially slower after chronic exposure. The metal is deposited in the lung in relatively insoluble forms. Soluble compounds are partly absorbed, but the extent of absorption in the respiratory tract has not been determined.

The absorption rate of vanadium compounds after ingestion depends on their solubility and chemical nature. Absorption of cationic vanadium is low, not exceeding 0.1–1% *(8)*. Skin is probably a minor route of absorption in man *(9)*.

Distribution

Absorbed vanadium is transported mainly in the plasma, bound to transferrin. The average value for the distribution of blood vanadium between plasma and cells in rats after an intravenous injection of 0.9–30 μg vanadium per kg was found to be 9 : 1 *(20)*. Pentavalent vanadium is reduced in erythrocytes to the tetravalent form. This reduction is a glutathione-dependent process *(21)*.

Byrne & Kosta *(19)* reported blood vanadium levels in healthy individuals of below 0.5 μg/litre. The analytical method used was neutron activation analysis with pre-separation of vanadium.

Vanadium is widely distributed in body tissues; principle organs of vanadium retention are kidneys, liver, testicles, spleen and bones. A major fraction of vanadium from cellular vanadium was found retained in nuclei *(22)*. In pregnant rats the injected vanadium was found in the fetus *(23)*.

Elimination

Vanadium is excreted mainly in the urine, but also in the faeces. Bile is probably not an important pathway for excretion into the faeces, but the existence of alternative routes for excretion into the gut (salivary excretion or direct transfer across the intestinal wall) has been suggested *(20)*. When 4.5–9.0 mg vanadium as diammonium oxytetravanadate was fed daily to 16 elderly persons, urinary excretion, although quite variable, amounted to a mean of 5.2% of the amount ingested *(24)*.

Health Effects

Nutritional studies have shown that vanadium is an essential element for the chick and rat *(3)*. Its deficiency may result in growth reduction, impairment of reproduction and disturbances in lipid metabolism. Vanadium is also essential for soil nitrogen-fixing microorganisms *(3)*. It may play a significant role in human nutrition *(2)*.

Biochemical, physiological and pharmacological properties of vanadium compounds have been reviewed *(1,25)*. It has been suggested that vanadium may be a regulatory agent for enzymatic activities in mammalian tissues. Vanadium is a potent inhibitor of many enzymes, while it stimulates adenylate cyclase. It has been shown to inhibit cholesterol biosynthesis and lower plasma cholesterol levels. Vanadium can also directly influence glucose metabolism *in vitro* and may play a role in its regulation. Lipid peroxidation of rat lung extracts, liver microsomes and mitochondria was induced by sulfite and accelerated by the presence of vanadium compounds.

Vanadium may play a physiological role as part of a control on levels of the endogenous antioxidant, glutathione. This may also be important with respect to toxic interactions of chemicals *(26)*.

Effects on experimental animals

The toxicity of vanadium has been found to be high when it is given parenterally, low when it is orally administered, and intermediate in the case of respiratory exposure. Toxicity is related to the valency of the vanadium compound (increasing with increasing valency). Vanadium is toxic both as a cation and as an anion *(16,27)*.

Severe acute exposure (tens of mg/m^3) is responsible for systemic effects. Most frequent findings in animal experiments were in the liver, kidneys, gonads and the nervous, haematological and cardiovascular systems *(2,3,16)*. However, systemic effects at very low levels of exposure were also reported *(28)*. When rats were continuously exposed to an aerosol of vanadium pentoxide at levels of 6 and 27 $\mu g/m^3$ (3.4–15 μg vanadium per m^3), neurotoxic effects appeared 30 days from the start of exposure. Measured chronaxy ratio between the flexor and extensor muscles of the tibia returned to normal about 18 days following cessation of exposure. Histopathological changes observed in the liver following the higher level of inhalation exposure (27 $\mu g/m^3$ for 70 days) included central vein congestion with scattered small haemorrhages, and granular degeneration of hepatocytes. The kidneys showed marked granular degeneration of the epithelium of the convoluted tubules. In the heart, myocardial vascular congestion was observed, with focal perivascular haemorrhages.

Numerous studies reported acute or chronic respiratory effects in laboratory animals following exposure to different vanadium compounds. Respiratory effects in rats at relatively low levels of vanadium pentoxide aerosol (3–5 mg/m^3) for 2 hours every other day for 3 months, included perivascular oedema, capillary congestion, haemorrhages, and in some cases desquamative bronchitis or pneumonia *(29)*. Similar effects were also observed at much lower levels. Rats exposed to vanadium pentoxide aerosol of 27 $\mu g/m^3$ for 70 days developed different respiratory effects, such as marked lung congestion, focal lung haemorrhages and extensive bronchitis *(28)*. Exposure to 2 μg vanadium pentoxide per m^3 was considered to be without effect.

Exposure to vanadium oxides may alter alveolar macrophage integrity and function to the detriment of pulmonary defence. Cytotoxicity, tested on rabbit alveolar macrophages *in vitro,* was directly related to solubility in the order $V_2O_5 > V_2O_3 > VO_2$. Dissolved vanadium pentoxide (6 $\mu g/ml$) also reduced phagocytosis *(3)*. The effect of vanadium compounds on the function of alveolar macrophages may result in an impairment of the lung's resistance to secondary bacterial infection.

It has been stated that $VOCl_2$ is a mutagen for *Escherichia coli* and $VOCl_2$, V_2O_5, NH_4VO_3 are capable of damaging DNA *(30)*. There is no information that vanadium compounds have embryotoxic, teratogenic and carcinogenic effects.

Effects on humans

Both acute and chronic poisonings have been described in workers engaged in the industrial production and use of vanadium. Most of the reported clinical symptoms reflect irritative effects of vanadium on the respiratory tract.

There is insufficient evidence that vanadium causes generalized systemic effects except at extremely high concentrations *(3)*. Several observers described only vague, general signs or symptoms and reported nervous disturbances, neurasthenic or vegetative symptoms, occasionally tremors, palpitation of the heart, high incidence of extrasystoles, changes in the blood picture (anaemia, leukopenia, punctatebasophilia of the erythrocytes), reduced level of cholesterol in the blood, etc. *(16)*.

Numerous studies have reported acute and chronic respiratory effects mainly due to exposure to vanadium pentoxide *(2,3,16)*. Most of the reported clinical symptoms reflect irritative effects of vanadium on the upper respiratory tract. Only at high concentrations (above 1 mg vanadium per m^3) were more serious effects of the lower respiratory tract observed, such as bronchitis and pneumonitis. Respiratory effects after acute or chronic exposure to low levels of vanadium are summarized in Table 1.

Acute changes in lung function in workers exposed to vanadium compounds during cleaning of boilers were reported (time-weighted average concentration of respirable dust 523 $\mu g/m^3$, 15.3% of vanadium, giving a value of 80 μg vanadium per m^3) *(31)*. In spite of the fact that the workers wore respirators, the changes in lung function developed within 24 hours and pre-exposure levels had not returned by the eighth day; however, the efficacy of the respirators was low (about 9% leakage).

Respiratory effects after acute exposure to vanadium pentoxide dust were also observed in a clinical study in which healthy volunteers were exposed to levels of 100, 200 and 1000 μg vanadium pentoxide per m^3 for 8 hours *(32)*. It was found that 98% of the particles were smaller than 5 μm. Coughing, which persisted for several days, was the most characteristic symptom at all three levels of exposure.

Chronic inhalation of vanadium pentoxide dusts in industry has resulted in rhinitis, pharyngitis, bronchitis, chronic productive cough, wheezing, shortness of breath and fatigue. Pneumonitis and bronchopneumonitis have also been observed *(2,3,16)*. It has been reported that vanadium workers are more susceptible to colds and other respiratory illnesses than others.

Respiratory effects after chronic exposure to vanadium pentoxide were described in refinery workers *(33)* at an average level of 300 μg vanadium per m^3 and at a maximum concentration of about 1 mg vanadium per m^3 (92% of the particles were smaller than 0.5 μm). In comparison to controls, a statistically significant higher incidence of symptoms, described as cough with sputum production, eye, nose and throat irritation, injected pharynx and green tongue were reported.

In workers exposed for several months to vanadium-containing dust in the range of 10–40 μg vanadium per m^3 (concentrations were previously higher, 200–500 μg vanadium per m^3), a macroscopic and microscopic survey of the upper respiratory tract of 63 males was performed *(34)*. The

Table 1. Respiratory effects after acute and chronic exposures to low levels of vanadium

Type of exposure	Vanadium compound	Concentration in $\mu g/m^3$		Symptoms	Reference
		Compound	Vanadium		
Acute					
Boiler cleaning	V_2O_5 V_2O_3	523	80	Changes in parameters of lung functions	*(31)*
Clinical study (experimental 8-hour exposure)	V_2O_5	1000	560	Respiratory irritation: persistent and frequent cough, expiratory wheezes	*(32)*
	V_2O_5	200	112	Persistent cough (7–10 days)	
	V_2O_5	100	56	Slight cough for 4 days	
Chronic					
Vanadium refinery	V_2O_5	536	300	Respiratory irritation: cough, sputum, nose and throat irritation, injected pharynx	*(33)*
Vanadium refinery	V_2O_5	18-71	10-40	Irritative changes of mucous membranes of upper respiratory tract	*(34)*
Vanadium processing	V_2O_5 V_2O_3	—	1.2-12.0	Respiratory irritation: injected pharynx	*(35)*

results were compared with a control group of workers exposed to inert dust only, and matched for age and smoking habits. Microscopic examination of nasal smears revealed a statistically significant increase in neutrophils, and biopsies of nasal mucosa showed significantly elevated numbers of plasma and round cells. The histological picture was characteristic for an irritating effect of vanadium dust on the mucous membranes seen earlier in exposed laboratory animals.

One study described a significant incidence (58%) of injected pharynx in workers engaged in the refining of vanadium pentoxide from soot generated by the combustion of heavy fuel-oil *(35)*. Concentrations of vanadium in the air at various locations in the work environment were all less than 400 $\mu g/m^3$, with mean values of 1.2–12.0 $\mu g/m^3$.

Vanadium compounds appear to be capable of inducing asthma bronchiale in previously nonatopic subjects and continuing manifestations of asthmatic symptoms were reported 8 weeks after subjects had left the industry (vanadium pentoxide refinery) *(36)*.

Some epidemiological data have shown positive correlations between the vanadium content of urban air and mortality from bronchitis, pneumonia, nephritis, cancer (other than lung cancer in males) *(37)* and "heart disease" *(38)*. However, these analyses did not take into account important, probably relevant, factors such as exposure to other chemicals, smoking habits, etc. *(16)*. In addition, most of these causes of excess mortality have not been reported in workers occupationally exposed to vanadium compounds *(3)*.

Evaluation of Human Health Risks

Exposure

The natural background level of vanadium in air has been reported to range from 0.02 to 1.9 ng/m^3 in Canada *(2)*. Vanadium concentrations recorded in rural areas varied from a few nanograms to tenths of a nanogram per m^3, and in urban areas from 50 to 200 ng/m^3. In big cities during winter, when high-vanadium fuel-oil was used for heating purposes, concentrations as high as 2000 ng/m^3 have been reported. Air pollution by industrial plants may be less than that caused by power stations and heating equipment.

The concentrations of vanadium in workplace air (0.01–60 mg/m^3) are much higher than in the general environment.

Health risk evaluation

The acute and chronic effects of vanadium exposure on the respiratory system of occupationally exposed workers should be regarded as the most significant factors when establishing air quality guidelines. Most of the clinical symptoms reported reflect irritative effects of vanadium on the upper respiratory tract, except at higher concentrations (above 1 mg vanadium per m^3), when more serious effects on the lower respiratory tract are observed. Clinical symptoms of acute exposure are reported *(31)* in workers exposed to concentrations ranging from 80 $\mu g/m^3$ to several mg vanadium per m^3, and in healthy volunteers *(32)* exposed to concentrations of 56–560 μg vanadium per m^3 (Table 1).

A study of occupationally exposed groups provides data reasonably consistent with those obtained from controlled acute human exposure experiments, suggesting that the lowest-observed-adverse-effect level for acute exposure can be considered to be 60 μg vanadium per m^3.

Chronic exposure to vanadium compounds revealed a continuum in the respiratory effects, ranging from slight changes in the upper respiratory tract, with irritation, coughing and injection of pharynx, detectable at 20 μg vanadium per m^3, to more serious effects such as chronic bronchitis and pneumonitis, which occurred at levels above 1 mg/m^3. Occupational studies illustrate the concentration-effect relationship at low levels of exposure *(33–35)*, showing increased prevalence of irritative symptoms of the upper respiratory tract; this suggests that 20 μg vanadium per m^3 can be regarded as the lowest-observed-adverse-effect level for chronic exposure (Table 1). There are no conclusive data on the health effects of exposure to airborne vanadium at present concentrations in the general population, and a susceptible subpopulation is not known. However, vanadium is a potent respiratory irritant, which would suggest that asthmatics should be considered a special group at risk.

There are no well documented animal data to support findings in human studies, although one study reported systemic and local respiratory effects in rats at levels of 3.4–15 μg/m^3 *(28)*.

Guidelines

Available data from occupational studies suggest that the lowest-observed-adverse-effect level of vanadium can be assumed to be 20 μg/m^3, based on chronic upper respiratory tract symptoms. Since the adverse nature of the observed effects on the upper respiratory tract were minimal at this concentration and a susceptible subpopulation has not been identified, a protection factor of 20 was selected. It is believed that below 1 μg/m^3 (averaging time 24 hours) environmental exposure to vanadium is not likely to have adverse effects on health.

The available evidence indicates that the current vanadium levels generally found in industrialized countries are not in the range associated with potentially harmful effects.

References

1. **Erdmann, E. et al.** Vanadate and its significance in biochemistry and pharmacology. *Biochemical pharmacology,* **33**: 945–950 (1984).
2. **Committee on Biologic Effects of Atmospheric Pollutants.** *Vanadium.* Washington, DC, National Academy of Sciences, 1974.
3. **Waters, M.D.** Toxicology of vanadium. *In:* Goyer, R.A. & Mehlman, M.A., ed. *Advances in modern toxicology. Vol. 2. Toxicology of trace elements.* New York, Wiley, 1977, pp. 147–189.
4. **Bengtsson, S. & Tyler, G.** *Vanadium in the environment.* London, University of London Monitoring and Assessment Research Centre, 1976 (MARC Report No. 2).

5. *Mineral commodity summaries.* Washington, DC, US Department of the Interior, Bureau of Mines, Division of Ferrous Metals, 1983.
6. **Galloway, J.N. et al., ed.** *Toxic substances in atmospheric deposition: a review and assessment.* 1980 (National Atmospheric Deposition Program Report NC-141).
7. **Anderson, D.** *Emission factors for trace substances.* Research Triangle Park, NC, US Environmental Protection Agency, 1973 (Report No. EPA-450/2-73-001).
8. **Davies, D.J.A. & Bennett, B.G.** *Exposure commitment assessments of environmental pollutants.* London, University of London Monitoring Assessment and Research Centre, 1983, Vol. 3 (MARC Report No. 30).
9. *Scientific and technical assessment report on vanadium.* Washington, DC, US Environmental Protection Agency, 1977 (Report No. EPA-600/6-77-002).
10. **Kretzschmar, J.D. et al.** Heavy metal levels in Belgium: a five-year survey. *Science of the total environment,* **14**: 85–97 (1980).
11. **Lahmann, E. et al.** *Heavy metals: identification of air quality and environmental problems in the European Community.* Luxembourg, Commission of the European Communities, 1986, Vol. 1 & 2 (Report No. EUR 10678 EN/I and EUR 10678 EN/II).
12. **Paznych, V.M.** *In:* [*Hygiene of population aggregates*]. Kiev, 1977, pp. 99–100 (Republican interagency collected volume) (in Russian).
13. **Aboulafia, J. et al.** Emissions de vanadium par les installations thermiques des raffineries de pétrole [Vanadium emissions from thermal installations of petroleum refineries]. *Pollution atmosphérique,* **101**: 13–20 (1984).
14. **Roščin, I.V.** [*Vanadium and its compounds*]. Moscow, Medicina, 1968, pp. 139–146 (in Russian).
15. **Williams, N.** Vanadium poisoning from cleaning oil-fired boilers. *British journal of industrial medicine,* **9**: 50–55 (1952).
16. **Vouk, V.** Vanadium. *In:* Friberg, L. et al., ed. *Handbook on the toxicology of metals.* Amsterdam, Elsevier-North Holland Biomedical Press, 1979, pp. 659–674.
17. **Myron, D.R. et al.** Vanadium content of selected foods as determined by flameless atomic absorption spectroscopy. *Journal of agricultural and food chemistry,* **25**: 297–300 (1977).
18. **Myron, D.R. et al.** Intake of nickel and vanadium by humans. A survey of selected diets. *American journal of clinical nutrition,* **31**: 527–531 (1978).
19. **Byrne, A.R. & Kosta, L.** Vanadium in foods and in human body fluids and tissues. *Science of the total environment,* **10**: 17–30 (1978).
20. **Sabbioni, E. et al.** Biliary excretion of vanadium in rats. *Toxicological European research,* **3**: 93–98 (1981).
21. **Hansen, T.V. et al.** The effect of chelating agents on vanadium distribution in the rat body and on uptake by human erythrocytes. *Archives of toxicology,* **50**: 195–202 (1982).
22. **Sabbioni, E. & Marafante, E.** Metabolic patterns of vanadium in the rat. *Bioinorganic chemistry,* **9**: 389–407 (1978).

23. **Söremark, R. et al.** Autoradiographic localization of V-48-labelled vanadium pentoxide (V_2O_5) in developing teeth and bones in rats. *Acta odontologica scandinavica,* **20**: 225–232 (1962).
24. **Tipton, I.H. et al.** Patterns of elemental excretion in long-term balance studies. *Health physics,* **16**: 455–462 (1969).
25. **Jandhyala, B.S. & Hom, G.L.** Physiological and pharmacological properties of vanadium. *Life sciences,* **33**: 1325–1340 (1983).
26. **Stacey, N.H. & Klaassen, C.D.** Inhibition of lipid peroxidation without prevention of cellular injury in isolated rat hepatocytes. *Toxicology and applied pharmacology,* **58**: 8–18 (1981).
27. **Venugopal, B. & Luckey, T.D.** *Metal toxicity in mammals. 2. Chemical toxicity of metals and metalloids.* New York, Plenum Press, 1978, pp. 220–226.
28. **Paznych, V.M.** [Maximum permissible concentration of vanadium pentoxide in the atmosphere]. *Gigiena i sanitarija,* **31**: 6–12 (1966) (in Russian).
29. **Roščin, I.V.** Vanadium. *In:* Izraelson, Z.I., ed. *Toxicology of the rare metals.* Jerusalem, Israel Program for Scientific Translations, 1967, pp. 52–59 (AEC-tr-6710).
30. **Kanematsu, N. & Kada, T.** Mutagenicity of metal compounds. *Mutation research,* **54**: 215–216 (1978).
31. **Lees, R.E.M.** Changes in lung function after exposure to vanadium compounds in fuel oil ash. *British journal of industrial medicine,* **37**: 253–256 (1980).
32. **Zenz, C. & Berg, B.A.** Human responses to controlled vanadium pentoxide exposure. *Archives of environmental health,* **14**: 709–712 (1967).
33. **Lewis, C.E.** The biological effects of vanadium. II. The signs and symptoms of occupational vanadium exposure. *AMA archives of industrial health,* **19**: 497–503 (1959).
34. **Kiviluoto, M. et al.** Effects of vanadium on the upper respiratory tract of workers in a vanadium factory. *Scandinavian journal of work, environment & health,* **5**: 50–58 (1979).
35. **Nishiyama, K. et al.** [A survey of workers on vanadium pentoxide]. *Shikoku igaku zasshi,* **31**: 389–393 (1977) (in Japanese).
36. **Musk, A.W. & Tees, J.G.** Asthma caused by occupational exposure to vanadium compounds. *Medical journal of Australia,* **1**: 183–184 (1982).
37. **Stocks, P.** On the relations between atmospheric pollution in urban and rural localities and mortality from cancer, bronchitis and pneumonia, with particular reference to 3,4-benzopyrene, beryllium, molybdenum, vanadium and arsenic. *British journal of cancer,* **14**: 397–418 (1960).
38. **Hickey, R.J. et al.** Relationship between air pollution and certain chronic disease death rates — multivariate statistical studies. *Archives of environmental health,* **15**: 728–738 (1967).

Part IV

Effects of inorganic substances on vegetation

32

The effects of nitrogen on vegetation

Atmospheres that are polluted with nitrogen oxides contain both nitric oxide (NO) and nitrogen dioxide (NO_2), often referred to together as NO_x. The quantitatively dominant nitrogen oxide in the atmosphere, produced by denitrification processes, is dinitrogen oxide (N_2O), which normally occurs at levels around $500 \mu g/m^3$ *(1)*. This gas is inert and insoluble in water. Dinitrogen oxide is normally not included when the abbreviation NO_x is used and it does not play a significant role in the context of tropospheric pollution by other oxides of nitrogen. In the stratosphere, however, dinitrogen oxide contributes to ozone depletion through photochemical reactions, and with time the upward trend of dinitrogen oxide levels, partly due to the increased use of agricultural fertilizers, may cause global climatic changes and increase the perceived amount of solar ultraviolet radiation.

Over half of the ammonia (NH_3) in the atmosphere is produced by the decomposition of the nitrogen-containing organic material arising from domestic animal husbandry. The remainder results from combustion processes or the production of agricultural fertilizers and amounted to a release of approximately 3.4×10^6 tonnes of nitrogen per year over Europe in 1977 *(2)*. Ammonia may have direct effects upon vegetation, but the ionic product in solution (ammonium) contributes significantly to nitrogen-based wet deposition and accounts for the bulk of reduced nitrogen returning to ecosystems. Ammonia has been suggested as an additional component contributing to the recent forest decline in Europe *(3)*.

Uptake of Nitrogen-based Air Pollutants by Vegetation

Wet and dry deposition of nitrogen oxides are important factors in the redistribution of nitrogen in the environment *(4)*, but the latter is more difficult to estimate than that of particulate nitrogen compounds in the atmosphere. Most understanding of mechanisms of dry deposition of pollutant gases has been obtained from work on sulfur dioxide and ozone. Van Aalst *(5)* supports the view that dry deposition of NO_x onto plants is a major component of NO_x dry deposition to all natural surfaces.

Total deposition of nitrogen-based atmospheric compounds is the sum of both dry and wet deposition. The deposition processes involved are strongly dependent upon climatic conditions and properties of the receiving surfaces. Determination of the total deposition may be done either by calculation (based on concentration levels and mathematical description of the deposition processes) or by direct measurements. The latter demand the application of a number of methods, together covering both ranges of dry and wet deposition rates. One approach consists of describing the accumulation in, and fluxes through, ecosystem compartments by analysing levels of nitrogen compounds in a time-series.

The effective dose of nitrogen oxides that a plant receives is determined by the rate of pollutant entry into the leaves. Consequently, there has been considerable interest in the net fluxes of nitrogen oxides into plants. Actual measurements of the net flux of nitrogen oxides into plants are scarce, although deposition velocities onto different surfaces and estimates of uptake by plant canopies have been measured in the field *(5–7)*. Nevertheless, some useful work has also been done using realistic levels found in urban atmospheres. For example, Freer-Smith *(8)* measured nitrogen dioxide deposition velocities of 2.89 mm/second and 3.18 mm/second onto poplar and birch trees respectively, using concentrations of less than $114 \mu g/m^3$ (0.06 ppm). Bengtson et al. *(7)*, using a range of concentrations likely to be experienced in the field, i.e. $95–510 \mu g/m^3$ (0.05–0.27 ppm), found that nitrogen dioxide fluxes were three times those of nitric oxide whereas Bennett et al. *(9)* quoted steady-state uptake rates for nitrogen dioxide and nitric oxide as 0.005 and 0.0003 ppm/minute per m^2 respectively — a 16-fold difference. Other work has shown that fluxes of nitrogen oxides are related to the exposure concentration of the gas and that this relationship is linear *(7,8)*.

The first resistance encountered by nitrogen oxides entering a leaf is the aerodynamic (boundary layer) resistance. Unstirred air layers on both sides of a leaf represent important barriers to the entry of nitrogen oxides and movement across such layers is by diffusion of gas molecules in response to differences in concentration. This boundary layer resistance varies with both wind speed and various leaf characteristics such as size, shape, orientation and development. As wind speed increases, resistance to the movement of gaseous molecules falls, uptake increases, and in this way wind speed can alter the rate of pollutant uptake *(10)*. Furthermore, the boundary layer is generally thinner at the edges of a leaf than at the centre, which might account for the increased pollutant damage often seen at leaf margins, especially on graminaceous species, where a steeper concentration gradient exists *(11)*.

Although epidermal cells occupy a much greater proportion of a leaf surface than do stomatal pores, the waxy cuticle covering them greatly increases the resistance to gases like sulfur dioxide and ozone. As a consequence, the cuticular resistance to them is generally very considerable, and much greater than that offered by the stomata and internal air spaces. Nevertheless, nitrogen oxides deposited on leaf and stem surfaces may dissociate in the water film and/or react with cuticular wax components, leading to damaged cuticular surfaces *(8)*. More recent unpublished studies have also shown that significant amounts of nitrogen dioxide may enter leaves by penetrating the cuticle, which means that closure of the stomata may reduce the dose of nitrogen dioxide received by a plant but not exclude damage or injury, unlike the case with sulfur dioxide or ozone. As a general rule, a plant responds less to nitrogen dioxide under conditions which cause stomatal closure, such as low light, humidity and nitrogen status *(12)*.

At high and acutely damaging concentrations, however, nitrogen dioxide can have a more harmful effect at night than a load of similar strength during the daylight period *(13)*. Raised concentrations of nitrites in chloroplasts can be encountered, despite comparatively low NO_x uptake. While rapid breakdown of the nitrite into ammonia ensues in chloroplasts during the day, as a result of sufficient production of adenosine triphosphate and $NADPH_2$ from photosynthesis, at night the activity of nitrite reductase is very sharply limited, owing *inter alia* to the lack of energy- and reduction-equivalents *(14,15)*.

Use of $^{15}NO_2$ has established that plants can remove nitrogen oxides from the air. However, Mansfield & Freer-Smith *(16)* point out that to enter a plant cell, a gaseous pollutant must pass through the extracellular water contained in the cell wall. The solubility of a gas is thus an important factor in determining the rate at which it is taken up. Zeevaart *(17)* suggested that nitrogen dioxide entering a leaf dissolves in the extracellular water to form nitrous and nitric acids which, by dissociation, form nitrite and nitrate ions. Later, Anderson & Mansfield *(18)* found that nitric oxide was more soluble in xylem sap than in distilled water. Since xylem sap is continuous with the extracellular water in a leaf, an enhanced solubility of nitric oxide is to be expected compared with that predicted by water solubility figures alone.

The effectiveness of plant metabolism in assimilating or transforming the dissociation products of nitrogen oxides (i.e. nitrite and nitrate ions) alters with the capacity of the leaf to take up nitrogen dioxide. The importance of mesophyll resistance in regulating nitrogen dioxide uptake has received little attention because of the difficulties involved in accurate measurement *(19,20)*. By deduction, Srivastava et al. *(21)* have implicated mesophyll resistance to nitrogen dioxide flux as the factor responsible for increased leaf tolerance over time to this pollutant gas. Furthermore, the effectiveness of the plant metabolism to assimilate or transform the products of the dissociation of nitrogen oxides could alter the leaf capacity for uptake of nitrogen oxides. This has been referred to as the residual internal resistance *(12)*.

Bennett et al. *(9)* found that nitrogen oxides were absorbed most efficiently by foliage near the top of the plant canopy, where light intensity and metabolic rates are highest. There is a body of evidence to show that the products of nitrogen oxides in plant leaves (nitrite and nitrate) are metabolized by normal mechanisms of nitrogen metabolism *(17,22)*. Consequently, they may act as supplementary foliar nitrogen fertilizers at low levels *(23)* and, when in excess, they may cause problems of toxicity and injury to plants. Amundsen & McClean *(24)* suggest that several woody species may be particularly susceptible to injury by nitrogen oxides because the leaves of such species lack the necessary enzymes to metabolize and detoxify the nitrate and nitrite formed by absorption of nitrogen oxides.

Effects on Plants and Plant Communities

Effects of low levels of NO_x alone

Most of the early studies on the effects of nitrogen oxides employed concentrations (greater than 1 ppm) that were much too high. However, recent investigations using lower doses, or doses near ambient levels, have been carried out to relate low-level fumigations more closely to field situations. For example, there have been some reports of the effects of lower doses of nitrogen dioxide on plant growth, often associated with work on pollutant mixtures *(9,11,20)*

Vegetation is recognized as a sink for atmospheric pollution, but some pollutants are phytotoxic *(9,25)* while some can be "beneficial" to plant growth *(18)*. Nitrogen oxides can act in different ways, depending on the dosage to plants. As a consequence, dose–response curves such as the one produced by Mansfield et al. *(26)* are valuable for understanding the effects of nitrogen oxides. There have been reports in the literature on the "beneficial" effects of nitrogen oxides on plant growth. There is much evidence that low concentrations of nitrogen dioxide below "toxicity thresholds" may stimulate growth, and such stimulation will not always be desirable for plants in the field. For example, susceptibility to insect attack or environmental stress such as frost may be increased. Mansfield et al. *(26)* demonstrated a stimulation of growth of *Phleum pratense* by a low dose of nitrogen dioxide (0.068 ppm for 20 weeks). Another grass, *Poa pratensis,* at first showed a similar increase when exposed to 0.062 ppm for 11 months (weekly means), although later in the exposure period polluted plants showed a growth reduction compared with controls *(27)*. Out of 6 broad-leaved trees exposed to a nitrogen dioxide concentration of 0.062 ppm for 150 days, only silver birch showed significant increases in growth after exposure, although other species exhibited only a slight stimulation in growth compared with controls *(27)*. However, Kress et al. *(28)* examined the effect of 28 days' exposure of American sycamore seedlings to a nitrogen dioxide concentration of 0.1 ppm and found no significant effects on growth throughout the exposure; however, polluted plants grew significantly better than controls after the exposure ceased. In conclusion, it appears that the response of plants to pollution by nitrogen oxides is species-dependent (or even variety-dependent) as well as dose-dependent.

Growth reductions caused by low concentrations of nitrogen oxides also vary between closely related species. Ashenden & Mansfield *(29)* found that the relative sensitivity of four grasses, *Lolium multiflorum, Phleum pratense, Dactylis glomerata* and *Poa pratensis,* was different. The last two species showed growth reductions at the end of the exposure period when exposed to 0.068 ppm for 20 weeks, whereas there was no significant effect on the growth of the first two species. During the 20 weeks of exposure, growth measurements were made and it was found that the response altered with the duration of the exposure and hence the dose *(25)*. The response of *Poa pratensis* to a nitrogen dioxide concentration of 190μg/m^3 (0.1 ppm) for 104 hours per week was further investigated by Whitmore & Freer-Smith *(27)*, who found a 55% reduction in the dry weight of polluted plants.

The amount of damage suffered by a plant varies according to a number of factors such as the concentration of the pollutant, length of exposure, plant age, edaphic factors, light and humidity. Symptoms are often divided into "invisible" (or hidden) injury and "visible" injury. The former type is where there is an overall reduction in growth but no obvious signs of damage; it is often associated with decreases in photosynthesis and transpiration. Actual visible injury by nitrogen oxides is extremely rare and produced only by very high concentrations, typified by chlorotic areas appearing on the leaves, with some necrosis.

Effects of nitrogen oxides in combination with other pollutants

The atmosphere often contains a mixture of pollutant gases, each of which is known to be phytotoxic. Some of these, like ozone, are secondary, in the sense that they are formed as a consequence of reactions involving emissions of primary pollutants such as nitric oxide and sulfur dioxide. Surveys in western Europe have shown that mixtures of nitrogen-based pollutants as well as sulfur dioxide and ozone often occur and that concentrations of NO_x may approach or exceed those of sulfur dioxide *(30)*. Ozone is more spasmodic in occurrence than the other two and is normally associated with the higher temperatures and light intensities of summer months, whereas concentrations of both sulfur dioxide and nitrogen-based air pollutants are highest in the winter. Comparisons between plants grown in untreated urban air and those grown in atmospheres from which all such pollutants were filtered have invariably shown that higher levels of toxicity were involved than could be found in experimental fumigations with single pollutants such as sulfur dioxide *(31)*. Recently a number of reviews of their combined damaging effects have appeared *(16)*, but the summarized responses demonstrate that the nature of the effect may be additive, synergistic (i.e. more than additive), or antagonistic.

In the case of fumigations using sulfur dioxide plus nitrogen dioxide, synergistic damage is normally detected. The consequences of this are better illustrated in terms of potential growth losses, demonstrated by a series of long-term mixed pollutant studies of grasses (see Table 1).

Foliar injury has often been caused by mixtures of sulfur dioxide and nitrogen dioxide at threshold concentrations much lower than for either pollutant alone, but the major consequence of pollutant mixtures is reflected in reductions of plant growth especially during the winter which are similar in magnitude to those found in comparative studies with unfiltered and filtered urban air.

The synergistic effects of nitrogen dioxide and sulfur dioxide at similar concentration ranges in damaging plants have been well established for short-term exposures but recent experimental evidence suggests that synergism of the same kind can also occur during long-term treatment *(33)*. Dose–response curves for $NO_2 + SO_2$ treatments for plants fumigated in

Table 1. Increase or reduction (relative to controls in clean air) in the growth of four grasses exposed for 140 days in winter to atmospheres containing nitrogen dioxide, sulfur dioxide or both

Plant and parameter	Increase or reduction (%) in growth after exposure to:		
	NO_2 (0.068 ppm)	SO_2 (0.068 ppm)	$NO_2 + SO_2$ (0.068 ppm) + (0.068 ppm)
Dactylis glomerata L.			
leaf area	+21	− 5	−72
dry weight of green leaves	− 7	−28	−83
dry weight of roots	−11	−37	−85
Poa pratensis L.			
leaf area	−17	−28	−84
dry weight of green leaves	−29	−39	−88
dry weight of roots	−17	−28	−91
Lolium multiflorum Lam.			
leaf area	+ 1	−22	−43
dry weight of green leaves	−10	−28	−65
dry weight of roots	+35	+ 7	−58
Phleum pratense L.			
leaf area	+30	−11	−82
dry weight of green leaves	+14	−25	−84
dry weight of roots	+ 1	−58	−92

Source: Ashenden, T.W. & Williams, I.A.D. *(25)* and Ashenden, T.W. *(32)*.

controlled environments have been established and suggest that the concentration of nitrogen dioxide above which injury occurs is between 0.02 and 0.07 ppm. Studies so far have been performed only on a few plant species but there is evidence that environmental conditions can greatly alter the response to $NO_2 + SO_2$. Thus concentrations that are damaging in winter may have little effect in summer or may even promote growth *(33)*.

Some recent experiments in which ozone has been added to mixtures of nitrogen dioxide and sulfur dioxide *(33)* suggest that the threshold for injury may be as low as 28.5 μg nitrogen dioxide per m^3 (0.015 ppm) when there are similar amounts of sulfur dioxide (40 $\mu g/m^3$ or 0.015 ppm) and ozone (60 $\mu g/m^3$ or 0.03 ppm).

Peak concentrations of nitrogen dioxide can have adverse effects on plants if sulfur dioxide is also present (bearing in mind that ozone levels are usually low during nitrogen dioxide peaks owing to scavenging of ozone by nitric oxide). A study by the Dutch National Health Council *(34)* led to the conclusion that sensitive plants are adequately protected from the adverse effects of nitrogen dioxide if the average concentration over four hours does not exceed 95 $\mu g/m^3$ (in the presence of similar concentrations of sulfur dioxide).

A number of studies have been undertaken to evaluate the possible physiological, biochemical and metabolic changes that may occur to account for these losses in plant productivity *(35)* due to exposure to mixed pollutants especially sulfur dioxide, ozone and nitrogen dioxide. Under the effect of nitrogen dioxide combined with sulfur dioxide or ozone, inactivation of the nitrite being formed in plants can be affected. While nitrite reductase activity commonly rises under the effect of nitrogen dioxide alone, the simultaneous presence of sulfur dioxide or ozone can limit or even prevent this substrate-related induction *(36,37)*. The loss of this potential nitrite inactivation mechanism must be regarded as one main reason for the increased harmful effect of nitrogen dioxide in conjunction with sulfur dioxide and/or ozone, while another is the formation of free radicals.

Wet deposition of nitrogen-based air pollutants and effects on sensitive ecosystems

In Europe it is likely that nitrogen-based air pollution contributes around 30% of the acidic wet precipitation. With heavy snowfall this contribution of nitrate can be particularly high when the snow melts. Because this often coincides with sensitive development stages of plants, wet deposition injury can be irreversible. Furthermore, nitrogen-based air pollutants can contribute to lower pH in poorly buffered soils and increased leaching of nutrients, release of metallic ions such as Al^{3+}, and root injury. Reduced growth and increased susceptibility to abiotic and biotic stresses may be the result *(33)*. In the forest ecosystems of Europe, nitrogen is often a factor limiting plant growth. Low levels of nitrogen input may enhance growth, but above these values the relationships between various plant nutrients may be disrupted, thereby adversely affecting the availability of nutrients, reducing plant growth and increasing susceptibility to abiotic and biotic stress factors.

Although deposition of nitrogen dioxide onto large bodies of water results in only minor transitory concentrations of nitrite in the surface film which are rapidly diluted by turbulent mixing, it is possible that in some oligotrophic lakes atmospheric nitrogen dioxide does result in enhanced nitrate concentrations, but these are likely to be smaller because nitrate will be largely retained by soils. However, when snows melt, nitrate falling on the catchment area may contribute in a significant manner to the concentration of nitrate in lakes.

The ecosystems most sensitive to nitrogen-based atmospheric compounds are those where plants with particularly high sensitivity to gases such as nitric oxide, nitrogen dioxide and ammonia predominate and are normally confined to oligotrophic conditions. Plant sensitivity to nitric oxide and nitrogen dioxide has probably been underestimated and future research should reveal to what extent these gases are involved in, for example, recent forest decline. Ammonia is also phytotoxic, but reaches levels sufficiently high to cause direct effects only locally. Ecosystems that are considered most threatened, in relation to eutrophication by nitrogen compounds, are the following.

Wetland ecosystem types

(*a*) Ombrotrophic[a] mires (e.g. raised bogs)

(*b*) Mires in granitic/gneissic regions

(*c*) Mires otherwise nitrogen-limited

Lakes

(*a*) Clear-water lakes with isoetid plant cover (*Lobelia dortmanna, Isoëte* spp., *Littorella uniflora,* etc.)

(*b*) Lakes with a high number of *Potamogeton* species, often dystrophic

Terrestrial ecosystems

(*a*) Heathlands with a high proportion of lichen cover

(*b*) Low meadow vegetation types on slopes and hills with different substrate characteristics, without added artificial fertilizer, and used for extensive grazing and haymaking

(*c*) Coniferous forests, especially those at higher altitudes

The effects observed in such oligotrophic systems are mainly due to increased competition from faster-growing plant species that are normally restricted in these ecosystems because of their low nitrogen nutrient content. Furthermore, increased mineralization rates caused by changes in carbon/nitrogen ratios may accentuate the eutrophication process and thereby

---

[a] Ombrotrophic = supplied with nutrients from rainfall only.

enhance the vegetational changes. This may ultimately lead to the disappearance of plant species originally characteristic of such ecosystems, with a loss not only of more species but also of irreplaceable genetic resources.

In the case of mires, a shift in the *Sphagnum* species composition is observed and, owing to increased mineralization, seed plants become more dominant. With respect to lakes, the effects are primarily observed as increased epiphytic growth and predominance of planktonic communities. Heathlands change to more species-poor vegetation with, for example, a strong dominance of *Empetrum nigrum,* while meadow vegetation is threatened by an increased establishment rate of nitrogen-demanding tall perennials such as *Cirsium, Carduus, Urtica* and *Chamaenerion* species.

The vegetational changes mentioned above have effects on animal populations as well. Examples of this are fish death due to oxygen depletion by plants and explosive attacks of bark beetles in damaged forest stands.

The major wetland plant ecosystems at risk from rising levels of nitrogen-based air pollutants are ombrotrophic mires composed of slow-growing species that have low nutrient requirements. The bryophytes are especially endangered. There can be little doubt that the present deposition of combined nitrogen from the atmosphere onto the Pennines region of the United Kingdom is optimal for the growth of *Sphagnum* species. Inhibition of *Sphagnum,* often the dominant plant of unpolluted ombrotrophic mires, occurs at a nitrogen dioxide concentration of $11.7 \mu g/m^3$, above which deposition of nitrogen dioxide may also contribute to ammonium toxicity. Thus, the magnitude of any effects of wet deposition of NO_x on sensitive ecosystems such as forests and ombrotrophic mires is uncertain and should not be considered in isolation from the effects of deposition of other pollutants, such as sulfate *(33).*

Input of ammonia and ammonium into forests may lead to enhanced soil acidification or, in the absence of nitrification (generally due to low pH values), to oversaturation with ammonia. Since plant roots cannot discriminate between different cations, the latter will lead to enhanced uptake of ammonium ions and reduced uptake of other cation nutrients (Ca^{++}, Mg^{++}, K^+), and cause tree damage through a deficiency of these nutrients.

Tolerance of Nitrogen-based Air Pollution

It is no longer in doubt that nitrogen-based pollution of the atmosphere can damage vegetation, especially in combination with other pollutants. By using normal pathways of nitrogen metabolism and natural means of adaptation, most plants attempt to detoxify and use the additional nitrogen from the atmosphere. In the longer term, these adjustments are often inadequate. The net costs to the plant in energy terms of repair and maintenance of detoxification are reflected in reduced growth and productivity. Those plant species that can tolerate and effectively scavenge free radicals (which are more likely in mixed fumigations) are likely to be more successful in the semiurban environment.

Evolution towards more tolerant species is undoubtedly taking place naturally. This is, of course, only valid for plants with a rapid generation sequence. In agriculture and horticulture, where a significant loss of productivity and amenity in certain species occurs, this process could be speeded up by collective breeding. Vegetation has a vast genetic potential for tolerance of environmental stresses. One of the most frequent limitations to plant growth besides lack of light and water is an inadequate nitrogen supply. A wide range of inherent adaptations and homeostatic readjustments take place in most plants to ensure maximum utilization of nitrogen under given circumstances. Some strains are better adapted than others and survive, while those less able to do so are eliminated. Resistance to or tolerance of pollutants is a well known phenomenon, especially in studies of plant sensitivity to sulfur dioxide.

The selection of tolerant strains can thus be extremely useful for agricultural purposes. However, to protect sensitive natural ecosystems, forests, and most agricultural crop species adequately against the adverse effects of nitrogen-based pollutants, the levels of these pollutants should be reduced to the proposed values specified below.

Evaluation of Ecological Effects

Nitrogen-containing compounds in the atmosphere are numerous; therefore, the evaluation of effects must take into account the various nitrogen-containing compounds as well as combinations with sulfur dioxide and ozone, which are more likely to occur together in Europe than elsewhere in the world.

Nitrogen dioxide is the most phytotoxic oxide of nitrogen. Foliar injury is often caused by mixtures of nitrogen dioxide with sulfur dioxide and/or ozone at threshold concentrations much lower than those for individual pollutants. However, the main consequence of pollutant mixtures is a reduction of plant growth.

On the basis of studies in which ozone has been added to nitrogen dioxide/sulfur dioxide, it can be concluded that, in order to protect sensitive plants from direct effects of nitrogen dioxide (in the presence of levels of sulfur dioxide and ozone not higher than $30 \mu g/m^3$ and $60 \mu g/m^3$, respectively), the nitrogen dioxide concentration should not exceed $30 \mu g/m^3$ as a yearly average of 24-hour mean values.

Limiting the one-year average level of a pollutant does not protect the environment effectively against peak values. Therefore, peak concentrations should also be limited. Sensitive plants are protected against the adverse effects of nitrogen dioxide if the average concentration over 4 hours does not exceed $95 \mu g/m^3$ (in the presence of similar concentrations of sulfur dioxide) *(34)*.

It is not possible to define no-effect levels of nitrogen-based air pollutants with respect to deposition, because even small changes in nitrogen fluxes in some ecosystems may induce major changes. Therefore, guidelines for total nitrogen deposition would be an effective approach for the protection of sensitive ecosystems. Unfortunately, methods for determining total nitrogen deposition are not widely available at present.

Oligotrophic ecosystems are threatened by the increasing deposition of nitrogen, and ombrotrophic mires are among the most sensitive ecosystems with regard to nitrogen-based air pollutants.

It has been demonstrated that, for the survival of the most sensitive wetland communities (ombrotrophic mires), total inputs of nitrogen from all sources should be below 3 g/m^2 per year. To achieve protection in these areas present rates of both wet and dry nitrogen deposition should be reduced *(33)*. It is difficult to relate atmospheric concentrations of different nitrogen-based atmospheric pollutants precisely to inputs of nitrogen into such systems, but it has been suggested that long-term levels of nitrogen dioxide should be below 10 μg/m^3 *(33)*. Levels of nitric oxide and ammonia should also be similarly restricted, although there are fewer data on which to base predictions.

Wet deposition of ammonium, nitrate and organic nitrogen may also contribute significantly to the eutrophication of these systems. If the approximate ratio of dry to wet deposition is taken to be 2 : 1 *(38)*, then inputs of nitrogen by wet deposition into ombrotrophic mires should be below 1 g/m^2 per year.

Guidelines

In the presence of levels of sulfur dioxide and ozone not higher than 30 μg/m^3 and 60 μg/m^3 respectively, the atmospheric concentration of nitrogen dioxide should be no higher than 30 μg/m^3 as a yearly average of 24-hour means and no higher than 95 μg/m^3 as a 4-hour average.

In order to protect sensitive ecosystems, the total nitrogen deposition should not exceed 3 g/m^2 per year.

References

1. **Bowen, H.J.M.** *Trace elements in the environment.* New York, Academic Press, 1982.
2. **Söderlund, R.** NO_x pollutants and ammonia emissions — a mass balance for the atmosphere over NW Europe. *Ambio,* **6**: 118–122 (1977).
3. **Nihlgård, B.** The ammonium hypothesis — an additional explanation to the forest dieback in Europe. *Ambio,* **14**: 2–8 (1985).
4. **Söderlund, R.** Dry and wet deposition of nitrogen compounds. *In:* Clark, F.E. & Rosswall, T., ed. *Terrestrial nitrogen cycles. Ecological bulletin (Stockholm),* **33**: 123–130 (1981).
5. **van Aalst, R.M.** Dry deposition of NO_x. *In:* Schneider, T. & Grant, L., ed. *Air pollution by nitrogen oxides.* Amsterdam, Elsevier, 1982, pp. 263–270.
6. **Fowler, D.** Dry deposition of SO_2 on agricultural crops. *Atmospheric environment,* **12**: 369–373 (1978).
7. **Bengtson, C. et al.** Deposition of nitrogen oxides to Scots pine (*Pinus sylvestris* L.). *In:* Drablos, D. & Tollan, A., ed. *Ecological impact of acid precipitation.* Oslo, SNSF, 1980, pp. 154–155.
8. **Freer-Smith, P.H.** *Chronic pollution injury to some tree species in response to SO_2 and NO_2 mixtures.* Thesis, University of Lancaster, 1983.

9. **Bennett, J.H. et al.** Acute effects of combination of sulfur dioxide and nitrogen dioxide on plants. *Environmental pollution,* **9**: 127–132 (1975).

10. **Ashenden, T.W. & Mansfield, T.A.** Influence of wind speed on the sensitivity of ryegrass to SO_2. *Journal of experimental botany,* **28**: 729–725 (1977).

11. **Taylor, O.C. & Eaton, F.M.** Suppression of plant growth by nitrogen dioxide. *Plant physiology,* **41**: 132–135 (1966).

12. **Law, R.M. & Mansfield, T.A.** Oxides of nitrogen and the greenhouse atmosphere. *In:* Unsworth, M.H. & Ormrod, D.P., ed. *Effects of gaseous air pollution in agriculture and horticulture.* London, Butterworth, 1982, pp. 93–112.

13. **van Haut, H. & Stratmann, H.** Experimentelle Untersuchungen über die Wirkung von Stickstoffdioxid auf Pflanzen [Experimental studies on the effect of nitrogen dioxide in plants]. *In: Schriftenreihe der Landesanstalt für Immissions- und Bodennutzungsschutz des Landes Nordrhein-Westfalen, No. 7.* Essen, Verlag W. Girardet, 1967, pp. 50–70.

14. **Kaji, M. et al.** Absorption of atmospheric NO_2 by plants and soils. VI. Transformation of NO_2 absorbed in the leaves and transfer of the nitrogen through the plants. *In: Studies on the effects of air pollutants on plants and mechanisms of phytotoxicity.* Tsukuba, Ibaraki, Japan, National Institute for Environmental Studies, 1980, pp. 51–58 (Research report No. 11).

15. **Yoneyama, T. & Sasakawa, H.** Transformation of atmospheric NO_2 absorbed in spinach leaves. *Plant cell physiology,* **20**: 263–266 (1979).

16. **Mansfield, T.A. & Freer-Smith, P.H.** Effects of urban air pollution on plant growth. *Biological review,* **56**: 343–368 (1981).

17. **Zeevaart, A.J.** Some effects of fumigating plants for short periods with NO_2. *Environmental pollution,* **11**: 97–107 (1976).

18. **Anderson, L.S. & Mansfield, T.A.** The effects of nitric oxide pollution on the growth of tomato. *Environmental pollution,* **20**: 113–121 (1979).

19. **Murray, A.J.S.** Light affects the deposition of NO_2 to the *Flacca* mutant of tomato without affecting the rate of transpiration. *New phytologist,* **98**: 447–450 (1984).

20. **Capron, T.M. & Mansfield, T.A.** Inhibition of net photosynthesis in tomato in air polluted with NO and NO_2. *Journal of experimental botany,* **27**: 1181–1186 (1977).

21. **Śrivastava, H.S. et al.** Inhibition of gas exchange in bean leaves by NO_2. *Canadian journal of botany,* **53**: 466–474 (1975).

22. **Durmišidze, S.V. & Nutsubidze, N.N.** Absorption and conversion of nitrogen dioxide by higher plants. *Doklady biochemistry,* **227**: 104–107 (1976).

23. **Faller, N.** Schwefeldioxid, Schwefelwasserstoff, nitrose Gase und Ammoniak als ausschliessliche S- bzw. N-Quellen der höheren Pflanzen [Sulfur dioxide, hydrogen sulfide, nitrous fumes and ammonia as the sole source of sulfur and nitrogen for higher plants]. *Zeitschrift für Pflanzenernährung, Düngung und Bodenkunde,* **131**: 120–130 (1972).

24. **Amundson, R.G. & MacLean, D.C.** Influence of oxides of nitrogen on crop growth and yield: an overview. *In:* Schneider, T. & Grant, C., ed. *Air pollution by nitrogen oxides.* Amsterdam, Elsevier, 1982, pp. 501–510.
25. **Ashenden, T.W. & Williams, I.A.D.** Growth reductions in *Lolium multiflorum* Lam. and *Phleum pratense* L. as a result of SO_2 and NO_2 pollution. *Environmental pollution,* **21**: 131–139 (1980).
26. **Mansfield, T.A. et al.** Effects of nitrogen oxides on plants: two case studies. *In:* Schneider, T. & Grant, I., ed. *Air pollution by nitrogen oxides.* Amsterdam, Elsevier, 1982, pp. 511–520.
27. **Whitmore, M.E. & Freer-Smith, P.H.** Growth effects of SO_2 and/or NO_2 on woody plants and grasses during spring and summer. *Nature,* **300**: 55–57 (1982).
28. **Kress, L.M. et al.** Growth impact of O_3, NO_2 and/or SO_2 on *Plantanus occidentalis. Agriculture and environment,* **7**: 265–274 (1982).
29. **Ashenden, T.W. & Mansfield, T.A.** Extreme pollution sensitivity of grasses when SO_2 and NO_2 are present in the atmosphere together. *Nature,* **273**: 142–143 (1978).
30. **Martin, A. & Barber, F.R.** *Sulphur dioxide, oxides of nitrogen and ozone measured continuously for two years at a rural site.* London, Central Electricity Generating Board, 1980 (Report No. Mid/SSD/80/0044/N).
31. **Bleasdale, J.K.A.** Effects of coal-smoke pollution gases on the growth of rye grass. *Environmental pollution,* **5**: 275–285 (1973).
32. **Ashenden, T.W.** The effect of long-term exposures to SO_2 and NO_2 pollution on the growth of *Dactylis glomerata* L. and *Poa pratensis* L. *Environmental pollution,* **18**: 249–258 (1979).
33. **Zierock, K.H. et al.** *Study on the need for an NO_2 long-term limit value for the protection of terrestrial and aquatic ecosystems.* Luxembourg, Commission of the European Communities, 1986 (Document No. EUR 10546EN).
34. **Gezondheidsraad.** *Advies inzake stikstofdioxid* [Guidelines for nitrogen dioxide]. The Hague, Ministry of Health and Environmental Hygiene, 1979.
35. **Ormrod, D.P.** Air pollutant interactions in mixtures. *In:* Unsworth, M.H. & Ormrod, D.P., ed. *Effects of gaseous air pollution in agriculture and horticulture.* London, Butterworth, 1982, pp. 307–331.
36. **Wellburn, A.R. et al.** Biochemical explanations of more than additive inhibitory effects of low atmospheric levels of sulphur dioxide plus nitrogen dioxide upon plants. *New phytologist,* **88**: 223–237 (1981).
37. **Robinson, D.C. & Wellburn A.R.** Light-induced changes in the quenching of 9-amino-acridine fluorescence by photosynthetic membranes due to atmospheric pollutants and their products. *Environmental pollution (Series A),* **32**: 109–120 (1983).
38. **Van Aalst, R.M. & Diederen, H.S.M.A.** *De rol van stikstofoxiden en ammoniak bij de depositie vanuit de lucht van bemestende en verzurende stoffen op de nederlandse bodem* [The role of oxides of nitrogen and ammonia deposited onto the Netherlands as a result of fertilizer application and acidification]. Delft, Hoofdgroep Maatschappelijke Technologie, 1983 (Report No. R83/42).

33

The effects of ozone and other photochemical oxidants on vegetation

Plants are especially sensitive to ozone and other photochemical oxidants *(1)*. Ozone and other photochemical oxidants may cause acute effects when high concentrations occur for short periods of time. However, chronic and subtle effects from long-lasting or repeated exposures are usually more important. The potential effects of these oxidants on long-living trees and relatively long-growing crops are considerable.

In general, exposures to elevated levels of ozone will occur over large areas, and at the same time levels of other photochemical oxidants will be higher. Thus, plants will be exposed to mixtures of several components of the photochemical smog, and possibly also other air pollutants (e.g. sulfur dioxide and acidic products).

Uptake of Photochemical Oxidants in Plants

The only route of exposure to photochemical oxidants is the air, and the amount of photochemical oxidants absorbed by the plants determines their response.

The uptake of oxidant gases from the ambient air into plants is largely determined by the normal gas exchange processes of plants via the stomata. Only the gas diffusing through the stomata and transported along the intercellular spaces of the leaves will arrive at the cellular surfaces where the first reactions may take place. The solubility of the gases in water is one of the most important properties in relation to further uptake and effects in plant cells *(2)*.

Environmental factors largely determine the gas uptake into the intercellular spaces of leaves. This physical process is dependent on the genetically determined properties of the plant and modified by environmental factors such as wind speed, temperature, relative humidity, light intensity, and other pollutants. In general, the factors that stimulate the turbulence of the air, together with those that enhance the opening of the stomata of the leaves, will increase the uptake of gases into the intercellular spaces.

Effects of Photochemical Oxidants on Plants and Ecosystems

Photochemical oxidants, especially ozone, damage the leaves and needles of sensitive plants largely by disrupting membrane integrity *(3)*, but metabolic processes such as photosynthesis are also affected *(2)*. Visible expressions such as leaf-yellowing, necrosis, defoliation and premature senescence may also become apparent. Characteristic ozone symptoms such as chlorotic and necrotic flecking have been observed in Europe in vines, potatoes, peas and beans. Clover species are among the most sensitive native plants in relation to ozone *(4)*. With Bel W3 tobacco being used as a biological indicator plant, ozone effects have been observed in many countries of the European Region. Band-forming necrosis on the underneath leaf surface of the small stinging-nettle is used as a biological indicator for peroxyacetyl nitrate *(5)*.

In addition to the visible, morphological effects, chronic and subtle effects on physiological processes such as photosynthesis and translocation of photosynthates may inhibit the production and distribution of carbohydrates in plants, decreasing the vitality of both leaves and roots and reducing growth and crop production. The plants may also have a reduced resistance to fungi, bacteria, viruses and insects, and also to climatic stresses (frost and drought). Ozone may also affect the quality of plant material used as food or fodder *(6)*.

At the ecosystems level, photochemical oxidants may have very severe impacts, not only on crops (seen as agro-ecosystems) but also on natural ecosystems, such as forests. Crops may be adversely influenced in regard to both quantity and quality, with obvious economic impacts. In natural ecosystems the inhibition of some sensitive species in competition with more tolerant species may result in the disappearance of the former.

Regarding the problem of forest decline in Europe, the effects of photochemical oxidants may be only part of the whole complex situation. Ozone, in combination with several air pollutants, may play an important role *(7)*. Thus, when investigating the many causes of acid precipitation, nutrient imbalances and other stresses, the almost ubiquitous ozone should not be overlooked.

Exposure–effect relationships for ozone and peroxyacetyl nitrate

The nature and intensity of the effects of photochemical oxidants on plants are dependent not only on exposure concentration and duration, but also on frequency of exposure and intervals between exposures; time of the day and year are also important. Also, the prevailing site conditions and the developmental stage of the plants have an influence *(8)*.

The characterization and representation of plant exposure to ozone have been, and continue to be, a major problem. Most studies have characterized the exposure by using mean ozone concentrations, although various averaging times have been used. Some studies have also used the cumulative ozone dose. Detailed enumerations of the effects of defined levels of ozone and peroxyacetyl nitrate are given by Guderian et al. *(8)*.

To determine the lowest concentrations of ozone that have an effect, Posthumus et al. *(9)* summarized data collected from the literature by Guderian & Rabe *(10)* (see Fig. 1). These data were derived largely from exposures of plants less than one year old and exposed in greenhouses or field chambers. The plant responses evaluated ranged from visible foliar injury to growth and yield reductions. When plants were exposed to ozone for a number of hours per day over a period of some weeks, the exposure time was calculated by adding the hours of real exposure. Each study was represented by a point in the coordinate system, indicating the lowest concentration and duration of exposure used in the study that caused an effect. The data sets do not necessarily establish a threshold that would prevent effects, because they reflect the limitations in measurement and

Fig. 1. Line of lowest concentrations of ozone that have an effect on plants

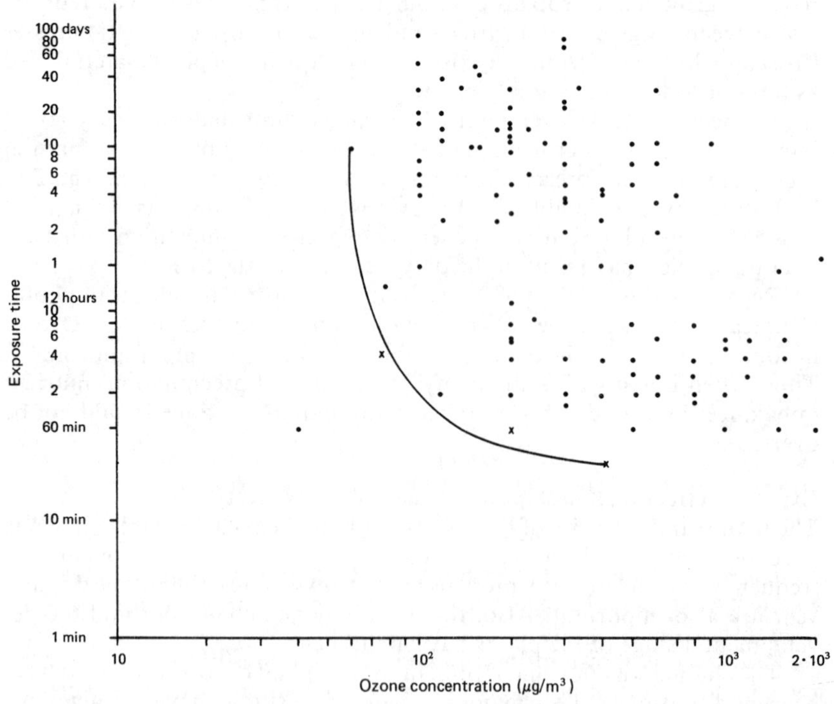

Source: Posthumus, A.C. et al. *(9)*.

experimental techniques. In addition, it is possible to establish only the injuriousness, not the harmlessness, of an exposure.

Despite extensive investigations, it is still possible that combinations of conditions that were previously thought to be harmless could be shown to have effects. The line in Fig. 1 represents the boundary of the range of lowest ozone concentrations and exposure durations that have an effect on a range of plant species. Exposures above and to the right of the line are likely to have an effect on susceptible plant species. It is unlikely that ozone exposures below and to the left of the line will cause a response. The no-adverse-effect levels proposed by Jacobson *(11)* are indicated in Fig. 1 by x. The line has simply been drawn through these points, not taking into account any one extreme value. In Table 1 the lowest (threshold) concentrations of ozone that have an effect are shown for three different exposure periods.

Table 1. Lowest (threshold) concentrations of ozone that have an effect

Period of exposure	Ozone concentration ($\mu g/m^3$)
1 hour	200
24 hours	65
100 days (growing season)	60

Most recent findings, especially those obtained in the context of the National Crop Loss Assessment Network in the USA *(12,13)*, indicate that subtle and chronic effects during the growing season largely fail to materialize, provided that the arithmetic mean of 60 $\mu g/m^3$ for 7 hours every day for the period of light during the growing period (100 days) is not exceeded.

Exposure doses of peroxyacetyl nitrate that have an effect have been considered by Jacobson *(11)* and Tonneijck *(14)* (see Fig. 2). Jacobson's study was based on 10 selected papers (line b). Tonneijck produced a line of lowest exposures that have an effect on the small stinging-nettle (*Urtica urens*), one of the most peroxyacetyl nitrate-sensitive species (line a). Only one study *(15)* has shown exposures that have an effect below this line.

The lowest (threshold) concentrations of peroxyacetyl nitrate that have an effect on the small stinging-nettle (*Urtica urens*) are given in Table 2 for several short-term exposures, calculated on the basis of line a in Fig. 2.

Peroxyacetyl nitrate primarily affects herbaceous crops, not trees *(8)*, and consequently it is not likely to be involved with forest decline. Furthermore, concentrations reported in Europe are not likely to pose a risk to vegetable crops. Long-term studies of peroxyacetyl nitrate effects have not

Fig. 2. Two lines of lowest concentrations of peroxyacetyl nitrate (PAN) that have an effect on plants

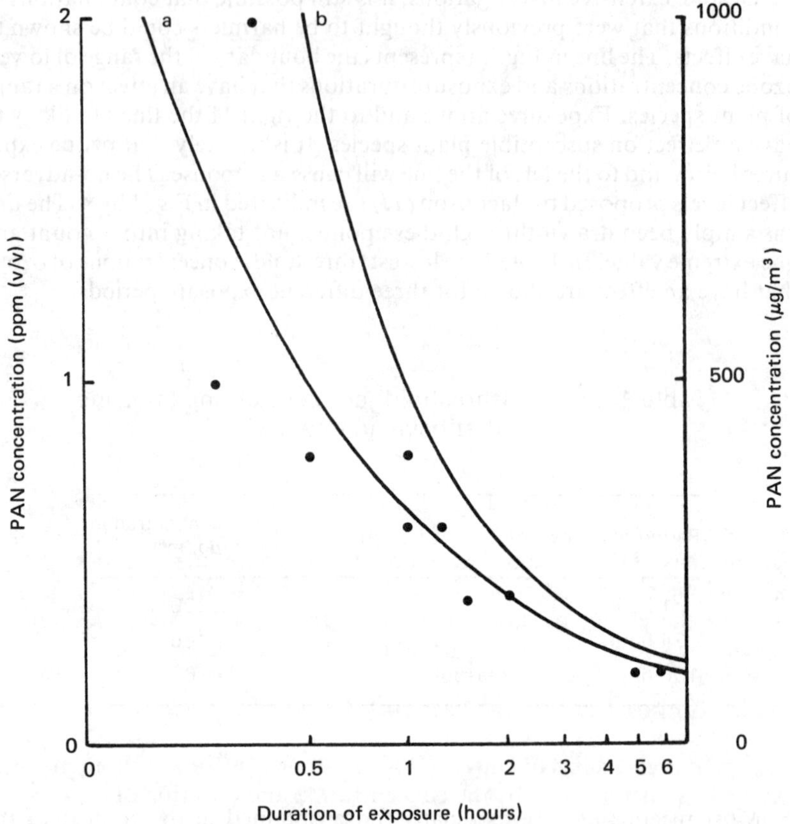

been conducted and studies have not been carried out on the effects of peroxyacetyl nitrate on growth or yield.

As stated above, the effects of photochemical oxidants on plants are strongly influenced by climatic and weather conditions. The data presented in Fig. 1 and 2 and Tables 1 and 2 relate to the exposure of the plants under varying climatic conditions. Thus, it can be postulated that the possible influences of many different environmental conditions are taken into account in this model.

Ozone in combination with other pollutants

The co-occurrence of ozone with sulfur dioxide and nitrogen dioxide and their potential effects, should be considered. Studies in the USA indicate that the presence of sulfur dioxide may increase the sensitivity of plants to leaf injury by ozone.

Table 2. Lowest (threshold) concentrations of peroxyacetyl nitrate (PAN) that have an effect

Period of exposure (hours)	PAN concentration ($\mu g/m^3$)
1	310
2	200
3	155
4	125
5	110
6	100
7	90
8	80

Research in the Federal Republic of Germany *(16)* and the Netherlands *(17)* has indicated that combinations of ozone, sulfur dioxide and nitrogen dioxide may result in a greater reduction in dry matter production than ozone alone or a combination of two pollutants. However, data in Europe showing the extent of co-occurrence of these pollutants are at present inadequate and do not allow further conclusions to be drawn.

Evaluation of Ecological Effects

The uptake and subsequent impacts of ozone and other photochemical oxidants are determined by the concentration and duration of exposure as well as by the genetic and environmentally conditioned sensitivity of the plants. The response is also influenced in large measure by environmental parameters, including wind speed, temperature, relative humidity, light intensity, elevation and soil moisture and chemistry, as well as the presence of other air pollutants. Interactions with fungus diseases and insect pests can also be important where ozone has predisposed the plant to infection and attack.

In addition to visible, morphological responses, impaired physiological processes may inhibit yield and reproduction, and reduce growth and quality at lower concentrations than those causing visible symptoms. In ecosystems such as forests, the more sensitive individuals and species may be eliminated, giving way to more tolerant (and sometimes less desirable) species, thus changing the sociobiology.

The time-concentration patterns of ozone exposure are critical to plant responses, as are site conditions and the stage of plant development. Data on exposure thresholds have been derived largely from exposures of plants less than one year old and exposed in greenhouses or field chambers. The plant responses evaluated ranged from visible injury to reductions in growth and yield. These data suggest that the lowest (threshold) concentrations of ozone that have an effect during the growing season are 200 $\mu g/m^3$ for 1 hour, 65 $\mu g/m^3$ for 24 hours and 60 $\mu g/m^3$ for the growing season (100 days). The attainment of mean ozone concentrations below these levels should be adequate to protect the most sensitive plant species and ecosystems. These concentrations are only slightly higher than, or may even overlap, background ozone concentrations. Nevertheless, the elevated ozone concentrations reported in Europe are harmful to plants and pose a risk to both agriculture and forests. Since the total exposure period is important to plant response, the life expectancy of the species becomes critical. Thus, the lowest concentration to have an effect on conifers with an 80–100-year rotation may be lower than that for annual, agricultural crops.

Peroxyacetyl nitrate primarily affects herbaceous crops, not trees, and consequently is not likely to be involved in forest decline. Furthermore, concentrations reported for Europe are not likely to pose a risk to agricultural crops. Data for peroxyacetyl nitrate indicate that the lowest concentrations to have an effect are 310 $\mu g/m^3$ for 1 hour and 80 $\mu g/m^3$ for 8 hours.

Guidelines

Concentrations of peroxyacetyl nitrate should stay below 300 $\mu g/m^3$ for 1 hour and 80 $\mu g/m^3$ for 8 hours.

The following guideline values for ozone are recommended: 200 $\mu g/m^3$ for 1 hour, 65 $\mu g/m^3$ for 24 hours and 60 $\mu g/m^3$ for the growing season (100 days).

References

1. **Grennfelt, P., ed.** *Ozone. Proceedings of an international workshop on the evaluation and assessment of the effects of photochemical oxidants on human health, agricultural crops, forestry, materials and visibility, Gothenburg, 29 February – 2 March 1984.* Gothenburg, Swedish Environmental Research Institute, 1984 (Document No. IVL-EM 1570).
2. **Tingey, D.T. & Taylor, G.E. Jr.** Variation in plant response to ozone: a conceptual model of physiological events. *In:* Unsworth, M.H. & Ormrod, D.P., ed. *Effects of gaseous air pollution in agriculture and horticulture.* London, Butterworth, 1982, pp. 27–42.
3. **Mudd, J.B.** Effects of oxidants on metabolic function. *In:* Unsworth, M.H. & Ormrod, D.P., ed. *Effects of gaseous air pollution in agriculture and horticulture.* London, Butterworth, 1982, pp. 189–203.
4. **Ashmore, M.R.** Effects of ozone on vegetation in the United Kingdom. *In:* Grennfelt, P., ed. *Ozone. Proceedings of an international workshop on the evaluation and assessment of the effects of photochemical oxidants on human health, agricultural crops, forestry, materials and visibility. Gothenburg, 29 February – 2 March 1984.* Gothenburg, Swedish Environmental Research Institute, 1984, pp. 92–104 (Document No. IVL-EM 1570).

5. **Posthumus, A.C.** Biological indicators of air pollution. *In:* Unsworth, M.H. & Ormrod, D.P., ed. *Effects of gaseous air pollution in agriculture and horticulture.* London, Butterworth, 1982, pp.27–42.
6. **Pell, E.J.** *Air pollution effects on food quality.* University Park, PA, Center for Air Environment Studies, Pennsylvania State University, 1982 (CAES report No. 645-82).
7. **Prinz, B. et al.** *Waldschäden in der Bundesrepublik Deutschland* [Forest damage in the Federal Republic of Germany]. Essen, Landesanstalt für Immissionsschutz des Landes Nordrhein-Westfalen, 1982 (LIS-Berichte Nr. 28).
8. **Guderian, R. et al.** Effects of photochemical oxidants on plants. *In:* Guderian, R., ed. *Air pollution by photochemical oxidants: formation, transport, control and effects on plants.* Berlin, Springer-Verlag, 1985, pp. 127–333.
9. **Posthumus, A.C. et al.** *Literatuur-vooronderzoek naar de belasting-effectrelaties voor planten ten aanzien van verschillende van kolengestookte installaties afkomstige luchtverontreinigingscomponenten* [A preliminary literature research into the burden–effect relationships for plants with regard to different air pollutants from coal-fired installations]. Wageningen, Instituut voor Planteziektenkundig Onderzoek, 1983 (Report No. R293).
10. **Guderian, R. & Rabe, R.** *Effects of photochemical oxidants on plants.* Brussels, Commission of the European Communities, 1982 (Final report No. U/81/519).
11. **Jacobson, J.S.** The effects of photochemical oxidants on vegetation. *VDI-Berichte,* **270**: 163–173 (1977).
12. **Heck, W.W. et al.** Ozone crop yield functions for loss assessment. *Journal of the Air Pollution Control Association,* **34**: 729–735, 810–817 (1984).
13. **Heck, W.W. et al.** Assessment of crop loss from ozone. *Journal of the Air Pollution Control Association,* **32**: 353–361 (1982).
14. **Tonneijck, A.E.G.** Urtica urens *L. as indicator plant for peroxyacetyl nitrate (PAN). Proceedings of the VIth World Congress on Air Quality.* Paris, International Union of Air Pollution Prevention Associations, 1983, Vol. 2, pp. 563–567.
15. **Nouchi, J.** [Effects of ozone and PAN concentrations and exposure duration on plant injury]. *Taiki osen gakkaischi,* **14**: 489–496 (1979) (in Japanese).
16. **Guderian, R. et al.** Wirkungen von Ozon, Schwefeldioxid und Stickstoffdioxid auf Fichte und Pappel bei unterschiedlicher Versorgung mit Magnesium und Kalzium sowie auf die Blattflechte *Hypogymnia physodes* [Effects of ozone, sulfur dioxide and nitrogen dioxide on spruce and poplar supplied with various rates of magnesium and calcium and on the lichen *Hypogymnia physodes*]. *VDI-Berichte,* **560**: 657–701 (1985).
17. **Mooi, J.** Wirkungen von SO_2, NO_2, O_3 und ihrer Mischungen auf Pappeln und einige andere Pflanzenarten [Effects of SO_2, NO_2, O_3 and their mixtures on poplars and certain other types of plant]. *Der Forst- und Holzwirt,* **39**(18): 438–444 (1984).

34

The effects of sulfur oxides on vegetation

In view of its degree of phytotoxicity and its wide distribution, sulfur dioxide (SO_2) is regarded in many parts of Europe as the most important air pollutant so far as plant damage is concerned.

Uptake of Sulfur Dioxide by Plants

Oxidized sulfur compounds can affect vegetation, directly by uptake through parts of the plants that are above ground, and indirectly via soil acidification. Direct effects on leaves are mainly determined by the dry deposition of sulfur dioxide. Like other gases, sulfur dioxide penetrates the mesophyll of leaves primarily through the stomata. The aperture of these pores is controlled by the prevailing environmental conditions, such as humidity, wind velocity, light and temperature. These external factors influence pollution uptake and, hence, the degree of injury received from a given burden of sulfur dioxide.

The absorption of gases is a consequence of the chemical potential gradient between the bulk air and the leaf interior. Flux is proportional to the pathway conductance and inversely related to the resistance to mass transfer along the diffusive pathway as the pollutant moves through the boundary layer, stomata and intercellular spaces and the liquid phases. Unlike the higher plants, the lower plants such as lichens and mosses do not have comparable outer cell layers, and this is one reason for their particular sensitivity to sulfur dioxide.

Effects on Plants and Plant Communities

The phytotoxic effects of sulfur dioxide are greatly influenced by the ability of plant tissue to convert dissolved sulfur dioxide into relatively nontoxic forms *(1)*. Sulfite (SO_3^{2-}) and hydrogen sulfite (HSO_3^-) are the major chemical species formed upon dissolution of sulfur dioxide in aqueous solutions. Plants can overcome phytotoxic effects from sulfite and hyrogen sulfite by converting them by enzymic and nonenzymic mechanisms, and sulfate

(SO_4^{2-}) thus accumulated is considerably less toxic than sulfite. Absorbed sulfur dioxide can be used by plants in the reductive sulfur cycle and thus serve as a fertilizer *(2)*.

Depending on the amount of sulfur dioxide taken up per unit of time, various kinds of biochemical and physiological effect take place in the plant tissue. These include the degradation of chlorophyll, reduced photosynthesis, raised respiration rates, and changes in protein metabolism, lipid and water balance and enzyme activity *(3–7)*. Chloroplasts have been regarded as the main sites of action *(8,9)*. The high pH value within plastids favours the formation of sulfite ions, which are commonly regarded as the toxic product of sulfur dioxide in aqueous solutions. High concentrations cause acute injury in the form of leaf necrosis, even after short periods of action, while long-term exposure to sulfur dioxide will result in chronic injury. Apart from visible effects, "subtle injury" should be mentioned; this may include reduced plant growth and greater susceptibility to climatic extremes or pathogens.

In plant communities, direct effects on species with different degrees of sensitivity and indirect effects in the form of changes in interspecies competition lead to a reduction in sensitive varieties, and this in turn can alter the structure and function of the whole community *(10)*. Parallel to this degradation, spontaneous or man-supported processes occur, in which either original, i.e. adaptively resistant members of the remaining community, or immigrants undergo secondary succession. These types of change naturally also have consequences for consumers and decomposers.

Factors Influencing Plant Responses

The reaction of a plant to a given air pollutant depends on the ambient exposure, the amount of pollutant diffusing into the leaves, and the autonomous and environmentally modified resistance of plants.

As with other phytotoxic gases, dose alone is not the decisive factor in the action of sulfur dioxide. For the same products of concentration (c) and time (t), the effect varies with concentration, duration of exposure, and frequency and sequence of impact.

Within certain ranges of concentration, a specific dose does not always induce the same magnitude of response. The extent of foliar injury increases progressively with increasing concentration when equal products of concentration and exposure duration are applied *(11)*. Therefore, peak concentrations are particularly hazardous.

Conversely, plants can recover in pollution-free periods. However, this recovery will not take place if the duration of a given impact is too long. Moreover, the pollution-free period must be sufficiently long to be effective *(12)*. At the present time, vegetation in Europe is principally threatened by the long-term effects of low concentrations of sulfur dioxide and the concurrent impact of other phytotoxic components.

In a given pollution situation, the diversity of the vegetation is one of the factors determining the kind and degree of damage caused. Individual plant species and varieties, and even individuals within a population, react with different degrees of sensitivity to stress resulting from air pollution.

In addition to legumes such as beans and clover, plant species highly sensitive to sulfur dioxide include berry-bearing shrubs and conifers such as white fir (*Abies alba*) and Norway spruce (*Picea abies*). Medium sensitivity is found in most grasses, including cereal species, and in Scots pine (*Pinus sylvestris*) and common beech (*Fagus sylvatica*) *(12)*. The black pine (*Pinus nigra austriaca*) and oak species have proved to be relatively resistant to sulfur dioxide *(13)*.

The genetically established degree of resistance alters according to the plant's stage of development and to external conditions of growth, such as soil and climate *(12)*.

Dose–response Relationships

When establishing and evaluating exposure–effect relationships, a distinction must be made between the effects of sulfur dioxide alone and in combination with other air pollutants. The results of field observations and fumigation experiments may form the basis for determining quantitative relationships. Data from several hundred experiments have been used to elucidate the quantitative relationships between sulfur dioxide concentrations and the effects on both annual and perennial plants.

Sulfur dioxide as a single component

Agricultural crops

It is generally accepted that the sulfur dioxide concentrations prevailing in most agricultural regions of Europe are unlikely to reduce cereal yield. However, there is evidence that some other species are affected by concentrations less than $100 \mu g/m^3$. Unfortunately, there are many discrepancies between various sulfur dioxide fumigation experiments on agricultural crops, which are probably due to interactions with environmental conditions and factors relating to the plants themselves. The lowest concentrations of sulfur dioxide shown to reduce the growth of higher plants by fumigation are $43 \mu g/m^3$ over 173 days for perennial ryegrass *(14)* and $55 \mu g/m^3$ over 28 days for tobacco and cucumber *(8)*. In the case of grass species, at least, the situation is complicated by evidence that both the magnitude and the direction of sulfur dioxide-induced growth effects can change markedly over a 12-month period. As a result of lowered physiological activity, plants can react in a more sensitive way to sulfur dioxide stress in winter than in summer *(15)*.

Experiments in which plants are grown in chambers ventilated with ambient and charcoal-filtered air have demonstrated a growth reduction caused by the former at sulfur dioxide concentrations generally lower than those shown to have reduced growth in controlled fumigations *(6,7)*. The lowest mean concentration of sulfur dioxide at which growth reduction has been recorded is $34 \mu g/m^3$ over 28 days in a filtration experiment with perennial ryegrass *(9)*. However, various other pollutants, notably nitrogen dioxide and ozone, will probably have contributed to the observed effects. This situation also probably applies to filtration studies in the United

Kingdom, where barley yield was reduced by ambient air containing mean sulfur dioxide levels as low as 50 μg/m³: in this case the presence of fluorides from neighbouring brickworks may also have been involved and the results cannot be extrapolated to agricultural areas in general *(16)*.

Forest trees

Currently, the International Union of Forest Research Organizations (IUFRO) recommends a sulfur dioxide annual mean of 25 μg/m³ as an acceptable level for the protection of trees, in areas where growth is poor owing to other environmental stresses (Table 1). This is based on field observations in Canada, Czechoslovakia and Finland *(18–21)*. For example, in Finland long-term exposure to sulfur dioxide concentrations between 20 and 40 μg/m³ resulted in elevated foliar sulfur levels and even chronic injury in the leaf tissue of conifers such as *Picea abies* and *Pinus sylvestris*, and in broad-leaf species such as *Betula pubescens (22)*. Effects of this kind were frequently observed in the springtime, being preceded by chronic injury during winter *(23,24)*. In Canada and Czechoslovakia the areas concerned are subject to contamination by other pollutants, particularly heavy metals. However, the extensive studies in Finland lend strong support to the IUFRO recommendation.

For areas with more equable growing conditions, IUFRO has recommended an annual mean of 50 μg/m³. This value is supported by several studies carried out under controlled as well as field conditions. Bonte et al. *(25)* showed growth reductions in two larch species after field exposure to 60 μg sulfur dioxide per m³ over two years. Exposure of four conifer species

Table 1. IUFRO air quality standards for sulfur dioxide (1978), confirmed in 1980

Period	Maximum level of sulfur dioxide (μg/m³) that allows: full production in most sites	full production and environmental protection[a]
Annual average	50	25
24-hour average	100[b]	50
97.5 percentile of 30-minute values in growing season	150	75

[a] Environmental protection against erosion and avalanches, and to ensure full production in higher regions of mountains, boreal zones, extreme sites, etc.

[b] The 24-hour average may be exceeded 12 times in a period of 6 months.

Source: Wentzel, K.F. *(17)*.

in central London (mean $SO_2 = 86\mu g/m^3$) resulted in greater growth over the first year than was seen in a rural site (mean $SO_2 = 30\mu g/m^3$), probably due to higher urban temperatures; however, after a further year in the field the two sulfur dioxide-sensitive spruce species showed substantial growth reductions in London, while the more tolerant pine species produced the same performance at both sites *(26)*.

Unfortunately, logistical constraints have so far restricted fumigation of trees to periods of up to three years only. Thus the significance of sulfur dioxide for the performance of a tree throughout its lifetime (80–250 years) remains unknown and can only be evaluated on the basis of incremental studies on mature trees at sites where some form of pollution record is available over a long period. Both studies point to the possibility of even lower sulfur dioxide concentrations reducing tree growth over a period of many years.

In a major investigation *(27)* around a point source of sulfur dioxide at Biersdorf in the Federal Republic of Germany, however, deciduous and coniferous tree species exposed in the field showed slightly reduced growth at annual mean concentrations of $50\mu g/m^3$ over four years. Other field studies are lacking. Future research to elucidate the significance of sulfur dioxide for tree growth over large areas of Europe will require more intensive sampling for tree performance, accompanied by systematic monitoring of pollutants, as well as the assessment of other environmental factors affecting tree growth, so that their relative importance can be assessed.

Besides the recommendation for annual average sulfur dioxide concentrations, IUFRO also recommends a 24-hour mean that should not be exceeded more than 12 times within a six-month period, as well as a 97.5 percentile of 30-minute values (Table 1). There is some doubt as to the significance of peak concentrations of sulfur dioxide occurring within overall means of $100\mu g/m^3$ or less: field observations have shown injury to conifers between 200 and $300\mu g/m^3$ for the 97.5 percentile *(28)*, while a controlled fumigation experiment over two years has demonstrated only small effects of peak concentrations up to $750\mu g/m^3$ within a $100\mu g/m^3$ mean, compared with the same mean at a constant concentration *(29)*. Thus, it appears that the IUFRO recommendations for peak concentrations are sufficient to protect forest trees.

Pollutant mixtures

Under European conditions, vegetation is usually exposed not only to sulfur dioxide but to a mixture of potential phytotoxic agents. Therefore, plant reactions represent the sum of the individual complex pollutant burden, which varies regionally according to emission sources. Because of their phytotoxic properties and wide distribution, the combined effects of sulfur dioxide, ozone, NO_x (nitric oxide and nitrogen dioxide) and acidic deposition prevail, with component-specific interactions varying regionally *(30)*. The special hazard to vegetation in the present complex pollution situations lies in the fact that in low ambient concentrations of pollutants, additive and synergistic effects are predominant *(30)*.

There is some indication that the recent decline of European forests is partly related to interactions of the above-mentioned components *(31)*. Since quantitative relationships, especially under field conditions, have so far been lacking, no clear-cut conclusions can yet be drawn for establishing air quality guidelines.

Interactions with other stresses

There is now abundant evidence concerning the impact of a range of environmental factors on sulfur dioxide-induced growth reduction, with conditions causing slow growth being particularly conducive to chronic injury. However, it is becoming clear that sulfur dioxide can predispose plants to injury by other environmental stresses *(23,32)*. Frost damage is known to be particularly important in this respect, while the frequently observed reductions in root/shoot ratio caused by sulfur dioxide will affect plants that are more susceptible to drought and possibly also affect mycorrhizae, thus disturbing nutrient uptake and water relations *(33)*. Furthermore, any kind of sulfur dioxide-induced reduction in vitality will lead to an increased susceptibility to plant pathogens *(34,35)*. However, these so-called secondary effects, which change the predisposition of host plants, are difficult to evaluate quantitatively, even though they are of great relevance under field conditions.

Total Deposition of Sulfur Oxides

The 1982 Stockholm Conference on Acidification of the Environment noted that a total annual sulfur deposition of 0.5 g/m^2 per year ($\approx$ 5 kg/ha) represents a potential threat to aquatic ecosystems with a low level of alkalinity *(36)*.

Wet deposition of sulfur compounds can affect terrestrial ecosystems in two different ways, either by direct impact on the above-ground parts of plants or by indirect impact, leading to changes in soil characteristics. Cation exchange capacity, base saturation and soil pH can be decreased after long-term exposure, resulting in decreased supply and uptake of plant nutrients, elevated concentrations of potential toxic metals and disturbances in the decomposer chain *(37)*. Forest soils overlying slowly weathering granite or porphyric bedrock are especially endangered (e.g. in Scandinavia and parts of the Federal Republic of Germany, such as the Black Forest and the Harz Mountains). The potential hazard has to be seen in the long-term impact of total acidic deposition in terrestrial ecosystems, to which sulfur compounds are the greatest contributors *(37)*.

Direct effects on the above-ground organs of plants can injure plant cuticles and exchange mechanisms in respect of organic and inorganic compounds *(37)*. As a result, increased leaching of chemical substances from plant organs is observed. Nutrient deficiency observed in areas with recent forest decline is frequently linked with leaching phenomena that are probably caused by combinations of acidic deposition and gaseous pollutants, mainly ozone *(38)*. However, the available data are so far insufficient to establish a guideline for protecting terrestrial ecosystems against acidic deposition.

Evaluation of Ecological Effects

Above a certain intensity, the effects on individual plants and plant communities described earlier impair the various functions of vegetation in the ecosystem as a producer, an economic resource and a gene reservoir. Lower yields and poorer quality lead to considerable economic losses in agriculture, forestry and horticulture. The extensive damage that is occurring, especially in forest ecosystems, increasingly threatens the ecological functions of vegetation, namely, maintaining the climatic and hydrological balance and protecting against erosion caused by wind and water. Disturbances in the natural regeneration of forest stands are the outward expression not only of reduced seed quality but also of impaired soil conditions following acid precipitation. Lastly, gene erosion occurs as a result of reduced expansion or the dying out of ecotypes or even species.

Trees with a long life-cycle suffer particularly from long-term exposure to sulfur dioxide, since here, too, subtle effects can accumulate over the years and decades to produce harmful overall effects, all the more so when other phytotoxic compounds such as NO_x, ozone and acid precipitation are also present. The dramatic deterioration in the health of forests in many parts of Europe has been related in particular to this complex type of air pollution.

There are no realistic chances of coping with the phenomenon of forest decline in Europe simply by breeding more resistant plant species, improving fertilization programmes or changing management practices. The main solution to the problem would seem to be to achieve a quick reduction in emissions of sulfur dioxide and other phytotoxic pollutants to a level that is still tolerable for vegetation.

Guidelines

The following guideline values for sulfur dioxide are recommended: annual average $30 \mu g/m^3$; 24-hour average $100 \mu g/m^3$.

This guideline, however, may not be sufficient in the case of extreme environmental conditions or if other ubiquitous pollutants are present. Although there are currently insufficient data on the combined impact of wet and dry deposition on terrestrial ecosystems to permit the establishment of an appropriate guideline, this should be seen as an ultimate goal for the protection of sensitive ecosystems.

References

1. **Malhotra, P.S. & Khan, A.A.** Biochemical and physiological impact of major pollutants. *In:* Treshow, M., ed. *Air pollution and plant life.* New York, Wiley, 1984, pp. 113–157.
2. **Faller, N.** *Schwefeldioxid aus der Luft als entscheidende Nährstoffquelle für Pflanzen* [Sulfur dioxide from air as a decisive source of nutrition for plants]. *Landwirtschaftliche Forschung,* **18**: 48–54 (1970).
3. **Black, V.J. & Unsworth, M.H.** Effects of low concentrations of sulfur dioxide on net photosynthesis and dark respiration of *Vicia faba. Journal of experimental botany,* **30**: 473–483 (1979).

4. **Brenninger, C. & Tranquilini, W.** Photosynthese, Transpiration und Spaltöffnungsverhalten verschiedener Holzarten nach Begasung mit Schwefeldioxid [Photosynthesis, transpiration and stoma behaviour of different kinds of tree after gassing with sulfur dioxide]. *European journal of forest pathology,* **13**: 228–238 (1983).
5. **Fowler, D. & Cape, J.N.** Air pollutants in agriculture and horticulture. *In:* Unsworth, M.H. & Ormrod, D.P., ed. *Effects of gaseous air pollution in agriculture and horticulture.* London, Butterworth, 1982, pp. 3–26.
6. **Mansfield, T.A. & Freer-Smith, P.** Effects of urban air pollution on plant growth. *Biological review,* **56**: 343–368 (1981).
7. **Bell, J.N.B.** Sulphur dioxide and the growth of grasses. *In:* Unsworth, M.H. & Ormrod, D.P., ed. *Effects of gaseous air pollution in agriculture and horticulture.* London, Butterworth, 1982, pp. 225–246.
8. **Mejstrik, V.** The influence of low SO_2 concentrations on growth reduction of *Nicotiana tabacum* L. cv. Samsum and *Cucumis sativus* L. cv. Unikat. *Environmental pollution (Series A),* **21**: 73–76 (1980).
9. **Awang, M.B.** *The effects of sulphur dioxide pollution on plant growth, with special reference to* Trifolium repens. Thesis, University of Sheffield, 1979.
10. **Guderian, R. & Knüppers, K.** Responses of plant communities to air pollution. *In: Proceedings of the Symposium on Effects of Air Pollutants on Mediterranean and Temperate Forest Ecosystems, June 22–27 1980.* Berkeley, 1980, pp. 187–199 (Pac. Southwest For. Range Exp. Stn., Gen. Tech. Rep. PSW-43).
11. **van Haut, H.** Die Analyse von Schwelfeldioxidwirkungen auf Pflanzen im Laboratoriumsversuch [The analysis of sulfur dioxide effects on plants in a laboratory experiment]. *Staub, Reinhaltung der Luft,* **21**: 52–56 (1961).
12. **Guderian, R.** *Air pollution. Phytotoxicity of acidic gases and its significance in air pollution control.* Berlin/Heidelberg/New York, Springer, 1977 (Ecological Studies No. 22).
13. **Wentzel, K.F.** Empfindlichkeit und Resistenzunterschiede der Pflanzen gegenüber Luftverunreinigung [Plant sensitivity and differences in resistance to air pollution]. *Forstarchiv,* **39**: 189–194 (1968).
14. **Bell, J.N.B. et al.** Studies of the effects of low levels of sulphur dioxide on the growth of *Lolium perenne* L. *New phytologist,* **83**: 627–643 (1979).
15. **Bell, J.N.B. & Clough, W.S.** Depression of yield in ryegrass exposed to sulfur dioxide. *Nature,* **241**: 47–49 (1973).
16. **Buckenham, A.H. et al.** The effect of aerial pollutants on the growth and yield of spring barley. *Annals of applied biology,* **100**: 179–187 (1982).
17. **Wentzel, K.F.** IUFRO studies on maximal SO_2 immissions standards to protect forests. *In:* Ulrich, B. & Pankrath, J., ed. *Effects of accumulation of air pollutants in forest ecosystems.* Dordrecht, D. Reidel, 1983, pp. 295–302.
18. **Huttunen, S.** *Ilman epäpuhtauksien leviäminen ja vaikutukset metsäympäristössä* [The spread and effects of air pollutants in a forest environment]. Oulu, Oulu University, 1980 (University Research Report, No. 12).

19. **Materna, J.** Luftverunreinigungen und Waldschäden [Air pollution and damage to forests]. *In: Symposium über Umweltschutz — eine internationale Aufgabe, Prag, 13.–15. März 1985* [Symposium on Environmental Protection — An International Responsibility, Prague, 13–15 March 1985]. Düsseldorf, Verein Deutscher Ingenieure, 1985 (in German and Czech).
20. **Materna, J. et al.** Výsledky měření koncentrací kysličníku siřičitého v lesích krušných hor [Results of measurements of SO_2 concentration in the Erzgebirge]. *Ochrana ovzdusi,* **6**: 84–92 (1969).
21. **Linzon, S.N.** Long-term chronic effects of sulphur dioxide on forest growth. *Aquilo series botanica,* **19**: 157–166 (1983).
22. *Rikkikomitean välimietintö I* [Interim report of the Sulphur Commission, No. I]. Helsinki, Government Printing Office, 1985 (Komiteamietintö 1985:35 [State Commission Reports, No. 1985:35]).
23. **Havas, P.J.** Injury in pines growing in the vicinity of a chemical processing plant in northern Finland. *Acta forestalia fennica,* **121**: 1–21 (1971).
24. **Davison, A.W. & Bailey, I.F.** SO_2 pollution reduces freezing resistance of ryegrass. *Nature,* **297**: 400–402 (1982).
25. **Bonte, J. et al.** *Etude des effets à long-terme d'une pollution chronique par* SO_2 [Study of the long-term effects of chronic SO_2 pollution]. Morlaas, Institut national de la recherche agronomique, 1981.
26. **Garsed, S.G. & Rutter, A.J.** Relative performance of conifer populations in various tests for sensitivity to SO_2, and the implications for selecting trees for planting in polluted areas. *New phytologist,* **92**: 349–367 (1982).
27. **Guderian, R. & Stratmann, H.** *Freilandsversuche zur Ermittlung von Schwefeldioxidwirkungen auf die Vegetation. III. Grenzwerte schädlicher* SO_2*-Immissionen für Obst- und Forstkulturen sowie für landwirtschaftliche und gärtnerische Pflanzenarten* [Field experiments to determine the effects of sulfur dioxide on vegetation. III. Threshold values of noxious SO_2 emissions for fruit and forestry crops and for agricultural and horticultural plant species]. Cologne, Westdeutscher Verlag, 1968 (Forschungsberichte des Landes Nordrhein-Westfalen, Nr. 1920).
28. **Stein, G. & Dässler, H.G.** Die forstliche Rauchschadengrossraumdiagnose im Erz- und Elbsandsteingebirge 1964–67. *Wissenschaftliche Zeitschrift — Technische Hochschule Dresden,* **10**: 1397 (1968).
29. **Garsed, S.G. & Rutter, A.J.** The effects of fluctuating concentrations of sulphur dioxide on the growth of *Pinus sylvestris* L. and *Picea sitchensis* (Bong.) Carr. *New phytologist,* **97**: 175–195 (1984).
30. **Lefohn, A.S. & Ormrod, D.P.** *A review and assessment of the effects of pollutant mixtures on vegetation — research recommendations.* Corvallis, OR, US Environmental Protection Agency, 1984 (Report No. EPA-600/3-84-037).
31. **Krause, G.H.M. et al.** Forest decline in Europe: development and possible causes. *Water, air, and soil pollution,* **31**: 647–668 (1986).
32. **Materna, J.** Impact of atmospheric pollution on natural ecosystems. *In:* Treshow, M., ed. *Air pollution and plant life.* New York, Wiley, 1983, pp. 397–416.

33. **Reich, P.B. et al.** Acid rain and ozone influence mycorrhizal infection in tree seedlings. *Journal of the Air Pollution Control Association,* **36**: 724–726 (1986).
34. **Dohmen, G.P. et al.** Air pollution increases *Aphis fabae* pest potential. *Nature,* **307**: 52–53 (1984).
35. **Laurence, J.A.** Effects of air pollutants on plant pathogen interactions. *Journal of plant diseases and protection,* **87**: 156–172 (1981).
36. *Ecological effects of acid deposition. Report and background papers 1982 Stockholm Conference on Acidification of the Environment. Expert meeting I.* Solna, National Swedish Environment Protection Board, 1982 (Report SNV pm 1636).
37. **Kommission Reinhaltung der Luft.** *Säurehaltige Niederschläge. Entstehung und Wirkungen auf terrestrische Ökosysteme.* Düsseldorf, Verein Deutscher Ingenieure, 1983.
38. **Krause, G.H.M. et al.** Forest effects in West Germany. *In:* Davis, D.D., ed. *Air pollution and the productivity of the forest. Proceedings of the Symposium, Washington, DC, 4–5 October 1983.* University Park, PA, Pennsylvania State University, 1983.

Annex 1

Tobacco smoking[a]

Usage and Trends

Smoking of tobacco is practised worldwide by hundreds of millions of people. In 1982, 6.7 million tonnes of tobacco were produced; annual per caput consumption in the USA ranged up to more than 3500 cigarettes. In developing countries, cigarette smoking is increasing, and many cigarettes and other products, including *bidis,* have very high tar (up to 55 mg per cigarette) and nicotine yields. In many developed countries, sizeable decreases in total consumption, sales and smoking rates have occurred. Generally, between one third and one half of men smoke, with some countries having notably higher rates. In most developed and some developing countries, about one third of women smoke, although in some countries fewer do.

Sales-weighted average tar and nicotine contents (as measured by standard laboratory methods) have declined significantly since the 1950s in some parts of the world. The chemical composition of smoke depends on (*a*) the type of tobacco; (*b*) cigarette design, including filtration, blend selection (e.g. reconstituted sheet, expanded tobacco), ventilation, paper and additives; and (*c*) the smoking pattern.

Tobacco is smoked principally in cigarettes, with pipes, cigars, *bidis* and other forms being used either to a minor extent or only in certain regions. Combustion of tobacco products delivers mainstream and sidestream smoke which differ in physicochemical nature. In addition, sidestream smoke contains greater amounts of identified carcinogens than mainstream smoke. Passive smoking is a universal phenomenon where smoking is common. The uptake of smoke constituents by smokers and by passive smokers has been studied in only a few countries, although extensive analysis of smoke shows cigarette smoking to be a major source of exposure to tobacco-specific nitroso compounds, polynuclear aromatic compounds, aromatic amines and some other carcinogens.

---

[a] This Annex is reproduced from *Tobacco smoking,* IARC Monographs on the Evaluation of the Carcinogenic Risk of Chemicals to Humans: Tobacco Smoking, Volume 38, 1986, pp. 309–314. It represents the views and expert opinions of an IARC Working Group that met in Lyon from 12 to 20 February 1985.

Carcinogenicity in Animals

Cigarette smoke has been tested for carcinogenicity in experimental animals by inhalation and by topical application of condensate and in other ways. Exposure of hamsters and rats to whole smoke results in the induction of malignant respiratory tract tumours. Cigarette smoke condensate induces skin cancers in mice and rabbits after application to the skin, and lung cancers in rats after intrapulmonary injection. Cigarette smoke contains many chemicals known to produce cancer in animals and/or humans.

More tumours occur in animals exposed to both cigarette smoke and 7,12-dimethylbenz[*a*]anthracene than to either one alone; the same is true for concomitant exposure to benzo[*a*]pyrene or radon daughters.

No study was available that was designed specifically to investigate the carcinogenicity of passive smoking to experimental animals.

Genetic Activity and Short-term Test Results

Tobacco smoke and smoke condensate are mutagenic and cause chromosomal damage in various test systems with multiple genetic endpoints. Exposure to these complex mixtures results in mutagenic urine in smokers and in increased chromosomal damage in the somatic cells of smokers compared to nonsmokers. Cigarette smoke condensate induces neoplastic transformation in mammalian cells *in vitro*.

Overall assessment of data from short-term tests on cigarette smoke

	Genetic activity			Cell transformation
	DNA damage	Mutation	Chromosomal effects	
Prokaryotes		+		
Fungi/Green plants		+	+[a]	
Insects		+		
Mammalian cells (*in vitro*)			+[a]	
Mammals (*in vivo*)			+	
Humans (*in vivo*)			+	

Degree of evidence in short-term tests for genetic activity: *Sufficient*

Cell transformation: No data

[a] Gas phase of smoke only.

Overall assessment of data from short-term tests on cigarette smoke condensate

	Genetic activity			Cell transformation
	DNA damage	Mutation	Chromosomal effects	
Prokaryotes		+		
Fungi/Green plants		+	+	
Insects		+		
Mammalian cells (*in vitro*)		+	+	+
Mammals (*in vivo*)			+	
Humans (*in vivo*)				
Degree of evidence in short-term tests for genetic activity: *Sufficient*				Cell transformation: Positive

Human Exposure

Smokers of cigarettes with low "tar" yields tend to inhale to a greater extent than do smokers of cigarettes with high "tar" yields, but, in general, their intake of smoke components is reduced.

Certain biochemical markers of smoke intake, e.g. cotinine in plasma, urine or saliva, are sufficiently sensitive and specific to identify passive smokers. Passive smokers who have been examined in western Europe and North America generally have levels between about 0.1% to 1% of these markers as compared to active smokers. The precise quantitative relationship between the measured levels of these markers and the intake of carcinogenic compounds in tobacco smoke is not known.

Approximately 80% of inhaled particles from cigarette mainstream smoke is deposited in the respiratory tract, the majority in the tracheobronchial region. Wide variation is found, however, among individuals. The distribution of particulate matter in the lung is similar in smokers of "high-" and "low-tar" cigarettes. The pattern of deposition of sidestream smoke is very different: the proportion deposited is smaller and is likely to occur mainly in the periphery of the lung.

Genetic Host Factors

Genetic polymorphism in microsomal monooxygenases exists in humans. Lung cancer patients with a history of smoking are more often extensive metabolizers of the drug debrisoquine or have high induced levels of aryl hydrocarbon hydroxylase than smokers without lung cancer. It remains to be established whether this association implies that individuals with such genotypes are at increased risk of tobacco smoke-associated cancer.

Cancer in Humans

Lung cancer is believed to be the most important cause of death from cancer in the world, with estimated total deaths in excess of one million annually. The major cause of the disease is tobacco smoking, primarily of cigarettes. Risk of lung cancer is particularly dependent on duration of smoking; therefore, the earlier the age at initiation of smoking, the greater the individual risk. Further, the longer the time period during which a major proportion of adults in a population have smoked, the greater the incidence and mortality from the disease in that population. Risk of lung cancer is also proportional to the numbers of cigarettes smoked, increasing with increasing cigarette usage. In populations with a long duration and heavy intensity of cigarette usage, the proportion of lung cancer attributable to smoking is of the order of 90%. This attributable proportion applies to men in most western populations; in populations in which women are increasingly using cigarettes, the attributable proportion in women is also approaching this level.

In smokers who have smoked for any length of time, the annual lung cancer risk incurred persists at approximately the same level after cessation of smoking, so that the increasing risk that would have been incurred by continuation of smoking is prevented.

Although cigarettes are the predominant cause of lung cancer, some increased risk also results from pipe and/or cigar smoking.

Smokers of other types of tobacco, particularly in Asia (e.g. of *bidis* in India), also appear to be at an increased risk of lung cancer. At present it is not possible to determine whether prolonged *bidi* smoking increases the risk of lung cancer to the same extent as does prolonged smoking of cigarettes.

Cigarettes appear to increase the risk of squamous-cell (epidermoid) and small-cell carcinomas of the lung to a greater extent than that of adenocarcinomas. However, each of these three main histological types of lung cancer is caused by tobacco smoking.

The risk of lung cancer associated with cigarette smoking is substantially increased in conjunction with high-dose exposures to radon daughters or asbestos.

Tobacco smoking (particularly of cigarettes) is an important cause of bladder cancer and cancer of the renal pelvis. The proportion of these diseases attributable to smoking in most countries with a history of prolonged cigarette usage is of the order of 50% in men and 25% in women. The relationships of risk with duration and intensity of smoking are similar to those for lung cancer, although the risks are lower. Pipe and/or cigar smoking probably also increases the risk of bladder cancer, but at lower levels than the risk caused by cigarette smoking.

Tobacco smoking is an important cause of oral, oropharyngeal, hypopharyngeal, laryngeal and oesophageal cancers. Pipe and/or cigar smoking appears to increase the risk of these cancers to approximately the same extent as cigarette smoking. The risks of these cancers associated with cigarette smoking are substantially increased in conjunction with high-dose exposure to alcohol. Tobacco smokers also appear to have increased risks for cancer of the lip.

Cigarette smoking is an important cause of pancreatic cancer and perhaps of renal adenocarcinoma. The proportion of these diseases that is attributable to smoking is not possible to quantify with the same accuracy as for lung cancer. The data now available on tobacco smoking and stomach and liver cancers do not permit a conclusion that the associations noted in some studies are causal.

Although the risk of cancer of the cervix is increased in tobacco smokers, it is not possible to conclude that the association is causal. Further, although in some studies a reduction in risk of endometrial cancer has been found in smokers as compared to nonsmokers it cannot be concluded that smoking protects against cancer at this site.

The cigarettes that are currently sold differ, in many countries, from those that were sold prior to the general recognition of the hazards associated with their use. When the newer cigarettes are smoked under standard laboratory conditions, the yield of some components — particularly of tar and nicotine — is, in consequence, reduced. It is difficult, however, to deduce from this how hazardous such cigarettes are likely to be as they tend to be smoked differently, and the differences observed with laboratory testing may not be reproduced when they are smoked by people. It is difficult, too, to detect their effect on a national scale, as the harmful effects of smoking accumulate over many years and the risk of developing cancer attributable to smoking depends on both recent and past exposure.

Nevertheless, the Group noted that:

1. Although smokers of "low tar"-level cigarettes tend to compensate for lower yields of nicotine and perhaps other smoke components, chiefly by changing the manner of smoking, they do not in general compensate fully for lower tar yields.

2. Case-control and cohort studies suggest that prolonged use of nonfilter and "high-tar" cigarettes is associated with greater lung cancer risks than prolonged use of filter and "low-tar" cigarettes.

3. In a few countries, in which smoking had been established for many years, a substantial reduction in mortality from lung cancer has been observed in young and middle-aged men, which is greatest in the youngest age groups. This has occurred at a time when the number of cigarettes smoked by young men in these countries has remained approximately constant. No substantial cause (or cofactor) has so far been identified that offers a

plausible explanation for the observed magnitude of the reduction of risk for lung cancer, other than changes in cigarette design which include reduction in tar content.

It was concluded that the risk of lung cancer associated with the types of cigarette commonly smoked before the middle 1950s is greater than that for modified cigarettes with "low-tar" levels now generally available in some countries.

The health benefits from the cessation of smoking, however, greatly exceed those to be expected from changes in cigarette composition.

Tobacco smoke affects not only people who smoke but also people who are exposed to the combustion products of other people's tobacco. The effects produced are not necessarily the same, as the constituents of smoke vary according to its source. Three main sources exist: mainstream smoke, sidestream smoke, and smoke exhaled to the general atmosphere by smokers. Smokers are exposed to all three to a greater extent than are nonsmokers. It follows that it is unlikely that any effects will be produced in passive smokers that are not produced to a greater extent in smokers and that types of effects that are not seen in smokers will not be seen in passive smokers. Examination of smoke from the different sources shows that all three types contain chemicals that are both carcinogenic and mutagenic. The amounts absorbed by passive smokers are, however, small, and effects are unlikely to be detectable unless exposure is substantial and very large numbers of people are observed. The observations on nonsmokers that have been made so far are compatible with either an increased risk from "passive" smoking or an absence of risk. Knowledge of the nature of sidestream and mainstream smoke, of the materials absorbed during "passive" smoking, and of the quantitative relationships between dose and effect that are commonly observed from exposure to carcinogens, however, leads to the conclusion that passive smoking gives rise to some risk of cancer.

Conclusion

There is *sufficient evidence* that inhalation of tobacco smoke as well as topical application of tobacco smoke condensate cause cancer in experimental animals.

There is *sufficient evidence* that tobacco smoke is carcinogenic to humans.

The occurrence of malignant tumours of the respiratory tract and of the upper digestive tract is causally related to the smoking of different forms of tobacco (cigarettes, cigars, pipes, *bidis*). The occurrence of malignant tumours of the bladder, renal pelvis and pancreas is causally related to the smoking of cigarettes.

Annex 2

Participants at WHO air quality guideline meetings

Planning Meeting on Air Quality Guidelines, Copenhagen, 28 February — 2 March 1984

Temporary advisers

Dr V. Bencko, Medical Faculty of Hygiene, Charles University, Prague, Czechoslovakia

Professor K. Biersteker, Agricultural University, Wageningen, Netherlands (*Vice-Chairman*)

Professor K.A. Bushtueva, Reference Centre on Air Pollution Control, Central Institute for Advanced Medical Education, Moscow, USSR

Dr R. Dolgner, Medizinisches Institut für Umwelthygiene an der Universität Düsseldorf, Federal Republic of Germany

Dr Mirka Fugas, Environmental Hygiene Laboratory, Institute for Medical Research and Occupational Health, Zagreb, Yugoslavia (*Chairman*)

Dr L. Grant, Environmental Criteria and Assessment Office, US Environmental Protection Agency, Research Triangle Park, USA

Dr A. Ileri, Environmental Health and Air Pollution Division, School of Public Health, Ankara, Turkey

Professor J.K. Piotrowski, Department of Toxicological Chemistry, Institute of Environmental Research and Bioanalysis, Lodz, Poland

Dr L.E. Reed, Industrial Air Pollution Inspectorate, Health and Safety Executive, London, United Kingdom (*Rapporteur*)

Dr M. Stupfel, INSERM, Le Vésinet, France

Mr R.E. Waller, Toxicology and Environmental Protection, Department of Health and Social Security, London, United Kingdom

Dr B.C.J. Zoeteman, Directorate-General of Environmental Protection, Ministry of Housing, Physical Planning and Environment, Leidschendam, Netherlands

Representatives of other organizations

Dr M.D. Gwynne, United Nations Environment Programme, Nairobi, Kenya

WHO Regional Office for Europe

Dr M. Benarie, Acting Regional Officer, Recognition and Control of Environmental Hazards

Mr W. Lewis, Consultant, Environmental Health Service

Dr R. Türck, Project Coordinator, Air Quality Guidelines

Mr J.I. Waddington, Director, Environmental Health Service

Working Group on Air Quality Guidelines for Noncarcinogenic Metals, Düsseldorf, 24–28 September 1984

Temporary advisers

Mr A.P.M. Blom, Directorate-General of Environmental Protection, Ministry of Housing, Physical Planning and Environment, Leidschendam, Netherlands

Dr A.C. Chamberlain, Newbury, Berkshire, United Kingdom

Dr M. Cikrt,[a] Centre of Industrial Hygiene and Occupational Diseases, Institute of Hygiene and Epidemiology, Prague, Czechoslovakia (*Chapter 31*)

Professor T.W. Clarkson,[a] Division of Toxicology, University of Rochester, School of Medicine, Rochester, USA (*Chapter 25*)

Dr U. Heinrich, Abteilung für Umwelthygiene, Fraunhofer-Institut für Toxikologie und Aerosolforschung, Hannover, Federal Republic of Germany

Dr I.E. Korneev, Department of Health — Environment, Institute of Hygiene of Water Transport, Moscow, USSR

Professor P. Mushak, Department of Pathology, University of North Carolina, Chapel Hill, USA (*Rapporteur*)

Professor J.K. Piotrowski, Department of Toxicological Chemistry, Institute of Environmental Research and Bioanalysis, Lodz, Poland

Professor M. Piscator,[a] Department of Environmental Hygiene, Karolinska Institute, Stockholm, Sweden (*Chapter 19*)

---

[a] Prepared first draft of, or made major contributions to, chapter indicated.

Dr M. Riolfatti, Istituto di Igiene, Università degli Studi, Padua, Italy

Professor D. Rondia, Laboratory of Environmental Toxicology, Institute of Chemistry, University of Liège, Belgium

Professor M. Sarić,[a] Institute of Medical Research and Occupational Health, Zagreb, Yugoslavia (*Chairman*) (*Chapter 24*)

Professor H.W. Schlipköter,[a] Medizinisches Institut für Umwelthygiene an der Universität Düsseldorf, Federal Republic of Germany (*Chapter 23*)

Representatives of other organizations

Dr B.G. Bennett, United Nations Environment Programme, Nairobi, Kenya

World Health Organization

Headquarters

Dr J. Parizek, International Programme on Chemical Safety

Regional Office for Europe

Dr R. Türck, Project Coordinator, Air Quality Guidelines

Working Group on Air Quality Guidelines for Inorganic Carcinogenic Air Pollutants, Copenhagen, 19–23 November 1984

Temporary advisers

Dr V. Bencko,[a] Medical Faculty of Hygiene, Charles University, Prague, Czechoslovakia (*Chapter 17*)

Dr B.G. Bennett, Monitoring and Assessment Research Centre (MARC), London, United Kingdom

Professor P.E. Enterline, Department of Biostatistics, Graduate School of Public Health, University of Pittsburgh, USA

Dr L. Fishbein, US Department of Human Services, Food and Drug Administration, National Center for Toxicological Research, Jefferson, USA (*Vice-Chairman*)

Professor M. Fischer,[a] Institut für Wasser-, Boden- und Lufthygiene des Bundesgesundheitsamtes, Berlin (West) (*Chapter 18*)

Dr C.A. van der Heijden, National Institute of Public Health and Environmental Hygiene, Bilthoven, Netherlands (*Rapporteur*)

[a] Prepared first draft of, or made major contributions to, chapter indicated.

Dr R.F. Hertel,[a] Fraunhofer-Institut für Toxikologie und Aerosolforschung, Hannover, Federal Republic of Germany (*Chapter 26*)

Dr K.R. Krijgsheld, Air Directorate, Directorate-General of Environmental Protection, Ministry of Housing, Physical Planning and Environment, Leidschendam, Netherlands

Dr S. Langård,[a] Telemark Central Hospital, Department of Occupational Medicine, Porsgrunn, Norway (*Chapter 21*)

Dr Susana Cerquiglini Monteriolo, Istituto Superiore di Sanità, Rome, Italy (*Chairman*)

Dr H. Muhle, Fraunhofer-Institut für Toxikologie und Aerosolforschung, Hannover, Federal Republic of Germany

Dr B.G. Pershagen, National Institute of Environmental Medicine, Stockholm, Sweden

Dr J.-P. Rigaut, Unité de recherches biomathématiques et biostatistiques, INSERM, Université de Paris, France

Professor T. Schramm, Zentralinstitut für Krebsforschung, Akademie der Wissenschaften der DDR, Berlin, German Democratic Republic

World Health Organization

International Agency for Research on Cancer

Dr H. Vainio, Unit of Carcinogen Identification and Evaluation

Regional Office for Europe

Dr J.O. Järvisalo, Regional Officer, Occupational Health

Dr R. Türck, Project Coordinator, Air Quality Guidelines

Consultation on Selection of Organic and Malodorous Substances for Inclusion in Guidelines for Air Quality, Delft, 10–11 December 1984

Temporary advisers

Professor L.J. Brasser, TNO Research Institute for Environmental Hygiene, Delft, Netherlands (*Chairman*)

Professor F. Kaloyanova-Simeonova, Institute of Hygiene and Occupational Health, Medical Academy, Sofia, Bulgaria

Dr M.A. Martens, Division of Toxicology, Institute of Hygiene and Epidemiology, Brussels, Belgium

[a] Prepared first draft of, or made major contributions to, chapter indicated.

Dr P.J.A. Rombout, Department for Inhalation Toxicology, National Institute of Public Health and Environmental Hygiene, Bilthoven, Netherlands

Mr R.E. Waller, Toxicology and Environmental Protection, Department of Health and Social Security, London, United Kingdom (*Rapporteur*)

Mr H.J.L.M. Wijnen, Air Directorate, Ministry of Housing, Physical Planning and Environment, Leidschendam, Netherlands

WHO Regional Office for Europe

Dr R. Türck, Project Coordinator, Air Quality Guidelines

Working Group on Establishing Guidelines for Organic Carcinogenic Air Pollutants, Athens, 4–8 March 1985

Temporary advisers

Professor R.E. Albert, Institute of Environmental Medicine, New York University Medical Center, USA

Mr A.P.M. Blom, Directorate-General of Environmental Protection, Ministry of Housing, Physical Planning and Environment, Leidschendam, Netherlands

Dr V.J. Feron, Department of Biological Toxicology, TNO-CIVO Toxicology and Nutrition Institute, Zeist, Netherlands

Dr I. Gut,[a] Institute of Hygiene and Epidemiology, Prague, Czechoslovakia (*Chapter 5*)

Dr C.A. van der Heijden, National Institute of Public Health and Environmental Hygiene, Bilthoven, Netherlands

Professor B. Holmberg, Unit of Occupational Toxicology, National Board of Occupational Safety and Health, Solna, Sweden

Dr P.-C. Jacquignon, Ministère de l'environnement, Mission des études et de la recherche, Neuilly sur Seine, France

Dr C. Maltoni, Istituto di Oncologia "Felice Addarii", Bologna, Italy

Professor D.B. Menzel, Laboratory of Environmental Toxicology and Pharmacology, Duke University Medical Center, Durham, USA (*Chairman*)

Dr W.J. Nicholson,[a] Environmental Sciences Laboratory, Mount Sinai School of Medicine, City University of New York, USA (*Chapter 16*)

---

[a] Prepared first draft of, or made major contributions to, chapter indicated.

Professor F. Pott,[a] Medizinisches Institut für Umwelthygiene an der Universität Düsseldorf, Federal Republic of Germany (*Chapter 11*)

Dr J.A. Sokal, Institute of Occupational Medicine, Department of Toxicity Evaluation, Lodz, Poland

Dr D. Steinhoff, Institut für Toxikologie, Bayer AG, Pharma-Forschungszentrum, Wuppertal, Federal Republic of Germany

Dr V.S. Turusov, Cancer Research Centre, USSR Academy of Medical Sciences, Moscow, USSR

Professor G. Ungvary, National Institute of Occupational Health, Department of Toxicology, Budapest, Hungary

Mr R.E. Waller, Toxicology and Environmental Protection, Department of Health and Social Security, London, United Kingdom (*Rapporteur*)

Observers

Mr M. Sabatakakis, Directorate of Sanitary and Environmental Protection, Ministry of Health and Welfare, Athens, Greece

Dr E. Velonakis, Ministry of Health and Welfare, Athens, Greece

Representatives of other organizations

Mr A. Price, Commission of the European Communities, Brussels, Belgium

World Health Organization

Headquarters

Professor M. Berlin,[a] WHO Scientist, Monitoring and Assessment Research Centre (MARC), London, United Kingdom (*Chapter 6*)

International Agency for Research on Cancer

Dr H. Vainio, Unit of Carcinogen Identification and Evaluation

Regional Office for Europe

Dr R. Türck, Project Coordinator, Air Quality Guidelines

---

[a] Prepared first draft of, or made major contributions to, chapter indicated.

Working Group on Air Quality Guidelines for Certain Organic Air Pollutants, Prague, 22–26 April 1985

Temporary advisers

Dr S.L. Avaliani, Laboratory of Toxicology — Air Pollutants, A.N. Sysin Institute of General and Community Hygiene, USSR Academy of Medical Sciences, Moscow, USSR

Professor U. Bauer, Chemisches und Lebensmittel-Untersuchungsamt, Bonn, Federal Republic of Germany

Dr F.A. Chandra, Department of Health and Social Security, London, United Kingdom (*Rapporteur*)

Dr M.M. Greenberg,[a] Environmental Criteria and Assessment Office, US Environmental Protection Agency, Research Triangle Park, USA (*Chapter 9, 13 & 14*)

Dr C.A. van der Heijden,[a] National Institute of Public Health and Environmental Hygiene, Bilthoven, Netherlands (*Chapter 8 & 15*)

Dr K.J. Hemminki,[a] Institute of Occupational Health, Helsinki, Finland (*Chapter 12*)

Dr J. Jäger, Institute of Hygiene and Epidemiology, Prague, Czechoslovakia

Professor E.M. Johnson, Department of Anatomy, Jefferson Medical College, Thomas Jefferson University, Philadelphia, USA (*Chairman*)

Dr K.R. Krijgsheld, Air Directorate, Directorate-General of Environmental Protection, Ministry of Housing, Physical Planning and Environment, Leidschendam, Netherlands

Dr R.J. Laib, Institut für Arbeitsphysiologie, Abteilung für Toxikologie und Arbeitsmedizin, Dortmund, Federal Republic of Germany

Dr J. Lener, Institute of Hygiene and Epidemiology, Prague, Czechoslovakia

Professor R. Masschelein, Section of Occupational and Insurance Medicine, School of Public Health, Catholic University of Leuven, Belgium

Dr B. Nikiforov, Institute of Hygiene and Occupational Health, Medical Academy, Sofia, Bulgaria

Professor F. Valic,[a] Andrija Štampar School of Public Health, Zagreb University, Yugoslavia (*Chapter 9 & 13*)

---

[a] Prepared first draft of, or made major contributions to, chapter indicated.

Dr P. Wolkoff, National Institute for Occupational Health, Hellerup, Denmark

Dr Grace Wood, Bureau of Chemical Hazards, Environmental Health Directorate, Health and Welfare Canada, Ottawa, Canada

Professor G. Ziglio, Istituto di Igiene, Università degli Studi, Milan, Italy

Observers

Dr V. Bencko, Medical Faculty of Hygiene, Charles University, Prague, Czechoslovakia

Dr M. Brezina, Research Institute for Preventive Medicine, Bratislava, Czechoslovakia

World Health Organization

International Agency for Research on Cancer

Mr J.D. Wilbourn, Unit of Carcinogen Identification and Evaluation

Regional Office for Europe

Dr R. Türck, Project Coordinator, Air Quality Guidelines

Consultation on Ecological Effects of Air Pollutants, Neukirchen, 24–28 June 1985

Temporary advisers

Dr J.N.B. Bell,[a] Centre for Environmental Technology, Imperial College, London, United Kingdom (*Chapter 34*)

Dr J.B. Bucher, Eidgenössische Anstalt für das forstliche Versuchswesen, Birmendorf, Switzerland

Professor E. Donaubauer, Forstliche Bundesversuchsanstalt, Vienna, Austria

Mr W. Goerke, Bundesministerium des Innern, Bonn, Federal Republic of Germany (*Chairman*)

Dr S.O. Huttunen,[a] Department of Botany, University of Oulu, Finland (*Chapter 34*)

Dr I. Johnsen, Institute of Plant Ecology, University of Copenhagen, Denmark (*Rapporteur*)

---

[a] Prepared first draft of, or made major contributions to, chapter indicated.

Dr G.H.M. Krause,[a] Landesanstalt für Immissionsschutz des Landes Nordrhein-Westfalen, Essen, Federal Republic of Germany (*Chapter 34*)

Dr K. Meijer, Air Directorate, Ministry of Housing, Physical Planning and Environmental Management, Leidschendam, Netherlands

Dr A.C. Posthumus,[a] Department of Phytotoxicology, Research Institute for Plant Protection, Wageningen, Netherlands (*Chapter 33*)

Dr D.T. Tingey, US Environmental Protection Agency, Corvallis, USA

Professor M. Treshow, University of Utah, Salt Lake City, USA

Dr A.R. Wellburn,[a] Department of Biological Sciences, University of Lancaster, United Kingdom (*Chapter 32*)

Dr H. Werner, Ministerium für Gesundheitswesen der Deutschen Demokratischen Republik, Berlin, German Democratic Republic

WHO Regional Office for Europe

Dr R. Türck, Project Coordinator, Air Quality Guidelines

Working Group on Indoor Air Quality — Radon and Formaldehyde, Dubrovnik, 26–30 August 1985

Temporary advisers

Mr A.P.M. Blom, Directorate-General of Environmental Protection, Ministry of Housing, Physical Planning and Environment, Leidschendam, Netherlands

Professor L.J. Brasser, TNO Research Institute for Environmental Hygiene, Delft, Netherlands (*Chairman*)

Mr O.I. Castrén, Laboratory for Natural Radiation, Finnish Centre for Radiation and Nuclear Safety, Helsinki, Finland

Dr I.N. Chernozemsky, Laboratory of Chemical Carcinogenesis and Testing, Institute of Oncology, Medical Academy, Sofia, Bulgaria

Mr K.D. Cliff, Aerosol Physics and Dosimetry Group, National Radiological Protection Board, Chilton, Didcot, United Kingdom

Dr Ildiko Farkas, National Institute of Hygiene, Budapest, Hungary

Dr V.J. Feron, Department of Biological Toxicology, TNO-CIVO Toxicology and Nutrition Institute, Zeist, Netherlands

Professor M.B. Festy, Laboratoire d'hygiène de la ville de Paris, France

Dr Mirka Fugas, Environmental Hygiene Laboratory, Institute for Medical Research and Occupational Health, Zagreb, Yugoslavia

[a] Prepared first draft of, or made major contributions to, chapter indicated.

Professor W. Jacobi, Institut für Strahlenschutz, Gesellschaft für Strahlen- und Umweltschutz, Neuherberg, Federal Republic of Germany

Dr L. Jech, Institute of Hygiene and Epidemiology, Prague, Czechoslovakia

Professor M.D. Lebowitz, The University of Arizona College of Medicine, Tucson, USA

Mr R.G. McGregor, Radiation Protection Bureau, Environmental Health Directorate, Health and Welfare Canada, Ottawa, Canada

Mrs M.E. Meek,[a] Environmental Health Directorate, Health and Welfare Canada, Ottawa, Canada (*Chapter 10*)

Dr A.V. Nero, Jr, Building Ventilation and Indoor Air Quality Program, Lawrence Berkeley Laboratory, University of California, Berkeley, USA

Dr V. Radmilovic, Sector for Sanitary Supervision and Preventive Health Care, Federal Committee for Labour, Health and Social Welfare, Belgrade, Yugoslavia

Dr P. Rudnai, National Institute of Hygiene, Budapest, Hungary

Professor B. Seifert,[a] Institut für Wasser-, Boden- und Lufthygiene des Bundesgesundheitsamtes, Berlin (West) (*Chapter 10*)

Professor J. Stolwijk,[a] Yale University School of Medicine, Department of Epidemiology and Public Health, New Haven, USA (*Rapporteur*) (*Chapter 10 & 29*)

Dr E. Stranden, National Institute of Radiation Hygiene, Osteraas, Norway

Dr Gun-Astri Swedjemark,[a] Environmental Laboratory, National Institute of Radiation Protection, Stockholm, Sweden (*Chapter 29*)

World Health Organization

International Agency for Research on Cancer

Mr J.D. Wilbourn, Unit of Carcinogen Identification and Evaluation

Regional Office for Europe

Dr M.J. Suess, Regional Officer, Recognition and Control of Environmental Hazards

Dr R. Türck, Project Coordinator, Air Quality Guidelines

[a] Prepared first draft of, or made major contributions to, chapter indicated.

Working Group on Air Quality Guidelines for Certain Malodorous Air Pollutants, Baden-Baden, 30 September — 4 October 1985

Temporary advisers

Dr Birgitta M. Berglund,[a] Department of Psychology, University of Stockholm, Sweden (*Chapter 7 & 22*)

Dr D. Djuric,[a] Institute of Occupational and Radiological Health, Belgrade, Yugoslavia (*Rapporteur*) (*Chapter 7*)

Dr M. Hangartner,[a] Institut für Hygiene und Arbeitsphysiologie, Zurich, Switzerland (*Chapter 7 & 22*)

Dr P.T. Jäppinen,[a] Enso — Gutzeit, Imatra, Finland (*Chapter 22*)

Professor E.P. Köster, Laboratory for Experimental Physics, University of Utrecht, Netherlands (*Chairman*)

Professor J. Stolwijk, Yale University School of Medicine, Department of Epidemiology and Public Health, New Haven, USA

Dr V. Thiele,[a] Landesanstalt für Immissionsschutz, Essen, Federal Republic of Germany (*Chapter 7 & 22*)

Mr H. Wijnen, Air Directorate, Ministry of Housing, Physical Planning and Environment, Leidschendam, Netherlands

Dr G. Winneke, Medizinisches Institut für Umwelthygiene an der Universität Düsseldorf, Federal Republic of Germany

Observer

Professor B. Seifert, Institut für Wasser-, Boden- und Lufthygiene des Bundesgesundheitsamtes, Berlin (West)

WHO Regional Office for Europe

Dr D. Kello, Consultant, Air Quality Guidelines

Dr R. Türck, Project Coordinator, Air Quality Guidelines

Working Group on Air Quality Guidelines for Major Urban Air Pollutants, Bilthoven, 14–19 October 1985

Temporary advisers

Professor J. Bignon, Clinique de pathologie respiratoire et environnement, Centre hospitalier intercommunal, Créteil, France

[a] Prepared first draft of, or made major contributions to, chapter indicated.

Mr A.P.M. Blom, Directorate-General of Environmental Protection, Ministry of Housing, Physical Planning and Environment, Leidschendam, Netherlands

Dr B. Brunekreef, Department of Environmental and Tropical Health, Agricultural University, Wageningen, Netherlands

Professor C. du V. Florey, Department of Community Medicine, Ninewells Hospital and Medical School, Dundee, United Kingdom

Professor B.D. Goldstein,[a] Department of Environmental and Community Medicine, Rutgers Medical School, Piscataway, USA (*Chapter 28*)

Dr Judith A. Graham,[a] Health Effects Research Laboratory, US Environmental Protection Agency, Research Triangle Park, USA (*Rapporteur*) (*Chapter 27*)

Professor M. Haider,[a] Institut für Umwelthygiene der Universität Wien, Austria (*Chapter 20*)

Dr A. Hasse, Umweltbundesamt, Berlin (West)

Professor B. Holma, Institute of Hygiene, Faculty of Medicine, University of Copenhagen, Denmark

Dr M.S. Islam, Medizinisches Institut für Umwelthygiene an der Universität Düsseldorf, Federal Republic of Germany

Professor W. Jedrychowski, Department of Epidemiology, Institute of Social Medicine, Medical School in Cracow, Poland

Professor T. Lindvall, National Institute of Environmental Medicine, Stockholm, Sweden (*Chairman*)

Professor M. Lippmann,[a] Institute of Environmental Medicine, New York University Medical Center, USA (*Chapter 30*)

Professor G. von Nieding,[a] Institut für Wasser-, Boden- und Lufthygiene des Bundesgesundheitsamtes, Berlin (West) (*Chapter 27*)

Professor G.F. Nordberg,[a] Department of Environmental Medicine, University of Umeå, Sweden (*Chapter 30*)

Dr L. Pelech, Institute of Hygiene and Epidemiology, Prague, Czechoslovakia

Dr P.J.A. Rombout,[a] Department for Inhalation Toxicology, National Institute of Public Health and Environmental Hygiene, Bilthoven, Netherlands (*Chapter 28*)

Professor D. Rondia, Laboratory of Environmental Toxicology, Institute of Chemistry, University of Liège, Belgium

Dr T. Schneider, National Institute of Public Health, Bilthoven, Netherlands

[a] Prepared first draft of, or made major contributions to, chapter indicated.

Dr C.A. Scibienski, Air Resources Board of California, Air Quality Standards Section, Sacramento, USA

Professor J. Spengler,[a] Harvard School of Public Health, Department of Environmental Science and Physiology, Boston, USA (*Chapter 30*)

Dr U. Thielebeule, Bezirks-Hygieneinspektion und -institut, Rostock, German Democratic Republic

Professor M. Wagner, Institut für Wasser-, Boden- und Lufthygiene des Bundesgesundheitsamtes, Berlin (West)

Mr R.E. Waller, Toxicology and Environmental Protection, Department of Health and Social Security, London, United Kingdom

Professor H. Zorn,[a] Institut für Arbeitsmedizin der Universität Tübingen, Federal Republic of Germany (*Chapter 20*)

Representatives of other organizations

Mr M. Wolf, Directorate-General for the Environment, Consumer Protection and Nuclear Safety, Commission of the European Communities, Brussels, Belgium

WHO Regional Office for Europe

Dr D. Kello, Consultant, Air Quality Guidelines

Dr R. Türck, Project Coordinator, Air Quality Guidelines

Informal Consultation on Air Quality Guidelines, Copenhagen, 17–21 March 1986

Temporary advisers

Professor B.D. Goldstein, Department of Environmental and Community Medicine, Rutgers Medical School, Piscataway, USA

Dr Judith A. Graham, Health Effects Research Laboratory, US Environmental Protection Agency, Research Triangle Park, USA (*Rapporteur*)

Professor R. Guderian, Institut für Angewandte Botanik, Universität Essen, Federal Republic of Germany

Dr V. Kodat, Department of Hygiene, Epidemiology and Microbiology, Ministry of Health of the Czech Socialist Republic, Prague, Czechoslovakia

Professor T. Lindvall, National Institute of Environmental Medicine, Stockholm, Sweden (*Chairman*)

---

[a] Prepared first draft of, or made major contributions to, chapter indicated.

Professor D.B. Menzel, Laboratory of Environmental Toxicology and Pharmacology, Duke University Medical Center, Durham, USA

Professor M. Piscator, Department of Environmental Hygiene, Karolinska Institute, Stockholm, Sweden

Dr J.A. Sokal, Institute of Occupational Medicine, Department of Toxicity Evaluation, Lodz, Poland

Professor J. Stolwijk, Yale University School of Medicine, Department of Epidemiology and Public Health, New Haven, USA

Professor F. Valic, Department of Occupational Health, Andrija Štampar School of Public Health, Zagreb University, Yugoslavia

WHO Regional Office for Europe

Dr J.O. Järvisalo, Regional Officer, Occupational Health

Dr D. Kello, Consultant, Air Quality Guidelines

Dr R. Türck, Project Coordinator, Air Quality Guidelines

Mr J.I. Waddington, Director, Environmental Health Service

Final Meeting on Air Quality Guidelines for the European Region, Copenhagen, 11–14 November 1986

Temporary advisers

Dr A. Economopoulos, Ministry of Planning, Housing and Environment, Environmental Pollution Control Project PERPA, Athens, Greece

Professor E. Oliveira Fernandes, Department of Mechanical Engineering, University of Porto, Portugal

Dr V.J. Feron, Department of Biological Toxicology, TNO-CIVO Toxicology and Nutrition Institute, Zeist, Netherlands

Mr W. Goerke, Bundesministerium für Umwelt, Naturschutz und Reaktorsicherheit, Bonn, Federal Republic of Germany

Professor B.D. Goldstein, Department of Environmental and Community Medicine, Rutgers Medical School, Piscataway, USA

Dr Judith A. Graham, Health Effects Research Laboratory, US Environmental Protection Agency, Research Triangle Park, USA

Professor R. Guderian, Institut für Angewandte Botanik, Universität Gesamthochschule Essen, Federal Republic of Germany

Dr A. Ileri, Environmental Health and Air Pollution Division, School of Public Health, Ankara, Turkey

Dr M. Jouan, Ministère des affaires sociales et de l'emploi, Direction générale de la santé, Paris, France

Professor F. Kaloyanova-Simeonova, Institute of Hygiene and Occupational Health, Medical Academy, Sofia, Bulgaria

Dr A.G.A.C. Knaap, National Institute of Public Health and Environmental Hygiene, Bilthoven, Netherlands

Dr V. Kodat, Department of Hygiene, Epidemiology and Microbiology, Ministry of Health of the Czech Socialist Republic, Prague, Czechoslovakia

Dr K.R. Krijsheld, Directorate-General of Environmental Protection, Ministry of Housing, Physical Planning and Environment, Leidschendam, Netherlands

Professor E. Lahmann, Institut für Wasser-, Boden- und Lufthygiene des Bundesgesundheitsamtes, Berlin (West)

Professor T. Lindvall, National Institute of Environmental Medicine, Stockholm, Sweden (*Chairman*)

Professor D.B. Menzel, Laboratory of Environmental Toxicology and Pharmacology, Duke University Medical Center, Durham, USA

Professor B. Paccagnella, Istituto di Igiene, Università degli Studi, Padua, Italy

Professor J.K. Piotrowski, Department of Toxicological Chemistry, Institute of Environmental Research and Bioanalysis, Lodz, Poland

Professor M. Piscator, Department of Environmental Hygiene, Karolinska Institute, Stockholm, Sweden

Professor D. Rondia, Laboratory of Environmental Toxicology, Institute of Chemistry, University of Liège, Belgium

Dr J.A. Sokal, Institute of Occupational Medicine, Department of Toxicity Evaluation, Lodz, Poland

Professor J. Stolwijk, Yale University School of Medicine, Department of Epidemiology and Public Health, New Haven, USA (*Vice-Chairman*)

Professor F. Valic, Andrija Štampar School of Public Health, Zagreb University, Yugoslavia

Mr R.E. Waller, Toxicology and Environmental Protection, Department of Health and Social Security, London, United Kingdom (*Rapporteur*)

Dr H. Werner, Ministerium für Gesundheitswesen der Deutschen Demokratischen Republik, Berlin, German Democratic Republic

Dr H.J. van de Wiel, National Institute of Public Health and Environmental Hygiene, Bilthoven, Netherlands

Representatives of other organizations

Mr E. Mackinlay, Commission of the European Communities, Brussels, Belgium

World Health Organization

International Agency for Research on Cancer

Dr E. Johnson, Unit of Analytical Epidemiology

Regional Office for Europe

Dr D. Kello, Consultant, Air Quality Guidelines

Dr R. Türck, Project Coordinator, Air Quality Guidelines

Mr J.I. Waddington, Director, Environmental Health Service